High Temperature Surface Engineering

Also from IOM Communications

B728 Materials for High Temperature Power Generation and Process Plant Applications

B726 Cyclic Oxidation of High Temperature Materials

B708 Advanced Heat Resistant Steel for Power Generation

B689 Advances in Turbine Materials, Design and Manufacturing

B723 Microstructural Stability of Creep Resistant Alloys for High Temperature Plant Applications

B667 Microstructural Development and Stability in High Chromium Ferritic Power Plant Steels

Forthcoming

B736 Parsons 2000

High Temperature Surface Engineering

Edited by

J. Nicholls and D. Rickerby

Proceedings of the
Sixth International Conference
in the Series
'Engineering the Surface'
23–25 September 1997
Edinburgh Conference Centre
Edinburgh, UK

Book 693
Published in 2000 by
IOM Communications Ltd
1 Carlton House Terrace
London SW1Y 5DB

IOM Communications Ltd
is a wholly-owned subsidiary of
The Institute of Materials

ISBN 1 86125 058 4

Typeset by IOM Communications Ltd

Printed and bound in UK at
The University Press, Cambridge

Contents

PREFACE

Over the last forty years a wide range of surface coatings have been developed to address the surface stability and thermal insulation of materials used in the gas turbine section of aero, industrial and land-based power generation equipment. *High Temperature Surface Engineering* reviews the surfacing technologies appropriate to oxidation, corrosion and thermal protection. Factors which underpin their choice for any given application are discussed in the proceedings. This highlights the importance of developing representative mechanical and physical test methods to elucidate coating degradation modes as an aid to establishing coating systems with improved engineering performance.

During the organisation of the conference and in the compiling of this book we have been privileged to work with many of the leading specialists in the field of High Temperature Surface Engineering and it is our hope that this book will be a valuable reference guide for Engineers and Material Scientists.

J. R. Nicholls
Cranfield University

D. S. Rickerby
Rolls–Royce plc

An Investigation of the Oxidation Resistance of Platinum Aluminide Coatings Produced by Either High or Low Activity Processes

G. FISHER, P. K. DATA, J. S. BURNELL-GRAY, W. Y. CHAN and R. WING*

Surface Engineering Research Group, University of Northumbria, UK
**Chromalloy United Kingdom Ltd, Derbyshire, UK*

ABSTRACT

Platinum aluminide coatings act as a remedy against certain aggressive conditions in which modern aero-gas turbines operate. High aluminium activity coatings, such as RT22LT†, provide excellent protection for turbine blades from oxidation and hot corrosive attack. This is due to the formation of an alumina-based scale which acts as an environmental barrier. The integrity of scale can however, be undermined by the diffusion of damaging elements from both within the coating and the substrate to the oxide/coating interface. High activity coatings are particularly susceptible to this type of degradation as they are formed by an inward diffusion process, resulting in substrate elements being present in the as-processed coating.

In contrast, as low activity coatings are formed by an outward diffusion process, the coating should be free from the potentially dangerous elements. This study investigated the oxidation resistance of a low aluminium activity platinum aluminide coating developed by Chromalloy United Kingdom Ltd. The ability of the coating to maintain a protective oxide was assessed, relative to RT22LT, after isothermal soaks at 1100°C, using SEM and XRD analysis to identify changes in coating morphology and degradation modes within the system. In comparison to RT22LT, the low activity coating was shown to have a superior performance, producing a complete and compact oxide scale and was less susceptible to both contamination and substrate elements.

1. INTRODUCTION

The adoption of surface coatings for the protection of turbine blades has arisen from a combination of the aggressive environment in which aero-gas turbines operate and the drive to develop more powerful and efficient engines. Diffusion coatings act to protect the turbine blade by promoting a complete and compact alumina-based scale, which acts as a barrier against the environment. The effectiveness of the coating relies on the maintenance of the scale, which can be undermined by both the environment and by damaging elements present within the component. This study assessed the oxidation

†Registered trademark of Chromalloy Gas Turbine Corporation

resistance at 1100°C of a development platinum aluminide diffusion coating, comparing its performance to that of a commercially available coating, to gain an understanding of the systems growth and degradation mechanisms.

2. EXPERIMENTAL

Several MarM002 DS superalloy pins – length 20 mm, diameter 6 mm – were coated with either the high activity platinum aluminide, RT22LT or the low activity development coating. The RT22LT coating was produced using a pack cementation process. For this process, the samples had an electroplated layer of platinum diffused into the substrates and were then cleaned and placed in an air tight container with a pack consisting of aluminium powder, a halide activator and an inert powder filler (usually a refractory oxide). The container is heated, in this case for 20h at 870°C, to promote diffusion. The halide activator will decompose and the halogen released will react with aluminium to form an aluminium halide which diffuses into the surface of the samples where it dissociates. The deposit on the surfaces diffused into the substrates, whilst the halide ions react with yet more aluminium and the cycle continues. The low activity coatings were produced using an 'out-of-pack' aluminising technique. After a platinum diffusion treatment, the samples were suspended above trays of aluminising powder. The system was heated, in this case for 6h at 1100°C, and the aluminium halide was transported over the sample surfaces by a carrier gas. The advantages of this system over pack cementation are that it enables the coating of more intricate shapes, offers the ability to mask off areas which are not to be coated and surface contamination is reduced as the components are not in contact with pack particles.[1,2] A schematic diagram of the production processes is shown in Fig. 1. The coated samples underwent isothermal soaks in laboratory air for 1 to 1000h at 1100°C. The as-processed and aged coatings were analysed using SEM, SIMS and XRD techniques.

3. ANALYSIS AND DISCUSSION

3.1 As-Processed Coatings

XRD analysis of the development coating in the as-processed condition confirmed that the system had a β phase structure, as expected for a low activity coating.[3] SEM analysis revealed that the coating had a two-layered structure commonly reported in the literature.[1,4] The outer layer being the β phase and the inner an interdiffusion zone, Fig. 2. The grain size of the coating was approximately 45μm. Analysis of RT22LT revealed that the coating had a three phased structure; again as reported in the literature, the outer phase being $PtAl_2$, the middle a β phase and the inner an interdiffusion zone, Fig. 3.[3] The grain size of the coating was approximately 20μm. Digimaps of both of the aluminides revealed the distribution of elements within the coatings. Of particular note was that elements, such as Ti, W and Hf were mostly confined to the interdiffusion zone of the development coating, whilst for RT22LT, they were prevalent throughout the outer layers of the coating, Fig. 4. The elemental distributions are

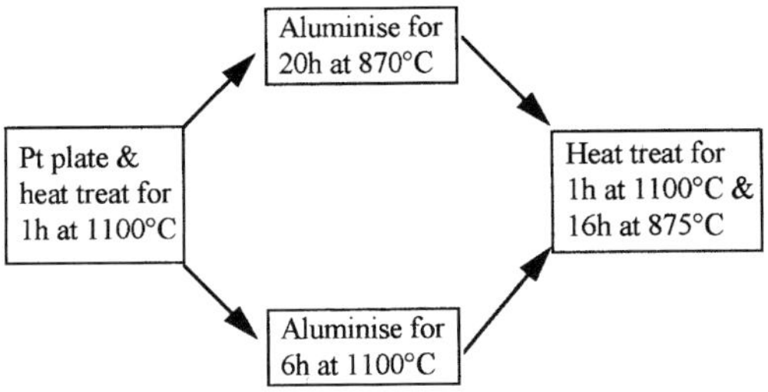

Fig. 1 Schematic diagram of the processing routes.

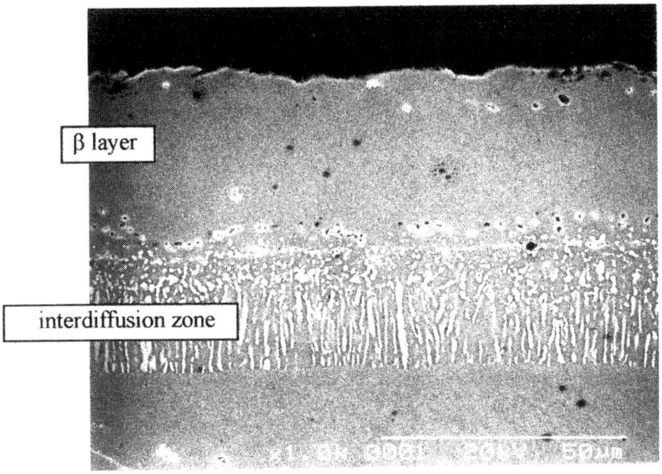

Fig. 2 Low activity coating, as processed, in section.

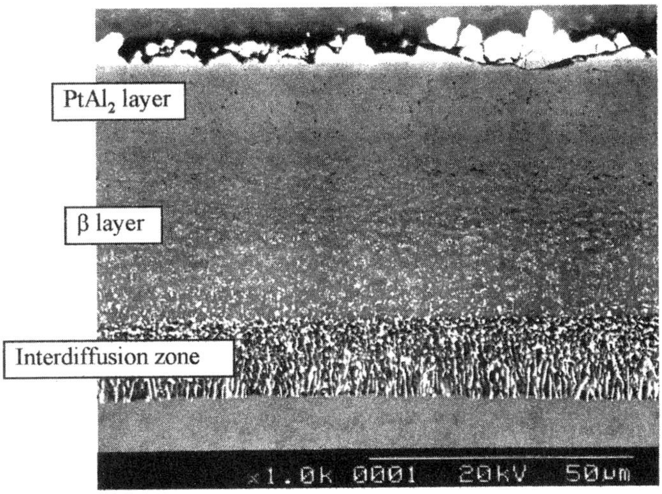

Fig. 3 High activity coating, as processed, in section.

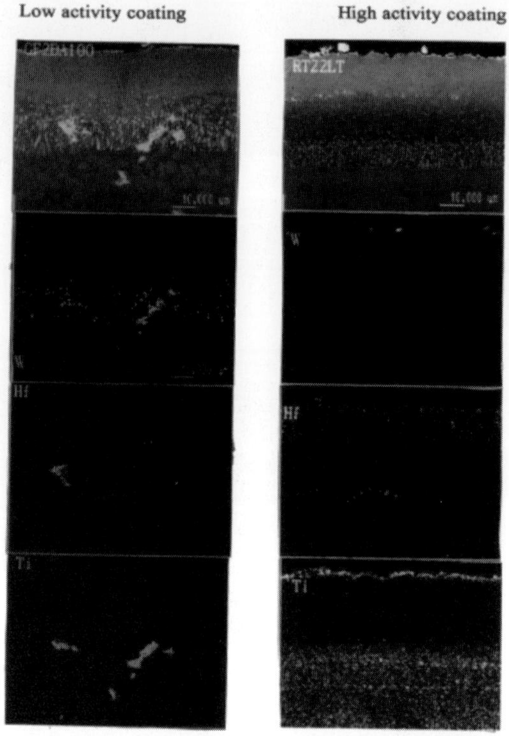

Fig. 4 Digimaps of high and low activity coatings, as processed, in section.

due to the production techniques employed. The high aluminium activity RT22LT coating was formed by the inward diffusion of aluminium into the substrate, whilst the low activity coating was formed by the outward diffusion of nickel. Therefore, the composition of high activity coatings relies, to a larger extent, on the composition of the substrate.

A consideration for these coatings was whether sulphur contamination had occurred during the aluminising process (from sulphur contaminates in the pack). Although sulphur is considered a detrimental element, there is some confusion over the actual effects of contamination. It has been argued that sulphur will segregate to the coating/oxide interface and interfaces within the oxide scale, such as grain boundaries. This segregation will alter the oxidation kinetics of the scale, promoting the outward diffusion of damaging elements and acting to embrittle the oxide.[5–8] Both of these factors will act to decrease the adhesion of the scale. Alternatively, Grabke *et al.*,[9] by using sputter profiling and Auger analysis, established that sulphur preferentially segregated to free surfaces produced by the growth of voids at the oxide/metal interface. The sulphur will promote further detachment by augmenting the effect of thermal and growth stresses, increasing void size and number and so decreasing the adhesion at the interface. It has also been argued however, that once sulphur has segregated to the voids it can play no further part in the degradation of the coating.[10,11]

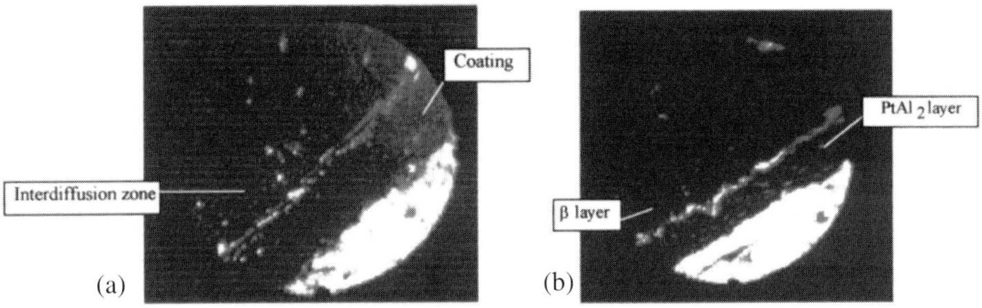

Fig. 5 SIMS analysis of (a) low activity coating, as processed, in section and (b) high activity coating, as processed, in section.

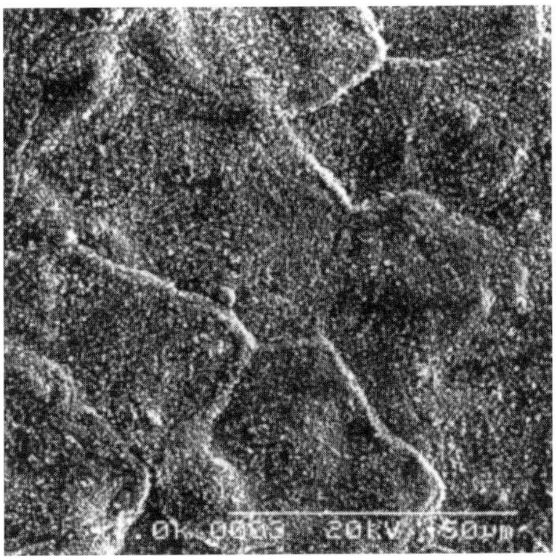

Fig. 6 Riged oxide formed by the low activity coating.

Therefore, as sulphur contamination is generally considered to be detrimental, the as processed coatings underwent SIMS analysis to determine the distribution of the element within each of the systems. The results are shown in Fig. 5. For RT22LT, sulphur contamination was concentrated at the interface between the $PtAl_2$ and the β layers. For the low activity coating, although the concentration of sulphur was of a similar level, the contaminant was found to be restricted to the interdiffusion zone. Therefore, the distance required for the sulphur to diffuse to the oxide interface was higher and so, potentially, the detrimental effect of the element was less.

3.2 Analysis of the Aged Coatings

After isothermal soaks of 5h at 1100°C, both coatings were found to have promoted alumina-based oxides. The scale on the low activity coating had a ridged morphology characteristic of an oxide grown by a β phase material,[12] Fig. 6. The oxide ridges or 'lace' structure were ringed by larger, more pronounced ridges that mirrored the grain boundaries of the coating. It has been proposed that the 'lace' structure is a conse-

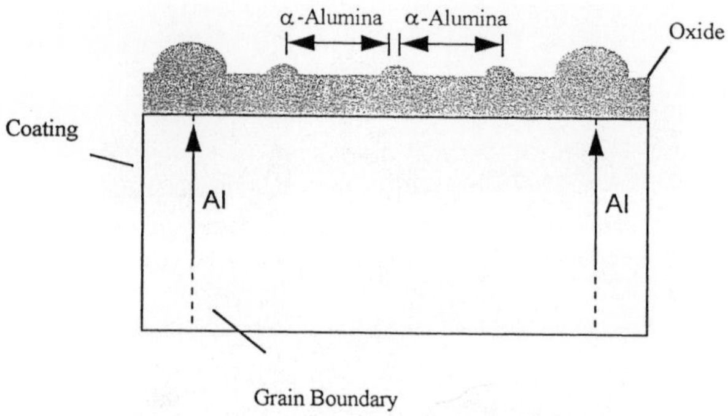

Fig. 7 Schematic diagram of the oxide growth mechanism for the low activity coating.

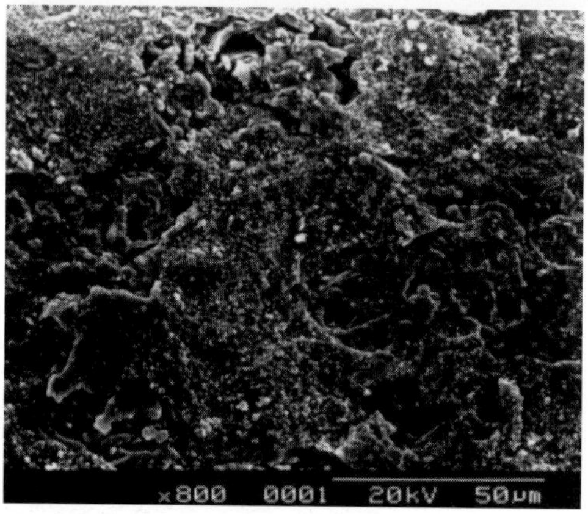

Fig. 8 Low activity coating, 1000h at 1100°C, surface.

quence of the transformation of intermediate aluminas to α-alumina.[12,13] During the initial stages of oxidation, cells of α-alumina will nucleate and grow laterally across the surface. This growth continues until the cells impinge, developing a ridge network. The larger ridges formed above the grain boundaries of the coating would suggest that the grain boundaries acted as high diffusion paths, promoting the outward diffusion of aluminium. Therefore, the oxide scale appears to form due to a combination of anion and cation transport. A schematic diagram of the oxide growth process is shown in Fig. 7. The oxide produced by the RT22LT coating was found to have a nodular morphology, without a ridged structure. This would suggest that the oxide was formed predominately by the inward diffusion of oxygen.

Up to 500h at 1100°C, the development coating appeared to be able to 'self-heal' damage incurred to the scale with the regrowth of alumina. However, as time at temperature was increased, spallations were observed within the scale, Fig. 8. In comparison, the RT22LT coating appeared to lose the ability to 'self-heal' damage after 250h at 1100°C and after 1000h, the level of oxide spallation was far higher than for the low activity coating, Fig. 9.

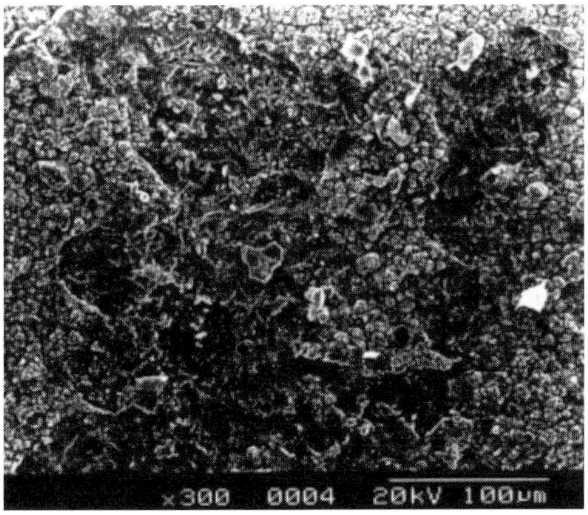

Fig. 9 High activity coating, 1000h at 1100°C, surface

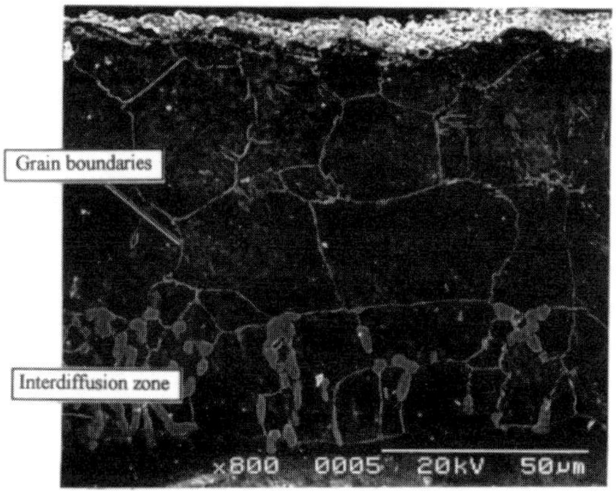

Fig. 10 High activity coating, 25h at 1100°C, in section.

SEM analysis of the coatings prepared in section revealed that after 25h at 1100°C elements, such as Ta, W and Ti, were outwardly diffusing from the interdiffusion zone along grain boundaries to the oxide interface, Fig. 10. As the ageing time was increased, the outward diffusion became more pronounced, with the formation of hafnia-rich pegs, enveloped in alumina and rich in tantalum and titanium, at the interface. These formed preferentially at grain boundaries, but also, to a lesser extent, across the oxide interface where the elements had diffused through the grains of the coating Fig. 11. At an optimum level, Hf is thought to be beneficial, due to such active element effects as pegging, slowing short-circuit transport etc.[3,11] However, excessive peg formation at the oxide interface appeared to have a detrimental effect. As oxidation progressed, the pegs grew outwards into the oxide and depleted the local area of aluminium. The stresses induced within the oxide led to the local spallation of the scale, Fig. 12.

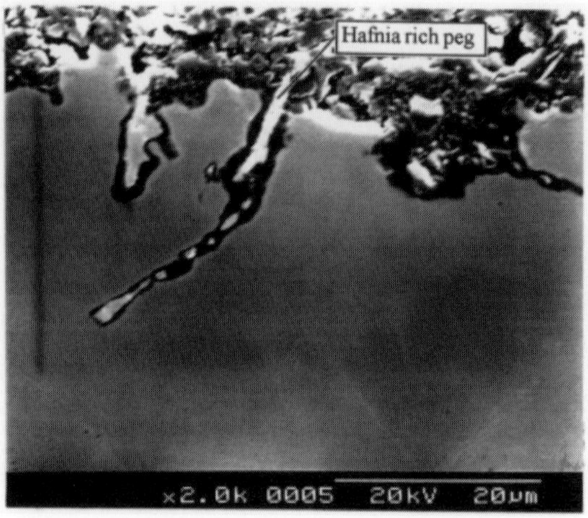

Fig. 11 Low coating activity, 750h at 1100°C, in section.

Fig. 12 Low activity coating, 750h at 1100°C, in section.

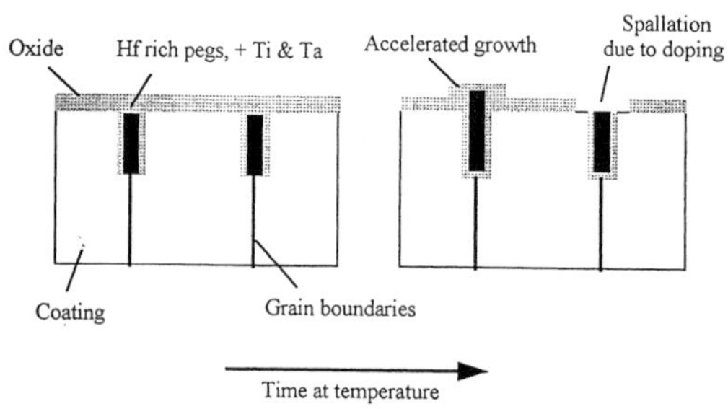

Fig. 13 Schematic diagram of failure modes.

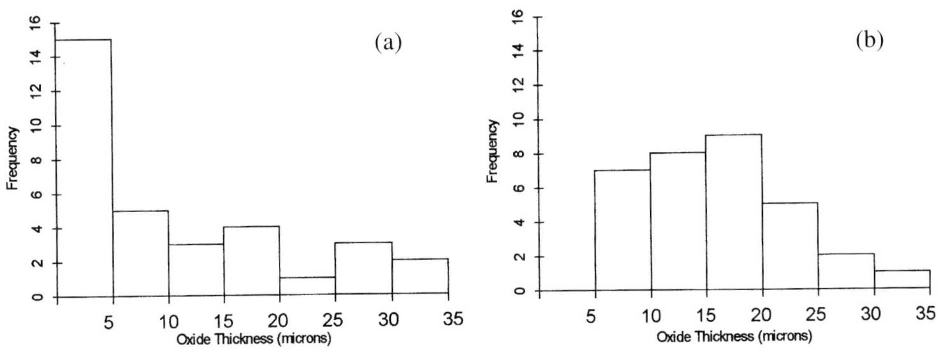

Fig. 14 Distrubution of oxide thickness, (a) low activity coating 1000h at 1100°C and (b) high activity coating 1000h at 1100°C.

Similarly, the incorporation of elements such as Ti within the scale acted as a life-limiting factor. The outward diffusion of these elements from the interdiffusion zone leads to their incorporation in the oxide scale, where they act to locally increase the oxide growth rate.[15] The increase in scale thickness will result in the formation of Kirkendall-type voids beneath the scale, leading to spallation. A schematic diagram of these failure modes is shown in Fig. 13.

Therefore, as these systems appear to fail due to the outward diffusion of detrimental elements, the factors that affect the life-time of the coatings are, the number of grain boundaries, the rate of diffusion through the grains of the coating and the composition of the coating (i.e. whether the coating, in the as processed condition, contains damaging elements and so decreasing the distance to the oxide interface). The superior performance of the low activity coating is due to its lower number of grain boundaries and cleaner composition. Both SEM and SIMS analyses revealed that the low activity coating, in the as processed condition, had restricted damaging elements, such as S, Ti and Hf, to the interdiffusion zone, whilst for the high activity coating, they were prevalent throughout the coating.

Due to the smaller grain size and therefore, higher number of grain boundaries within RT22LT, the aged coating was more susceptible to the formation of pegs and the incorporation of Ti etc. into the scale. This was highlighted by both the observed levels of spallation and oxide thickness measurements, which demonstrated that after 1000h at 1100^{0}C, the development coating had produced a thinner, more compact oxide with less internal oxidation, Fig. 14. Similarly, measurements of the coating thickness remaining after 1000h at 1100^{0}C revealed that the low activity coating was degrading on an even front, with few areas in poor condition. In comparison, the high activity coating had a more distributed thickness, with a higher proportion of the coating in a poor condition, Fig. 15.

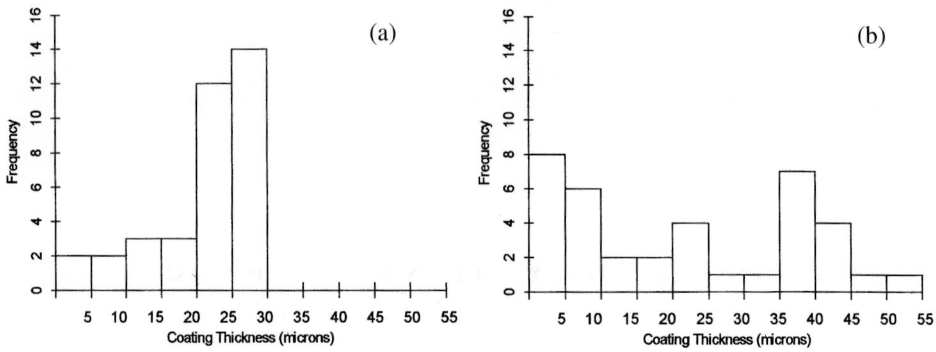

Fig. 15 Distribution of coating thickness, (a) low activity coating 1000h at 1100°C and (b) high activity coating 1000 at 1100°C.

4. CONCLUSIONS

1. Degradation mechanisms identified for these coating systems include the excessive build-up of hafnia-rich pegs and the incorporation of elements such as Ti within the alumina scale.
2. Factors that affect the adhesion of an oxide scale grown by platinum aluminide/MarM002 systems include the number of grain boundaries, the diffusion rate of damaging elements through the grains of the coating and the composition of the coating.
3. The low activity coating had superior oxidation performance at 1100°C due to its larger grain size and that, in the as processed condition, damaging elements were confined to the interdiffusion zone.

ACKNOWLEDGEMENTS

This study forms part of an EPSRC funded programme. We wish to acknowledge the help and support of Chromalloy United Kingdom Ltd, Rolls-Royce plc and Johnson Matthey.

REFERENCES

1. R. STRIEFF: 'Protection of materials by advanced high temperature coatings', *Journal de Phisique IV,* 1993, **3**, 172.
2. C. DURET and R. PICHOIR: 'Protective coatings for high temperature materials: CVD and pack cementation processes', *Coatings for high temperature applications,* E. Lang ed., Applied Science Publishers, 1984, 33.
3. J. SCHAEFER, G. M. KIM, G. H. MEIER and F. S. PETTIT: 'The effects of precious metals on the oxidation and hot corrosion of coatings', *The role of active elements in the oxidation behaviour of high temperature metals and alloys,* E.

Lang ed., Elsevier Applied Science, 1991, 231.

4. G. W. GOWARD and D. H. BOONE: 'Mechanisms of the formation of diffusion aluminide coatings on nickel-base superalloys', *Oxidation of Metals,* 1971, **3**(5), 475.
5. G. MOULIN, G. BEN ABDERRAZIK, J. M. ROUSSELET and A. M. HUNTZ: 'Influence of sulphur on oxide scale adherence to Ni-Cr alloys', *Mat Sci and Tech,* 1988, 4, 479.
6. T. T. HUANG, R. RICHTER, Y. L. CHANG and E. PFENDER: 'Formation of aluminium oxide scales in sulphur containing high temperature environments', *Met Trans,* 1985, **16A,** 2051.
7. D. G. LEES: 'On the reasons for the effects of dispersions of stable oxides and additions of reactive elements on the adhesion and growth mechanisms of chromia and alumina scales - The sulphur effect', *Oxidation of Metals,* 1987, **27**(1/2), 75.
8. J. L. SMIALEK: 'The effect of sulphur content on alumina scale adhesion', *Microscopy of Oxidation, Conference Proceedings 26–28/3/90,* M. J. Bennett and G. W. Lorimer eds, Institute of Materials, 1991, 258.
9. H. J. GRABKE, G. KURBATOV and H. J. SCHMUTZLER: 'Segregation beneath oxide scales', *Oxidation of Metals,* 1995, **43**(1/2), 97.
10. K. L. LUTURA and C. L. BRIANT: 'Mechanism of adhesion of alumina on MCrAlY alloys', *Oxidation of Metals,* 1986, **26**(5/6), 397.
11. F. H. STOTT: 'Principles of growth and adhesion of oxide scales', *The role of active elements in the oxidation behaviour of high temperature metals and alloys,* E. Lang ed., Elsevier Applied Science, 1991, 3.
12. R. PRESCOTT, D. F. MITCHELL and M. I. GRAHAM: 'A study of the growth of alumina scales using high resolution imaging SIMS', *Corrosion Science,* 1994, **50**(1), 62.
13. G. C. RYBICKI and J. L. SMIALEK: 'Effect of the θ–α–Al_2O_3 transformation on the oxidation behaviour of β–NiAl + Zr', *Oxidation of Metals,* 1989, **31**(3/4), 275.
14. P. FOX and G. J. TATLOCK: 'Effect of tantalum additions on the oxidation of overlay coated superalloys', *Mat Sci and Tech,* 1989, **5**(8), 816.

Characterisation of Aluminide Coatings Formed on Nickel-Base Superalloys by Vapour Aluminising

A. B. SMITH, A. KEMPSTER* AND J. SMITH*

Institute of Polymer Technology and Materials Engineering, Loughborough University, Leicestershire LE11 3TU

**Diffusion Alloys Limited, Hatfield, Hertfordshire, AL9 5JW*

ABSTRACT

A production-scale vapour aluminising process has been developed which facilitates the protective aluminide coating of the surface of internal cooling channels/passages in nickel-base superalloy turbine blades and vanes, without recourse to reduced pressure or pressure pulsing. Initial studies confirmed that aluminide coatings, thus formed, were present both at the entrance to and exit from these channels. Explicit restraints, however, precluded the necessary detailed determination of the characteristics of the vapour aluminide coating and hence the elucidation of any variation in these characteristics along the length of the cooling channels in an actual blade/vane. To circumvent this problem, a previously devised experimental reactor set-up has been utilised to simulate the vapour aluminising of the internal channels in a blade/vane. Small, rectangular Inconel 738 samples were suspended in three strategic locations in the reactor coating chamber, representing the entrance to, the mid-point of and the exit from the cooling channels. After the vapour aluminising process, the coating formed on each of these samples was subjected to the required detailed characterisation employing the following techniques: X-ray diffraction, profilometry, scanning electron microscopy, optical microscopy, microhardness testing and Auger electron spectroscopy. The results thus obtained are presented and discussed and any variation in the characteristics of the vapour aluminide coating with increasing sample distance from the source powder mixture identified.

INTRODUCTION

Aluminide diffusion coatings are routinely applied to nickel-base superalloy blades and vanes used in the hot sections of gas turbines to enhance their resistance to high temperature oxidation and hot corrosion. Over the past 40 years the main coating process employed for this purpose has been pack cementation (aluminising),[1] in which, basically, the parts to be coated are embedded in an appropriate reactive powder mixture and the whole subjected to an apposite thermal cycle. Although pack aluminising has many attractive features as regards coating the external surfaces of blades and vanes, the necessity for intimate part/pack powder contact means that it is not readily applicable to coating their internal surfaces, such as the walls of cooling channels/ passages.[2] Whilst large channels can be coated by filling them with fine pack powder, this approach is not feasible when the cooling channels are small and of complex geometry, partly because of the difficulty of reliably introducing and removing the powder.[1] Although conventional chemical vapour deposition may be utilised to circumvent this problem,[3] the most widely employed alternative process is vapour aluminising.

Vapour aluminising resembles the pack process in the sense that a reactive powder mixture is employed, but differs from it in that the parts to be coated are physically separated from this mixture. Volatile aluminium halides are generated in the powder mixture when at temperature. To stimulate aluminide coating formation on the walls of cooling channels in blades/vanes, either a flow of inert gas may be used to transport the halides through the channels,[4] or the pressure in the coating chamber may be reduced[5] or reduced and pulsed.[6] Surprisingly, given the current pre-eminence of vapour aluminising for this application, there appears to be a relative paucity of detailed published information on the characteristics of such aluminide coatings and their variation along cooling channels. This is addressed in the present work. A production-scale vapour aluminising process has been developed which facilitates the aluminide diffusion coating of the walls of internal cooling channels in nickel-base superalloy turbine blades/vanes, without recourse to either reduced pressure or pressure pulsing. It has been confirmed that aluminide coatings thus formed are present both at the entrance to and exit from these channels. The purpose of the present work was to determine, in some detail, the characteristics of the aluminide coatings and to elucidate any variation in these characteristics along the length of the cooling channels.

EXPERIMENTAL WORK

The prohibitive cost of nickel-base superalloy turbine blades/vanes, in concert with the complex geometry of their internal cooling channels and the non-planar nature of the walls of these channels, effectively precluded the use of an actual blade or vane in the present work. A previously devised experimental reactor set-up[7] was therefore utilised to simulate the vapour aluminising of the internal channels in a blade/vane. This is shown schematically in Fig. 1. The length of Inconel tube constituting the reactor chamber is consonant with the total length of the cooling channels in a typical blade/vane; the small holes in the nickel foil cap on the end of the tube are analogous to the channel exit holes. The powder mixture employed was located beneath the Inconel tube, the whole assembly being contained within a larger, resistively heated chamber under a H_2 atmosphere. H_2 from a separate source was used as carrier gas, entering the vapour aluminising assembly below the powder mixture. Inconel 738 (nominal composition, wt%: 0.17% C, 16% Cr, 1.7% Mo, 0.9% Nb, 3.4% Ti, 3.4% Al, 0.01% B, 0.1% Zr, 0.5% max. Fe, 2.6% W, 1.7% Ta, 8.5% Co, 61% Ni) was selected for the substrate material. Samples of final size ~13 mm square by 3 mm thick were initially milled from as-cast material and then surface ground on all their faces. A hole was spark eroded in the corner of each sample. As indicated in Fig. 1, two samples were suspended by wire in each of three strategic locations in the Inconel tube, identified as

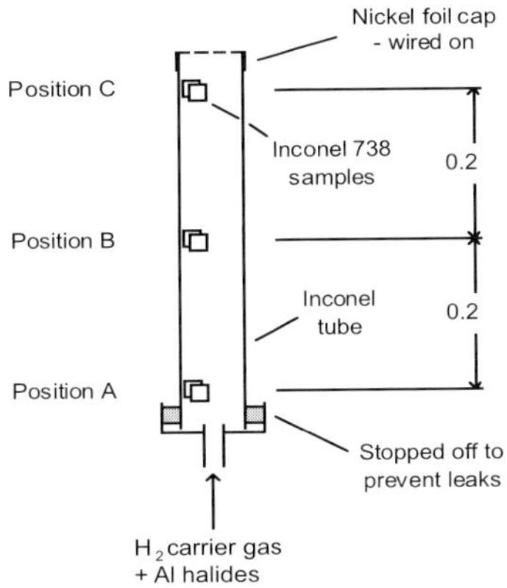

Fig. 1 Schematic diagram of experimental reactor set-up.

positions A, B and C, representing the entrance to, the mid-point of and the exit from the cooling channels respectively. The vapour aluminising process was carried out at 1050 °C for 4.5 hours at atmospheric pressure, with a H_2 carrier gas flow rate of 10 l min^{-1}. It is germane to stress that an initial study,[7] using this set-up, had confirmed that the thicknesses of aluminide coatings formed on three similar sets of samples were in harmony with those determined at the corresponding cooling channel locations in an actual turbine blade. A detailed characterisation of the vapour aluminide coatings formed on the Inconel 738 samples in the present work was then carried out.

X-ray diffraction, using a diffractometer and Cu Kα radiation, was first applied to identify the structure of the vapour aluminide coating on the two large faces of all six of the Inconel 738 samples. X-ray spectra were acquired between 2θ values of 2° and 149°. Profilometry was then employed to assess the surface roughness of the six vapour aluminised samples, plus two of the uncoated Inconel 738 samples for comparison. On the two large faces of each sample, two roughness average (R_a) readings were taken on five evenly divided sampling lengths parallel to a sample edge and on five similar lengths at 90°. All the vapour aluminised samples were then given an evaporated gold coating and the nature of the surface of the coating on the two large faces of each determined by direct observation in a scanning electron microscope. One of the position A, position B and position C vapour aluminised samples was then metallurgically sectioned and half of each mounted in a proprietary compound, followed by lapping on cast iron plates, vibratory polishing using 1 μm diamond compound and finish polishing using a suspension of 0.05 μm γ-alumina powder in water. To elucidate coating microstructures, a metallographic examination of each as-polished sample was first carried out using Nomarski differential interference contrast. This was followed by etching in 1% HF–20% HNO_3 and re-examination using conventional bright field illumination. The peripheral variation in total thickness of the vapour aluminide coating on each polished and etched sample was then measured optically.

Microhardness tests were next employed to determine through-thickness hardness contours for the coating on the remaining half of the sectioned position A, B and C vapour aluminised samples. These tests were performed on lapped and vibratory polished low-angle (5.6°) taper sections through one of the large faces of each sample, which 'mechanically magnified' coating thickness by a factor of 10. A Vickers indentor was used under a load of 25 g applied for a time of 5 seconds. Finally, composition-depth profiles for the remaining position A, B and C vapour aluminised samples, extending from the surface of the coating on one of the large faces of each sample into the Inconel 738 substrate, were obtained using Auger electron spectroscopy (AES). In order to avoid the problems associated with the determination of AES composition-depth profiles by the normal sequential ion bombardment method, the ball-cratering technique was employed; each profile being obtained by point-to-point AES analysis down the sloping side of the resulting crater wall. Quantification was achieved using experimentally derived relative sensitivity factors.

RESULTS AND DISCUSSION

It is well-established that the coatings formed on nickel-base superalloys during pack aluminising are primarily composed of intermetallic phases found in the aluminium-nickel binary system.[8,9] Reference to the aluminium-nickel equilibrium phase diagram[10] revealed that, at the vapour aluminising temperature of 1050 °C employed in the present work, this binary system contains the following intermetallic phases: δ-Ni_2Al_3, β-NiAl and γ'-Ni_3Al. Consequently the X-ray diffraction patterns determined for the vapour aluminide coating on the two large faces of each of the six Inconel 738 samples in the present study were compared with those for Ni_2Al_3, NiAl and Ni_3Al from the ASTM powder diffraction file (card numbers 14–648, 20–19 and 9–97 respectively). As exemplified by Fig. 2, which shows representative X-ray diffraction patterns determined for the coatings on the position A, position B and position C vapour aluminised samples, this comparison indicated, in every case, the presence only of β-NiAl phase. However, it also exposed the existence of several additional low-angle reflections which could not be identified. As can be seen from Fig. 2, these reflections decreased in both number and intensity in going from position A, through B to C. A concurrent, relatively small, but sustained increase in coating (110) preferred orientation is also evident from Fig. 2. Having ascertained that the vapour aluminide coating formed on each of the six Inconel 738 samples consisted predominantly of β-NiAl, coating lattice parameters were calculated from a plot of a_{hkl} versus the Nelson and Riley function of Bragg angle[11] to correct for possible errors due to absorption. As can be seen from the values quoted in Fig. 2, which are reasonably typical, the lattice parameter of the coatings was thus established to increase in going from position A, through B to C, indicative, it is thought, of a change in vapour aluminide coating composition.

Figure 3 shows typical optical sections through the surface of the as-polished position A, position B and position C vapour aluminised samples, obtained from their metallographic examination using Nomarski differential interference contrast. Complementary sections observed under conventional bright field illumination after etch-

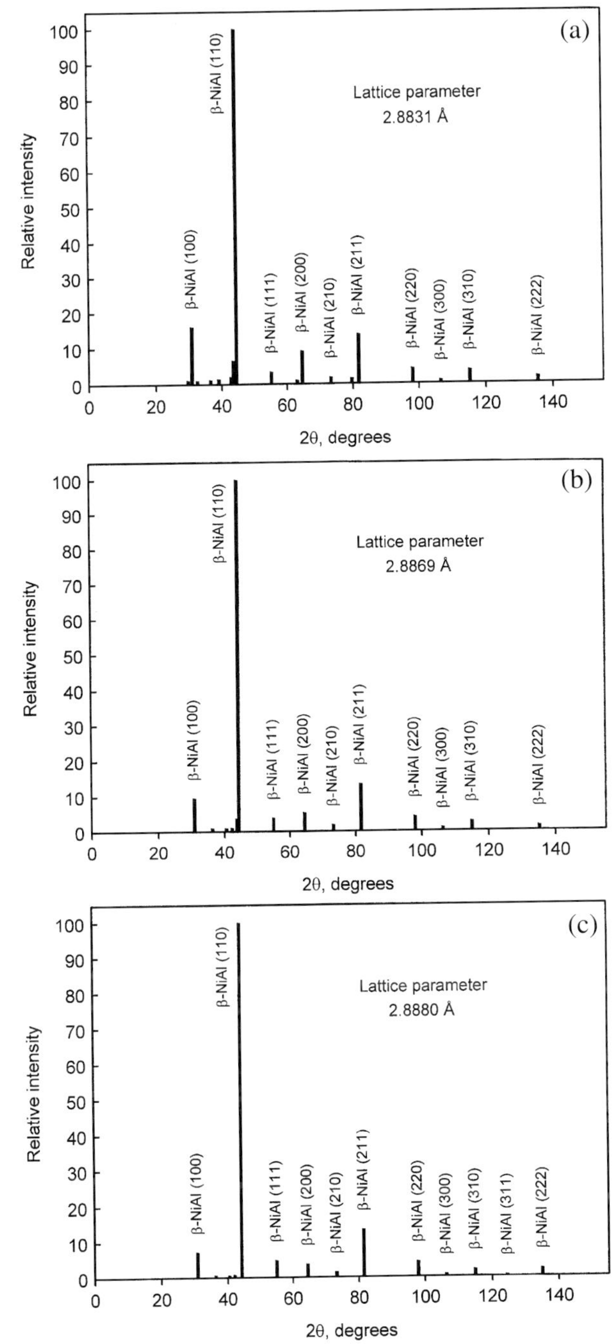

Fig. 2 X-ray diffraction patterns and lattice parameter values for vapour aluminide coatings on Inconel 738 samples – Cu Kα radiation. (a) position A sample, (b) position B sample, (c) position C sample.

ing in 1% HF–20% HNO_3 are presented in Fig. 4. It is clearly evident from these sections, particularly those in Fig. 3, that the microstructure of the vapour aluminide coating formed on each Inconel 738 sample in the present work is fundamentally of the same form, essentially consisting of an inner and an outer zone. The former contains a high density of second phase particles. Three regions can be distinguished in the outer coating zone: a preponderant central region rich in second phase particles, and an upper region adjacent to the coating surface and a lower region adjacent to the inner coating zone both with substantially less second phase particles. These three regions are also demarcated by their etching response and grain size and shape (Fig. 4). Portentously, as evinced by Figs 3 and 4(c), relatively large and smaller substrate carbides (authenticated as such by energy-dispersive X-ray analysis and microhardness tests) were clearly seen to be present in the outer coating zone. These were, however, only detected in the lower and central regions of this zone, never in its upper region. Both Figs 3 and 4 show that in going from position A, through B to C, there is a decrease in the number of second phase particles in the outer coating zone central region.

It is suggested that the second phase particles in the outer coating zone central region may be the source of the additional low-angle reflections present on the X-ray diffraction patterns determined for the vapour aluminide coatings on the Inconel 738 samples (Fig. 2). Using equations and data given by Cullity[11] it has been simplistically calculated that the effective depth of X-ray penetration over the 2θ range where these additional reflections occur extends from ~15 μm to ~23 μm from the coating surface. Reference to Figs 3 and 4 clearly shows that this covers much of the outer coating zone central region on each of the vapour aluminised samples. If the proposed hypothesis is sound, the noted decrease in both the number and intensity of these additional reflections in going from position A, through B to C, can be simply rationalised in terms of the observed contemporary decrease in the number of second phase particles in the outer coating zone central region.

The microstructure of the vapour aluminide coatings formed on the Inconel 738 samples in the present study diverges substantially from those observed for commensurate coatings produced by other vapour aluminising processes.[2,12] Furthermore, it differs from the archetypal inward and outward diffusion (pack) aluminide coating microstructures classified in the seminal work of Goward and Boone,[8] but there is no contradiction here as the latter represent extremes of microstructure and intermediate ones are known to exist.[13]

It is also qualitatively evident from the optical sections through the surface of the vapour aluminised samples from the present work (Figs 3 and 4) that there is a decrease in total coating thickness in going from position A, through B to C. This is quantitatively corroborated in Fig. 5, which shows the measured, not insubstantial, peripheral variation in total thickness of the vapour aluminide coatings formed on the Inconel 738 samples. Figures 3, 4 and 5 all signal that the decrease in total coating thickness in going from position B to C is greater than that in going from position A to B. This is corroborated by overall average total coating thicknesses for the position A,

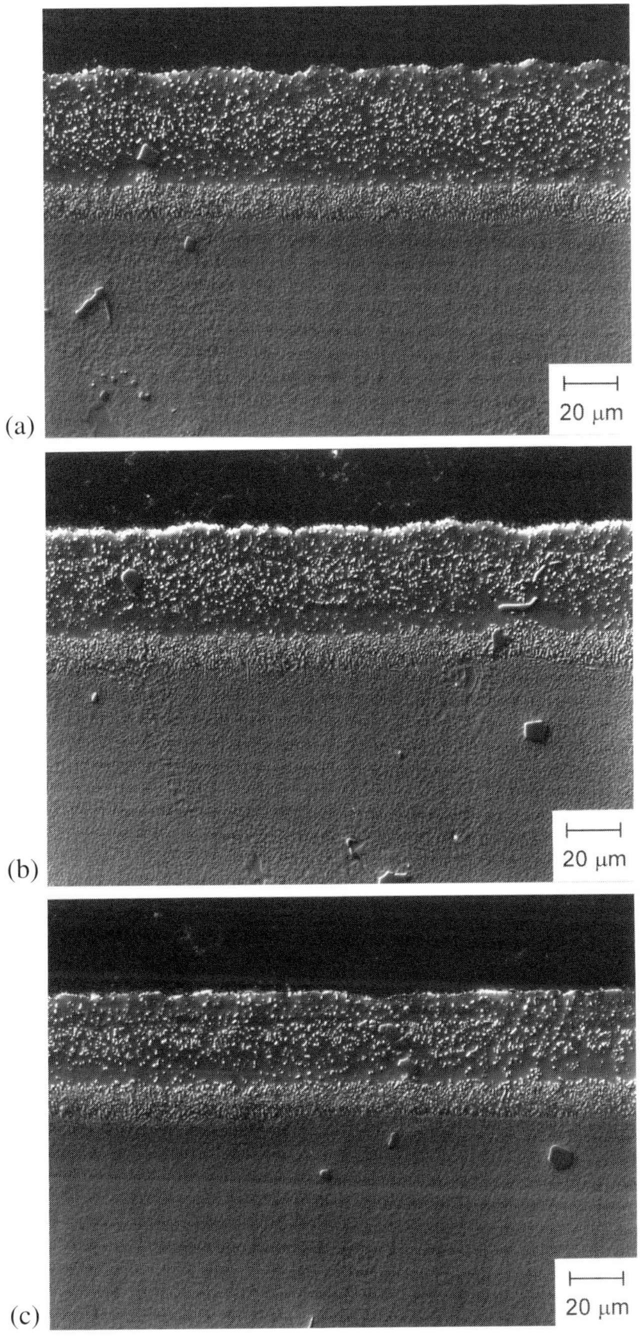

Fig. 3 Optical sections through surface of vapour aluminised Inconel 738 samples, as-polished, Nomarski differential interference contrast. (a) position A sample, (b) position B sample, (c) position C sample.

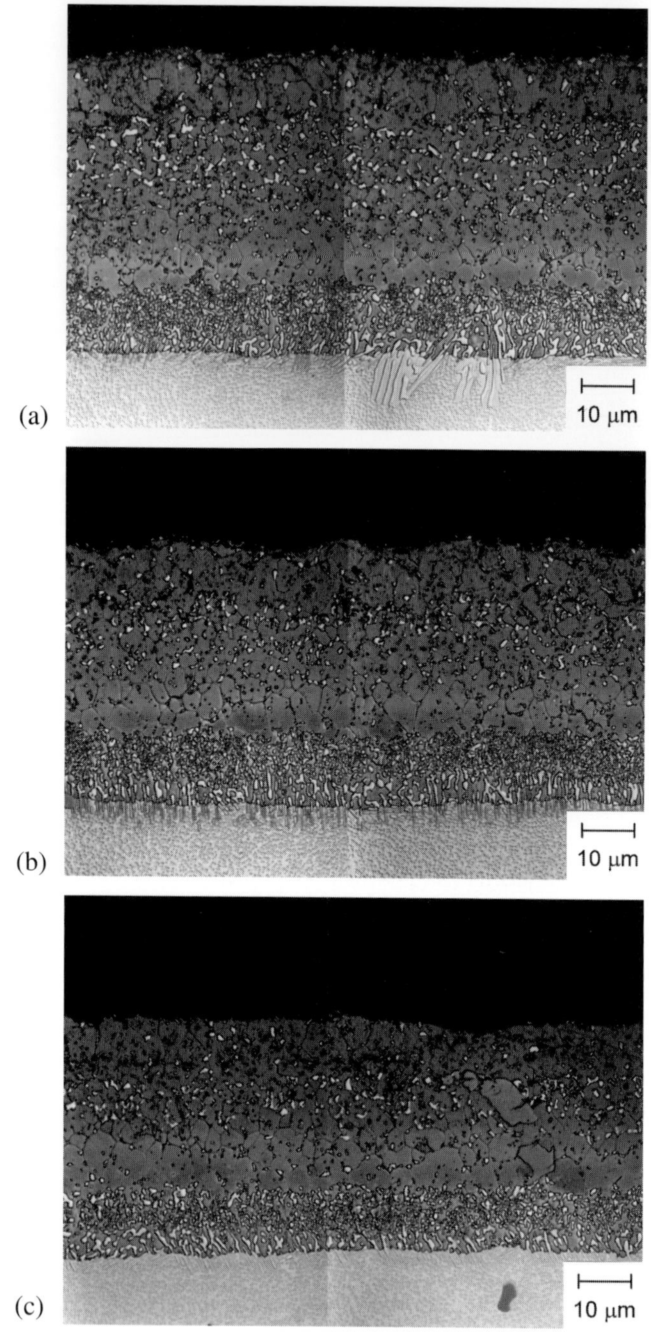

Fig. 4 Optical sections through surface of vapour aluminised Inconel 738 samples, etched in 1% HF–20% HNO_3, bright field illumination. (a) position A sample, (b) position B sample, (c) position C sample.

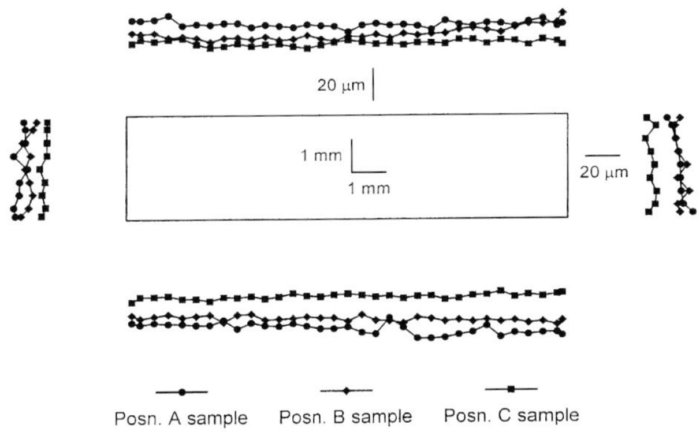

Fig. 5 Peripheral variation in total thickness of vapour aluminide coatings on Inconel 738 samples.

position B and position C vapour aluminised samples, calculated from the 84 data points for each sample in Fig. 5, of 61.4 μm, 56.2 μm and 45.9 μm respectively. These effects are attributed simply to decreasing aluminising potential with increasing sample distance from the source powder mixture. An analogous trend has been noted in previous studies concerning other vapour aluminising processes[2,4–6] involving inert carrier gas flow, reduced coating chamber pressure and reduced and pulsed chamber pressure.

Figure 6 shows the nature of the surface of the vapour aluminide coatings formed on the Inconel 738 samples in the present work, as determined by scanning electron microscopy. Whilst there appears to be no material variation in the intrinsic surface structure of the coatings in going from position A, through B to C, there does seem to be a diminution in surface roughness. The latter observation is sustained, with one exception, by the mean surface R_a values determined for the vapour aluminised samples, shown in Fig. 7. It is thought likely that the anomaly is attributable to uncharacteristically poor surface finish of that particular Inconel 738 sample prior to vapour aluminising.

The AES composition-depth profiles determined for the position A, B and C aluminised samples are presented in Fig. 8. Uncertainties in the analyses are anticipated to be within the range ± 2 to 10% of the values shown, depending on the signal to noise ratio and the particular element involved. For clarity, some of the minor (in terms of amount present) elements have been omitted from the profiles. The total content of tungsten plus tantalum is exhibited because of overlap of the Auger peaks for these elements. It is clear from Fig. 8 that generally the same features are displayed by each composition-depth profile. The plurality of the outer coating zone can be seen to consist primarily of nickel plus aluminium, with titanium and, more saliently, chromium also present in each case. Manifestly, the level of chromium is generally well in excess of its quoted solubility limit in β-NiAl of 3 to 5 wt%.[14] It is thought likely that

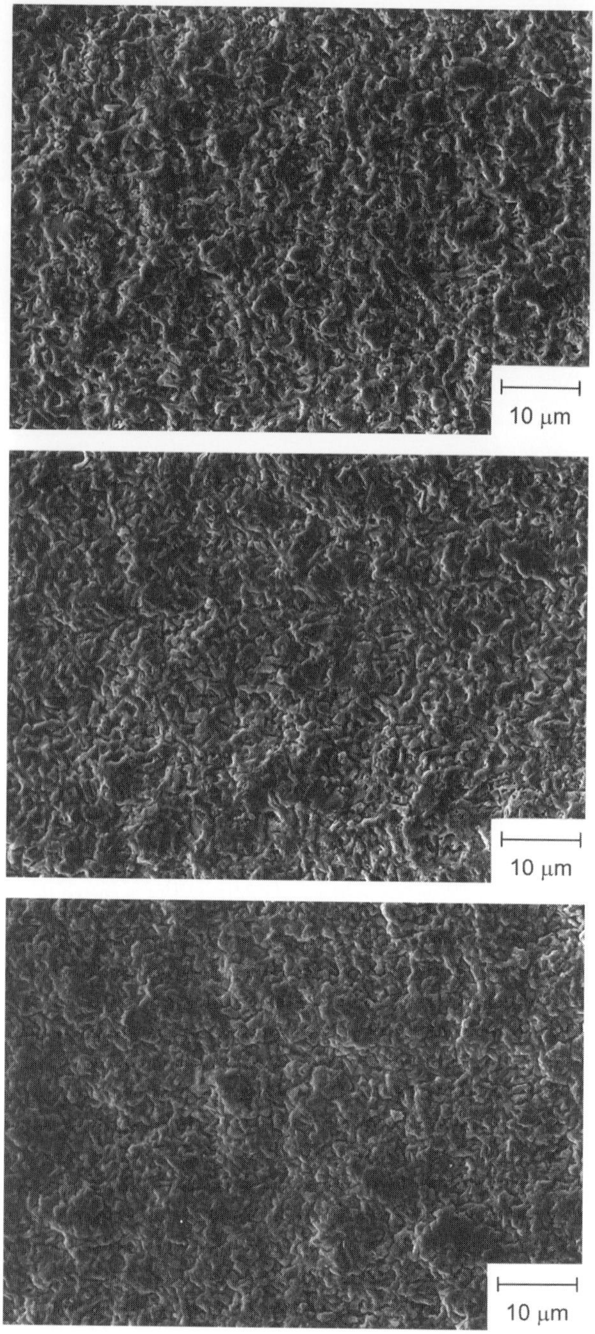

(a) (b) (c)

Fig. 6 Surface of vapour aluminide coatings on Inconel 738 samples. (a) position A sample, (b) position B sample, (c) position C sample.

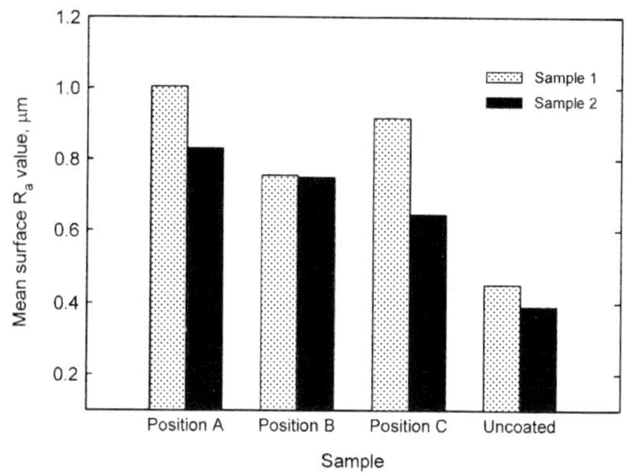

Fig. 7 Mean surface roughness average (R_a) values for vapour aluminised Inconel 738 samples.

the observed presence of the second phase particles in the outer coating zone (Figs 3 and 4), particularly in its central region, is pursuant to this. Consentient with the X-ray diffraction results (Fig. 2), it is evident from Fig. 8 that in going from sample position A, through B to C, there is a reduction in the aluminium content of the outer coating zone, but that all three outer coating zone compositions remain indicative of β-NiAl phase. Quoting genuinely characteristic compositions is difficult. However, for the portion of the outer coating zone on each sample where the coating nickel and aluminium contents are relatively constant, the following are reasonably representative average wt% levels: position A ~ 58.5% Ni, 28.5% Al; position B ~ 64% Ni, 27% Al; position C ~ 65% Ni, 25.5% Al. This decrease in aluminium content, as with total coating thickness, is attributable simply to decreasing aluminising potential with increasing sample distance from the source powder mixture. Coating nickel content may be observed from Fig. 8 to decrease sharply, accompanied by a rapid rise in chromium content, at the inner coating zone on each vapour aluminised sample. This is considered to be indicative of its formation by outward diffusion of nickel causing transformation of the $\gamma + \gamma'$ Inconel 738 substrate matrix in this location into either a hypo-stoichiometric β-NiAl or γ'-Ni_3Al matrix containing a variety of excess phases. Given the composition of Inconel 738, reference to work by Fleetwood[15] suggests the former to be most likely.

Figure 9 shows the through-coating thickness microhardness contours determined for the position A, B and C vapour aluminised samples in the present study. All are clearly of the same form. They exhibit a precipitous increase from the Inconel 738 substrate hardness of ~ 400 $HV_{0.025}$ to a maximum of ~ 850 $HV_{0.025}$ in the inner coating zone. This is followed, in the outer coating zone, by a gradual decrease to a minimum value and a further increase to essentially a microhardness plateau extending to/close to the coating surface. For the position A and B samples, this plateau is at a hardness of ~ 600 $HV_{0.025}$ but falls to ~ 500 $HV_{0.025}$ in the case of the position C sample. These values lie within the hardness range of 500 to 700 HV quoted for nickel-aluminide coatings with aluminium contents of 25 to 30 wt%.[16]

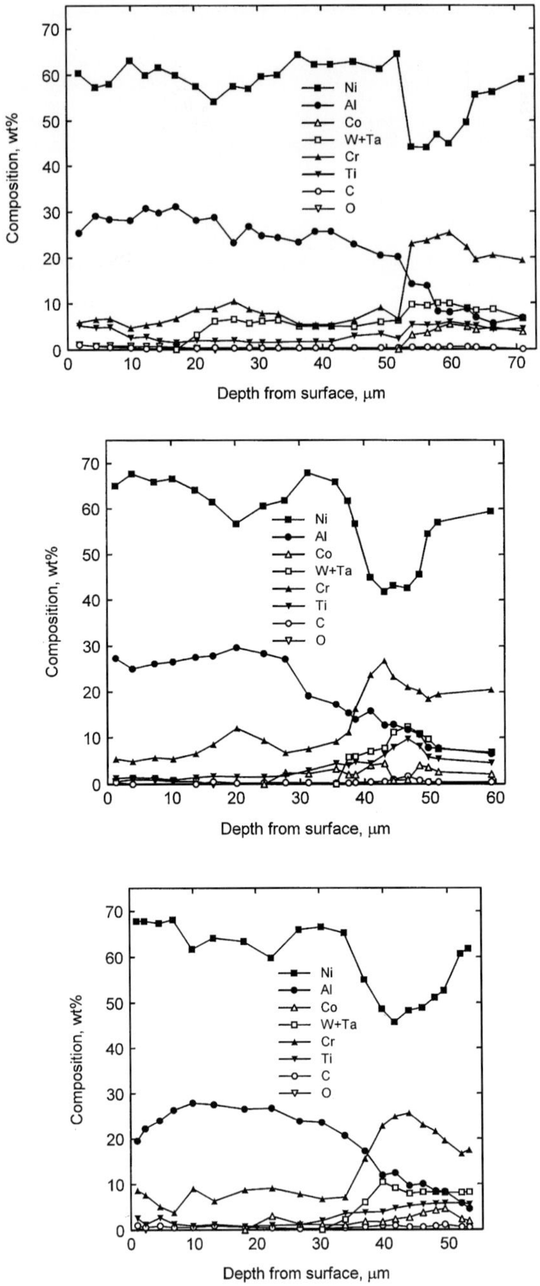

Fig. 8 Auger electron spectroscopy composition-depth profiles through surface of vapour aluminised Inconel 738 samples. (a) position A sample, (b) position B sample, (c) position C sample.

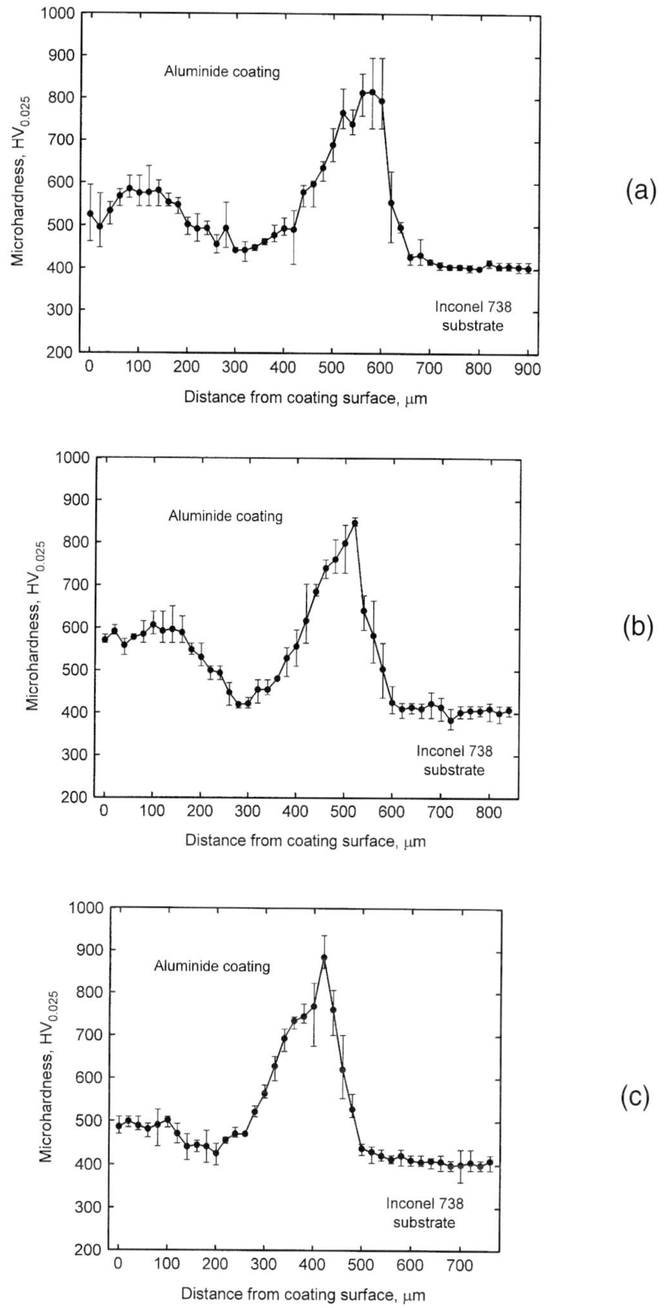

Fig. 9 Through-coating thickness microhardness contours for vapour aluminised Inconel 738 samples determined on low-angle (5.6°) taper sections. (a) position A sample, (b) position B sample, (c) position C sample.

MAJOR CONCLUSIONS

1. The vapour aluminide coatings formed on the Inconel 738 samples in the present work consist primarily of β-NiAl. Their microstructure diverges substantially from those observed for commensurate coatings formed by other vapour aluminising processes. It also differs from the archetypal inward and outward diffusion (pack) aluminide coating microstructures classified in the seminal work in this area, but there is no contradiction here as the latter represent extremes of microstructure and intermediate ones are known to exist.
2. Several of the characteristics of the vapour aluminide coatings formed on the Inconel 738 samples in the present work display no substantive change with increasing sample distance from the source powder mixture and hence, by implication, increasing length along a turbine blade/vane cooling channel. Some others, chiefly total coating thickness and outer coating zone aluminium content, do exhibit reckonable variation, in both cited cases an attenuation. This does not, however, exceed tolerable limits. It is attributed simply to decreasing aluminising potential with increasing distance from the source powder mixture.

ACKNOWLEDGMENTS

The authors wish to express their thanks to the following: the Directors of Diffusion Alloys Limited for their support and permission to publish this paper, the Head of IPTME, Loughborough University, for the provision of research facilities, Mr. G. W. Critchlow of the Institute of Surface Science and Technology, Loughborough University, for performing the AES analyses and Ms. W. H. Chua of Singapore Polytechnic for her assistance with the other experimental work.

REFERENCES

1. J. E. RESTALL and M. I. WOOD: 'Alternative Processes and Treatments', *Materials Science and Technology*, 1986, **2**, 225–231.
2. R. S. PARZUCHOWSKI: 'Gas Phase Deposition of Aluminium on Nickel Alloys', *Thin Solid Films*, 1977, **45**, 349–355.
3. J. S. SMITH and D. H. BOONE: 'Platinum Modified Aluminides – Present Status', *American Society of Mechanical Engineers Paper 90-GT-319*, 1–10.
4. R. S. BENDEN and R. S. PARZUCHOWSKI: 'Gas Phase Deposition of Aluminium using a Complex Aluminium Halide of an Alkali Metal or an Alkaline Earth Metal as an Activator', *Patent US 4132816*, 2 January 1979.
5. L. VANDENBULCKE, G. LEPRINCE and B. NCIRI: 'Low-Pressure Gas-Phase Pack Cementation Coating of Complex-Shaped Alloy Surfaces', *Materials Science and Engineering*, 1989, **A121**, 379–386.
6. J. E. RESTALL and C. HAYMAN: 'Improvements in or Relating to the Diffusion Coating of Metal or Other Articles', *Patent GB 1549845*, 8 August 1979.

7. Y. K. CHIA: 'Out-of-Pack Aluminide Coatings on Nickel-Base Superalloys', MSc thesis, Loughborough University, Great Britain, 1996.
8. G. W. GOWARD and D. H. BOONE: 'Mechanisms of Formation of Diffusion Aluminide Coatings on Nickel-Base Superalloys', *Oxidation of Metals*, 1971, **3**(5), 475–495.
9. G. F. SLATTERY: 'Microstructural Aspects of Aluminised Coatings on Nickel-Base Alloys', *Metals Technology*, 1983, **10**, 41–51.
10. M. HANSEN and K. ANDERKO: *Constitution of Binary Alloys*, 2nd Edition, McGraw-Hill, New York, 1958.
11. B. D. CULLITY: *Elements of X-Ray Diffraction*, 2nd Edition, Addison Wesley, London, 1978.
12. G. GAUJÉ and R. MORBIOLI: 'Vapor Phase Aluminizing to Protect Turbine Airfoils', *High Temperature Protective Coatings*, S. C. Singhal ed., The Metallurgical Society of AIME, Warrendale, Pa, 1983.
13. G. W. GOWARD and L. W. CANNON: 'Pack Cementation Coatings for Superalloys: a Review of History, Theory and Practice', *Transactions of the American Society of Mechanical Engineers*, 1988, **110**, 150–154.
14. D. CHATTERJI, R. C. DEVRIES and G. ROMEO: 'Protection of Superalloys for Turbine Applications', *Advances in Corrosion Science and Technology*, Vol. 6, M. G. Fontana and R. W. Staehle eds, Plenum Press, New York, 1976.
15. M. J. FLEETWOOD: 'Influence of Nickel-Base Alloy Composition on the Behaviour of Protective Aluminide Coatings', *Journal of the Institute of Metals*, 1970, **98**, 1–7.
16. G. W. MEETHAM: 'Use of Protective Coatings in Aero Gas Turbine Engines', *Materials Science and Technology*, 1986, **2**, 290–294.

Plasma Sprayed Coatings for Improved Corrosion Resistance in Aggressive Sulphur Containing Environments

J. L. JIMÉNEZ,*† D. J. BAXTER,* J. F. NORTON*
and F. J. PÉREZ†

**EC, JRC, Institute for Advanced Materials, 1755 ZG Petten, NL*
†Universidad Complutense de Madrid, Avda. Complutense S/N, 28040 Madrid, SP

ABSTRACT

The microstructures of vacuum plasma sprayed (VPS) Al/Si and air plasma sprayed (APS) Ni/Cr/Si coatings on a model Fe–12%Cr alloy and on the commercial alloys HCM 12 and Sanicro 28, and their corrosion behaviour in a simulated non-equilibrium coal gasifier heat exchanger atmosphere (H_2–64CO–3.8CO_2–0.8H_2S) at 600°C are described. The VPS coatings were compact and rich in Ni-aluminide, whereas the APS coatings, characteristic of the APS process, were porous and contained internal oxides. The 80 to 150 μm thick coatings eliminated corrosion attack of the base alloys within 1000 hours of exposure at 600°C. The rate of corrosion attack was comparable to that of highly alloyed, and therefore more expensive, bulk materials sometimes considered for these applications. The long term effects of coating/base alloy interdiffusion were demonstrated following a test in an inert atmosphere.

INTRODUCTION

In fossil fuel-fired power plants, materials are exposed to sulphidising, low-oxygen activity gaseous atmospheres in which rapid corrosive degradation can take place. While highly alloyed and therefore expensive materials can be used to reduce the rate of corrosion, the option of coating low-alloy steels may be preferred. In any case, the surfaces of components need to contain substantial amounts of at least two of the elements, Cr, Al and Si in order to resist rapid corrosive degradation. The selection of coating process is important in respect not only of ensuring adequate corrosion resistance, but of being capable of application to large components both in initial plant construction and during periods of repair. The air plasma spraying process is capable of meeting the commercial requirements. The goal of this work was to evaluate coatings applied to three alloys using a commercial powder with a chemical composition (50Ni–49.5Cr–0.5Si) capable of providing corrosion resistance similar to that of highly alloyed materials. Since coating life is a function of both the rate of wastage due to corrosion and the effects of interaction with the substrate alloy, both corrosion behaviour and coating/alloy interactions were evaluated. A second, Al/Si-rich coating applied by VPS followed by a diffusion heat treatment, was used for comparison purposes.

EXPERIMENTAL

Materials and Sample Preparation

The alloys selected (Table 1) were HCM 12, a commercial alloy manufactured by Sumitomo, Japan, and approved in ASME codes for power plant applications and commonly used at 350°C, a model Fe–12%Cr–2.65%Si alloy (manufactured by INCO Alloys, UK) whose composition was intended to provide better corrosion resistance than that of HCM 12 and Sanicro 28, another commercial alloy which has shown acceptable behaviour in certain heat exchanger applications in coal gasification plants. Two reasons are given for coating three materials: (i) to evaluate the coatings with respect to both corrosion resistance and interaction with the base alloy and (ii) in case of failure of the coating, the subsequent corrosion behaviour of the three different alloys would be studied.

Test pieces had a diameter of 6 mm, length 40 mm and domed ends (3 mm radius). Prior to coating all test pieces were shot blasted and ultrasonically degreased in ethanol. Two coatings were selected: (i) onto all alloys, 50/50 NiCr (actual composition Ni–49.3Cr–1.65Si–0.49Mn) from a -75+38 μm powder, air plasma sprayed to a thickness of 250 μm and surface finished by centre-less grinding to 150 μm, Fig. 1(a). Its average porosity content was 2.8%, while the oxides represented 9.1% of the coating volume; (ii) onto Sanicro 28, Al–12%Si vacuum plasma sprayed followed by a diffusion treatment of 12 hours at 900°C, 30 minutes at 1100°C, followed by rapid cooling. The purpose of this thermal treatment was to favour the interdiffusion of the aluminium into the base alloy. The resulting coating thickness ranged from 80–140 μm, Fig. 2(a).

Table 1 Chemical analysis of the coated alloys (wt%).

Alloy	Fe	Cr	Si	Ti	Mn	Mo	Ni	S	C
Model	Bal.	12.2	2.65	<0.01	1.24			0.004	0.13
HCM 12	Bal.	12.25	0.2		0.48	1		0.001	0.1
Sanicro 28	Bal.	27	0.7		2	3.5	31	0.015	0.02

Both coatings were exposed at 1000°C for 1000 hours in argon. The purpose of this test was to study the effect of temperature alone on the coating–substrate interactions. The test resulted in interdiffusion between the base alloy and the coating, as well as a microstructure coarsening in the case of the VPS coating, Fig. 2(b). Although argon provided an inert atmosphere to the reaction chamber, the remaining air and humidity

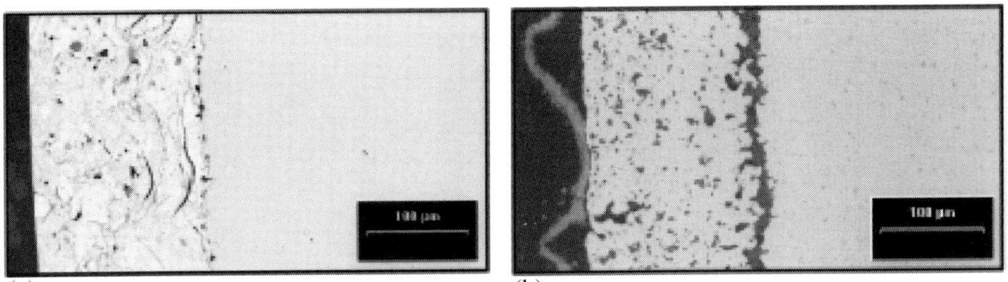

(a) (b)

Fig. 1 SEM micrographs of the model alloy APS coated (a) model alloy APS coated and subsequently surface ground, (b) APS coated model alloy after 1000 h in Ar at 1000°C.

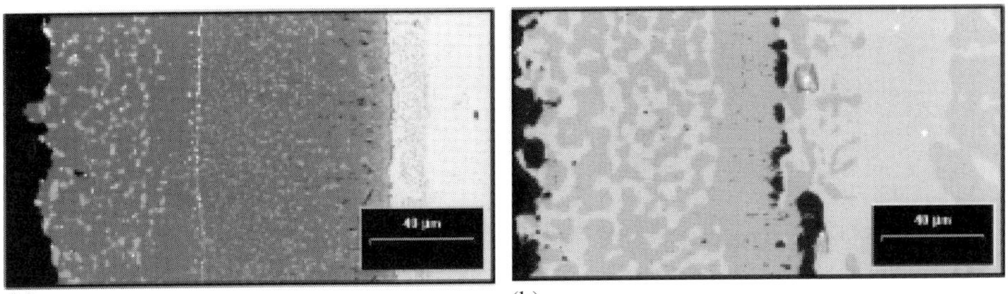

(a) (b)

Fig. 2 SEM micrographs of VPS coated Sanicro 28 (a) VPS coated Sanicro 28 after initial diffusion treatment, (b) VPS coated Sanicro 28 after 1000 h in Ar at 1000°C.

in the ceramic tubes in the rig might have been responsible for the small amount of oxidation observed on the coatings. Oxidation also took place at the VPS coating–substrate interface, where coarse alumina grains developed. APS coatings also suffered oxidation of the bulk coating and formed an outer oxide that spalled on cooling at the end of the 1000 hour exposure. Within the coating, where the oxygen partial pressure is probably several orders of magnitude lower, silicon has segregated to form silicon oxide as a secondary phase. In addition, an inner silicon layer located between the alloy and the coating was formed. Due to the low silicon content in the coating, it is more likely that the silicon forming the SiO_2 layer at the coating–2.65%Si Model alloy interface originates in the substrate. EDX analysis showed that iron had readily diffused into the coating while nickel from the coating had diffused into the alloy, to a depth of at least 87 μm from the original alloy–coating interface. The exposure to a higher temperature was intended to give an indication of potential changes in the coatings occurring over a much longer exposure at 600°C.

Test Environment

Materials were exposed to an environment typical of a dry-feed entrained slagging gasifier. The CO-rich gas produced here is typically at equilibrium at 1400°C, but when it leaves the gasifier it comes into contact with heat exchanger tubes at a much

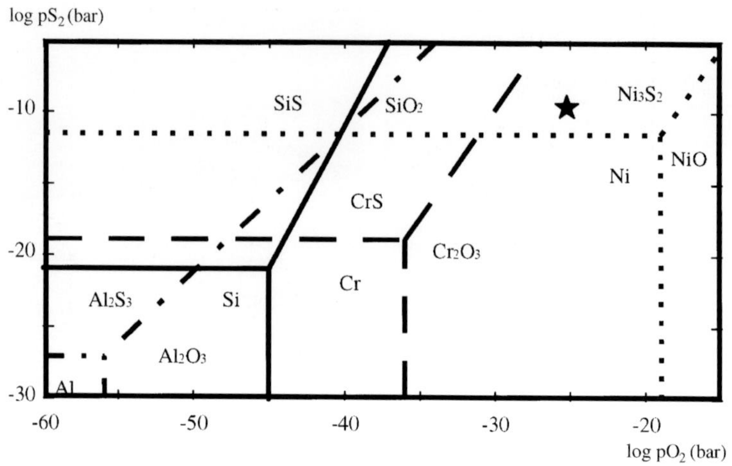

Fig. 3 Phase stability diagram for the elements forming the coatings, at 600°C.

lower temperature (250°C approximately), resulting in an intermediate temperature at the tubes surface of about 600°C. This rapid quenching results in the gas no longer being in thermodynamic equilibrium. To reach these conditions in the laboratory, a gas mixture with a composition 64%CO–3.8%CO_2–0.8%H_2S–H_2 balance was selected. Isothermal tests were conducted at a temperature of 600°C and a pressure of 1.5 bar. A high flow rate of 30 to 35 litres per hour (residence time in the furnace hot zone was one minute) was used in order that the gas mixture did not reach thermodynamic equilibrium at 600°C. The position of the gas relative to stable oxide/sulphide phases is shown on the phase stability diagram for the elements forming the coatings, Fig. 3. The apparent oxygen and sulphur partial pressures of the non-equilibrium gas were calculated following a procedure suggested by Perkins.[1]

Equipment and Experimental Technique

Test pieces were measured, ultrasonically degreased in acetone, and weighed; then placed in individual alumina crucibles and loaded into an alumina reaction chamber of a vertical, top-loading autoclave, described elsewhere,[2] at room temperature. After ensuring the correct gas composition and flow rate, the rig was heated up to 600°C, at an average rate of approximately 7°C per minute. The recorded time for the test was started when the required temperature was reached. Testing was interrupted by switching off the furnace and cooling (0.8°C per minute) in stagnant gas, after accumulated periods of 50, 100, 250, 500, 750 and 1000 hours for mass-change measurements. Metal loss measurements were made after times of 100, 500 and 1000 hours. Two measuring techniques were followed for the latter, Fig. 4: one referred to as 'line', where the measurement was taken directly on a line fixed in the microscope's screen, and the other referred to as 'band' where only the highest value inside a 40 μm wide

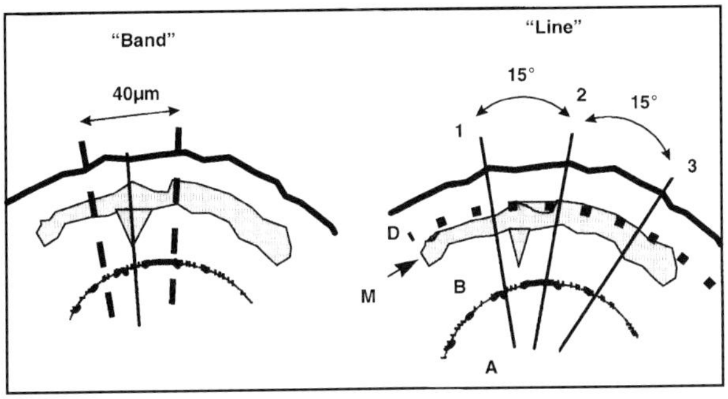

Fig. 4 Schematic representation of the measuring methods followed when recording metal loss data. A – alloy; B – remaining sound coating; M – inner corrosion scale, metal loss; D – outer scale.

band was recorded. Measurements were repeated at 15° intervals around transverse sections of the cylindrical test pieces. Three cross-sections enabled 72 individual measurements per original test piece to be made. Detailed surface and cross-sectional structural analysis using XRD, SEM/EDX, EPMA and optical microscopy were also carried out on the test pieces exposed for 100, 500 and 1000 hours.

RESULTS

Corrosion Kinetics

Weight change measurements indicate a substantial reduction in corrosion rate when the materials are coated, with the VPS Al/Si coating generally performing better than the plasma sprayed, Ni/Cr/Si coating, Fig. 5. Sanicro 28 experienced a weight gain of 21.8 $mg{\cdot}cm^{-2}$ after 1000 hours, but gains of only 8.4 $mg{\cdot}cm^{-2}$ when APS coated, and 2.3 $mg{\cdot}cm^{-2}$ when VPS coated. The model alloy gained 41.2 $mg{\cdot}cm^{-2}$ after 1000 hours, but only 9.1 $mg{\cdot}cm^{-2}$ when APS coated. The corrosion rates of all APS coatings are comparable to more highly alloyed materials (e.g. MA956) sometimes considered for these applications. In addition, parabolic rates were followed by the coated materials with two clear stages: an initial stage where corrosion attack occurs quickly, and a second, steady stage in which a diffusion barrier has developed and protects the material from severe corrosion. Very good reproducibility of weight change data was observed.

Structural Analysis

Surface examination

Using XRD analysis a very intensive peak of Ni_3S_2 was detected and also significant amounts of $NiCr_2S_4$ and Cr_2S_3 on the surface of the 50/SONiCr coatings after exposure for 1000 hours to the mixed gas. Cr_2O_3 was also detected after removing the outermost corrosion products.

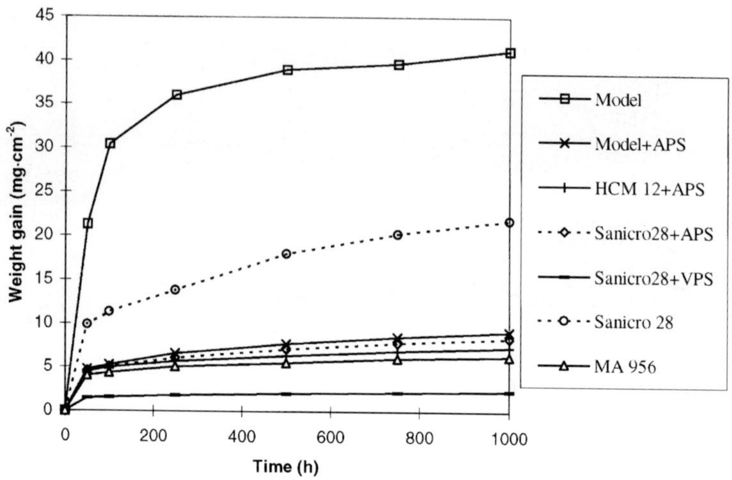

Fig. 5 Corrosion kinetics for some alloys and coatings after 1000 hours in the test environment.

SEM analysis shows that the APS coating surface was covered by a porous layer which was composed mainly of large grains, Fig. 6. Two main compositions were determined by EDX analysis: grains rich in chromium with an elongated structure, and grains rich in nickel showing a more irregular, sometimes spongy, porous structure.

After 100 hours, the Al/Si VPS coating was covered by two types of structure, Fig. 7:

(i) A non-homogeneous, porous layer formed by a network of aluminium oxide in which chromium sulphides are included and particles rich in silicon. On top of this layer appear some nodules of iron-rich sulphides that are a consequence of the relatively easy outwards diffusion of iron.

(ii) A homogeneous, very porous layer formed mainly from aluminium, with some Cr, Fe, Si, O, and very small amounts of sulphur and nickel also present, mainly indicating the unmodified coating surface.

After 1000 hours a completely different surface had developed, Fig. 8. Iron has diffused outwards forming an external FeS layer which completely covers the coating surface.

Cross-sectional examinations

Upon exposure in the active gas, APS coatings form a complex corrosion product. This consists of a double layered sulphide scale on the metal surface and penetration of sulphur, forming sulphide pegs within the metal, Fig. 9. The double layer consists of (i) an outer, porous scale composed of a mixture of nickel and chromium sulphides;

Fig. 6 Scanning electron micrographs of the surface of 50/50 NiCr APS coating 1000 hours exposure to the gas environment.

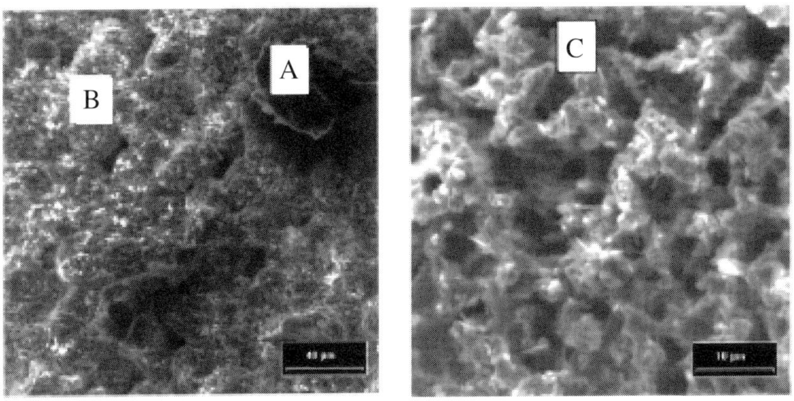

Fig. 7 Scanning electron micrographs of the surface of Al/Si VPS coating after 100 hours of exposure to the gas environment.

Fig. 8 Scanning electron micrographs of the surface of Al/Si VPS coating after 1000 hours of exposure to the gas environment.

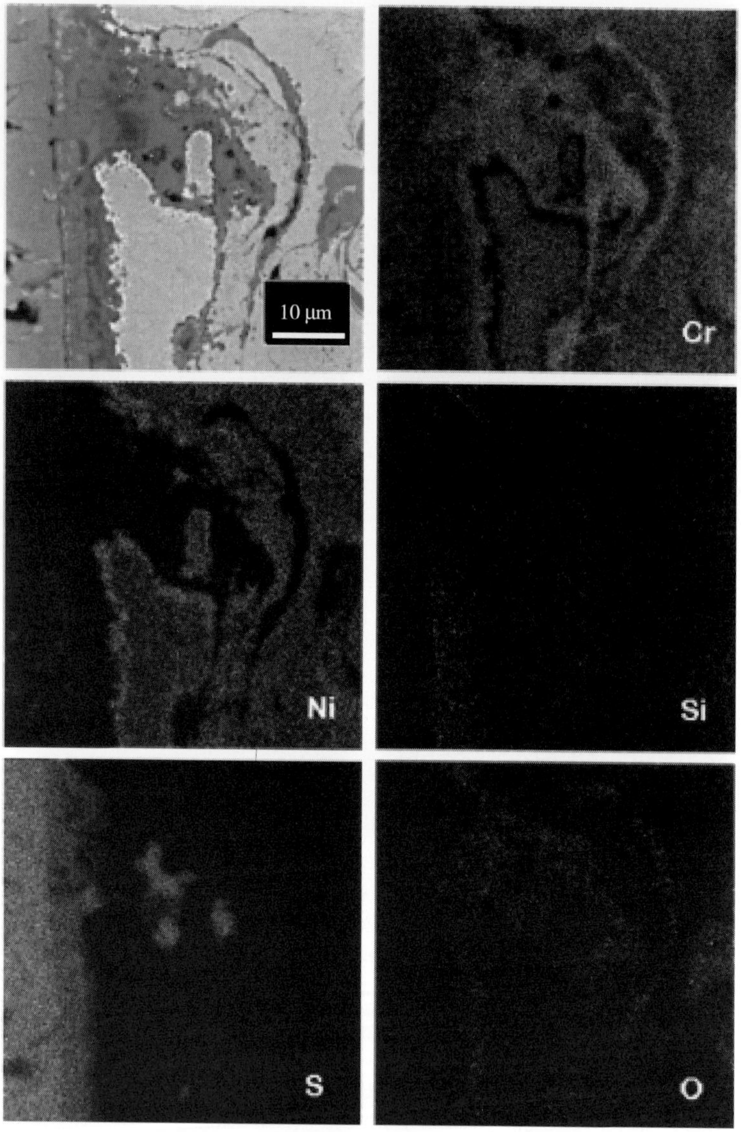

Fig. 9 SEM X-ray mapping on 50/50 NiCr APS after exposure to the gas environment for 100 hours.

(ii) an inner, compact layer where silicon concentrates. The latter can be further divided into two parts: an outermost part rich in sulphur and chromium, and an innermost part rich in chromium and oxygen. In the metal immediately beneath the corrosion product was a chromium depleted zone approximately 1–2 μm wide as shown in the X-ray map.

Non-uniform corrosion attack resulted in the formation of pegs, containing mainly chromium, sulphur and some oxygen. They contained a very small concentration of nickel, similar to that for the chromium-rich oxides in the bulk coating. With time of exposure the sulphur penetrated deeper into the coating. Sulphur was detected at a depth of 85 μm from the original coating surface after 1000 hours of exposure.

Al/Si VPS coatings form a single but multi-component corrosion layer, consisting mainly of an Al–Cr-rich sulphide in the region near to the scale–gas interface, with Al-rich oxide forming a particulate network adjacent to the original coating surface, Fig. 10. Metallic particles, rich in silicon and molybdenum, were distributed throughout the scale. For exposure times less than 500 hours, the corrosion products exhibited very poor adherence to the metal substrate and as a result detailed examination of morphology could only be carried out after long test times. The base elements in the Sanicro 28 alloy, iron, nickel and chromium, which were present in the coating due to the diffusion heat treatment participated to only a very limited extent in surface corrosion product formation. The overall attack was uniform and no sulphur in the coating was observed.

Metal Loss

For section loss the following measurements were made:

- Loss of section, given by the maximum depth of corrosion attack.
- Remaining sound coating (i.e. coating unaffected by corrosion).

Probability plots (e.g. Fig. 11) display the distribution of the measurements. Measurements were made using the two different methods, 'line' and 'band'. By using the band method the chance of missing small 'pits' is reduced and consequently the measurement of section loss is greater than when the line method is used, Fig. 11, unless corrosion is very uniform in which case the two data sets in Fig. 11 would overlap. Measurements of the corroded VPS coatings, which were attacked uniformly, resulted in the data obtained by the line and band methods being superimposed. Owing to the poor adherence of the corrosion product on the VPS coatings after less than 500 hours of exposure, the measured remaining sound coating thickness was subtracted from the original coating thickness, in order to estimate the section loss of the coating. Non-uniform attack of the APS coatings resulted in the deviation from a linear distribution at high values of metal loss, Fig. 11. Here, data obtained using the band method indicates a comparatively higher section loss than that obtained using the line measurements.

Table 2 summarises all section loss data. All three APS coatings show the same range of results, regardless of the material coated. If the mean values of the 1000 hours data are simply plotted and fitted to linear rates in the manner often used in many past investigations an average metal loss of 0.169 mm per year results from the extrapolation. This means that a coating of at least 170 μm would be required for a service life of 1 year, and an equivalent of almost 2 mm would be required for a

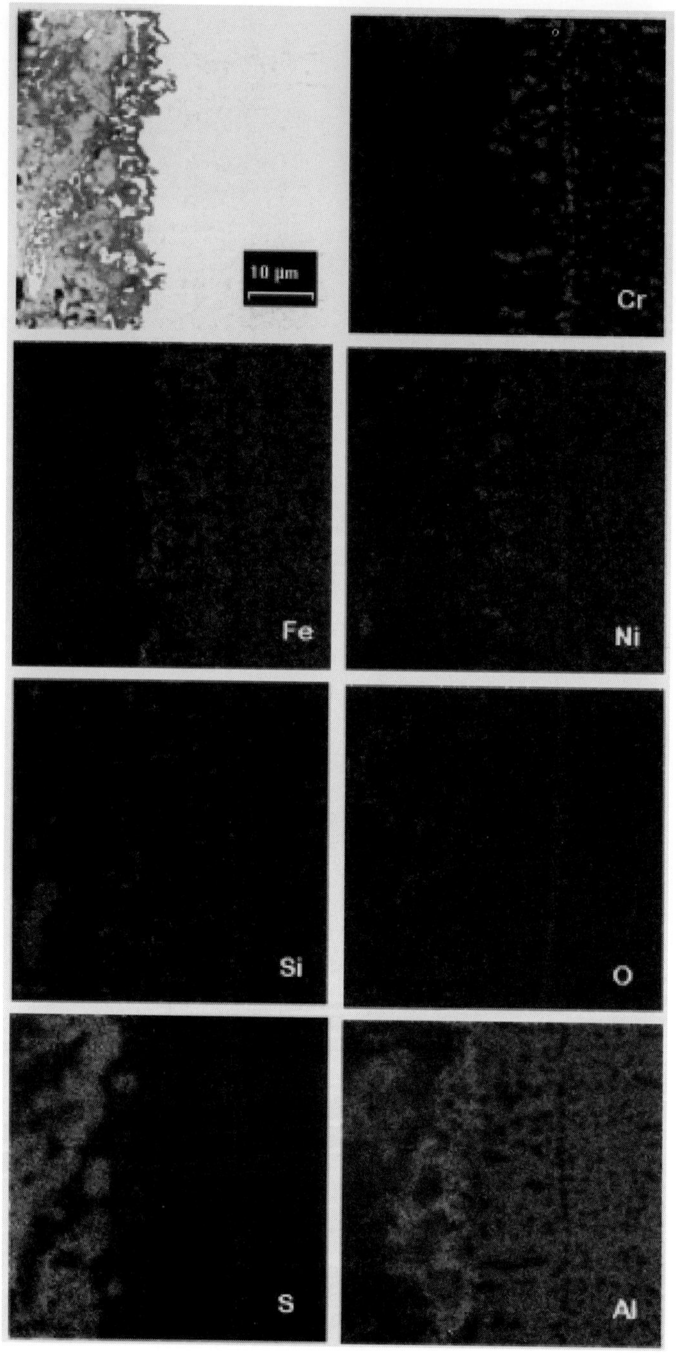

Fig. 10 SEM X-ray mapping on Al/Si VPS after exposure to the gas environment for 1000 hours.

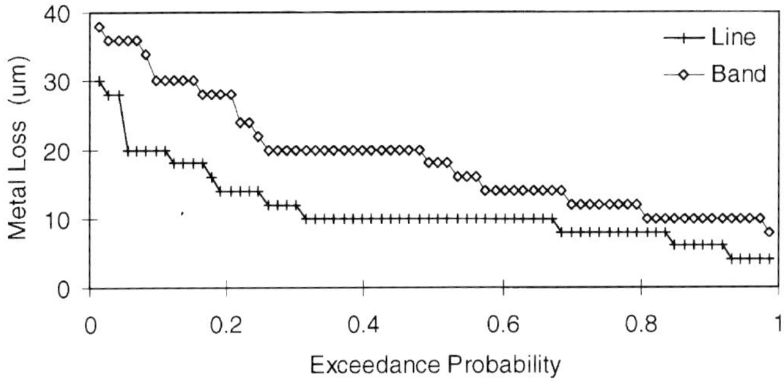

Fig. 11 Exceedance probability plot for the section loss of 50/50 NiCr applied by APS on to the model alloy, after 1000 hours of experiments.

service life of 10 years. Having collected an extensive set of data at different times of exposure it is possible, however, to re-evaluate corrosion kinetics using section loss data in place of the usual weight change data, Fig. 12.

Table 2 Statistic values for the metal loss (mean, maximum and standard deviation) and remaining coating (R.C.) measurements.

		Line			Band			
	t (h)	Mean	St. dev	R. C.	Mean	Max	St. dev	R. C.
Model + APS	100	7.5	30	133	11.8	38	6.6	129
	500	9.2	6.3	137	13.8	48	6.5	133
	1000	19	9.6	126	26	60	12.9	122
HCM 12 + APS	100	7.5	3.4	159.7	11.3	46	5.9	155.8
	500	9.6	4.7	137	12.7	60	5.8	135
	1000	11.3	5.4	131	18.8	52	8	125
Sanicro 28 + APS	100	8.2	5.1	143.7	11.6	44	5.7	142.3
	500	8.7	3.5	123	12.5	42	4.4	119
	1000	14.1	8.2	149	18.2	60	9.4	147
Sanicro 28 + VPS	100	–	23.2	72	–	–	–	–
	500	–	39.5	98.8	–	25	–	–
	1000	–	38.3	98.6	–	35	–	–

Unfortunately, insufficient data points have so far been established in this investigation (section loss at only 100, 500 and 1000 hours) to permit reliable curve fitting to be carried out. Nevertheless, a sound estimate of scatter around each data point is available, which makes estimation of the range of section loss expected to be found for each material possible. Further analysis of the data using extreme value statistical methods, as proposed by Nicholls,[3] would produce additional information on material life assessment.

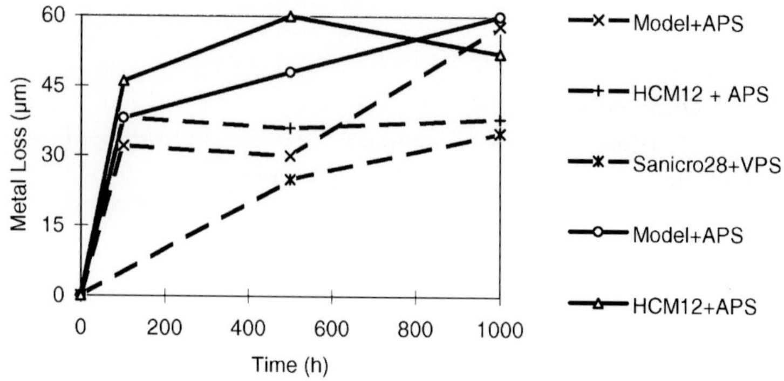

Fig. 12 Maximum metal loss points found after 100, 500 and 1000 hours of test exposure. Dashed lines reproduced values collected from a set of 72 recorded measurements while solid lines show data assumed as absolute maxima.

Conclusions

High temperature exposure modifies the microstructure of both coatings; coarsening and interdiffusion were observed.

Kinetics assessed by weight change had excellent reproducibility and very likely parabolic laws are followed. This implies diffusion controlled mechanisms operate and protection against corrosion is achieved with time. On the other hand, these data alone do not reflect the occurrence of pitting on APS coatings.

Section loss data show a considerable scatter and the absence of more data points (collected at different times of exposure) does not allow reliable curve fitting. Longer term experiments in the future will permit kinetics of metal loss to be better described.

50/50 NiCr APS coatings are as corrosion resistant as high alloy monolitic materials previously tested under the same conditions. Al/Si VPS coatings provide even better protection, although application of coatings by VPS might not be as practical under plant conditions as that of APS coatings.

Oxides formed during the APS process enable sulphur penetration into the coating.

ACKNOWLEDGEMENTS

The authors acknowledge the collaboration of S. Canetoli and P. Puschek of the Corrosion Group and also colleagues within the Structural Analysis Group under the direction of P. Moretto who contributed to the microstructural analysis, i.e. K. Schuster (SEM), P. D. Frampton (optical metallography) and R. Van Nieuwenhuyzen (image processing). This work was carried out within the European Commission's Research and Development Programme.

REFERENCES

1. R. A. PERKINS, G. MORSE and W. C. COONS: *EPRI report AP-2518*, August 1982.

2. J. J. BARNES: *The Influence of Sulphur on the High-Temperature Caburization of Fe-Ni-Cr alloys*, Ph.D. 1983.
3. J. R. NICHOLS and P. HANCOCK: 'Prediction of High Temperature Performance using Statistical Analysis Techniques', *Plant Corrosion,* J. E. Strutt and J. R. Nicholls eds, Ellis Horwood, 1987.

The Role of Defects on the Mechanical Integrity of Alumina Scales on MA 956

V. GUTTMANN, F. HUKELMANN and G. BORCHARDT*
European Commission, Directorate General JRC, Joint Research Centre, Institute for Advanced Materials, Petten, The Netherlands
**Technical University Clausthal, Clausthal-Zellerfeld, Germany*

ABSTRACT

The present work deals with the mechanical integrity of Al_2O_3 scales which grow at high temperatures on the Fe-based ODS-type alloy MA 956. The main objectives concern the assessment of the formation and evaluation of scale defects and their influence on mechanically induced scale failure. The experiments were carried out at 1100°C for a maximum exposure time of 5200 h under isothermal creep strain.

The results revealed the formation of a continuous well adherent scale of principally Al_2O_3 which obeyed sub-parabolic growth kinetics. Defects appearing in the scale include compositional defects as Y(Al)- and Ti-oxides and physical defects in the form of pores of different sizes. All defects became more pronounced for longer exposure times.

Scale failure was triggered by chain-like pore arrangements which caused intergranular scale fracture. An average value of $K_{IC} = 0.78$ MPa $m^{1/2}$ was calculated on the basis of composite defects which resulted from interactions between single defects in close proximity.

INTRODUCTION

The oxidation/corrosion resistance of service relevant materials strongly influences their applicability at elevated temperatures. Protection is generally achieved by the development and maintenance of appropriate scales such as Cr_2O_3 or Al_2O_3. An important aspect in terms of protection concerns the mechanical integrity of scales, i.e. their resistance against initiation and growth of cracks under thermal and mechanical stresses. As crack sources, inherent defects such as pores and microcracks play a crucial role. Therefore attention has to be paid to the generation and evolution of these defects and their specific influence on failure mechanisms.

The present work deals with the mechanical stability of the Al_2O_3 scale forming on the Fe-based ODS alloy MA 956 at high temperature under simultaneous alloy deformation. Emphasis was devoted to the characterisation of scale inherent defects and their relevance on scale failure.

EXPERIMENTAL

The alloy with a nominal composition shown in Table 1 was delivered by INCO Alloys Ltd. (Hereford, UK) as 2 mm thick sheet, from which creep samples with a gauge length of 45 mm and a width of 6 mm and reference samples of 10 x 15 mm were machined by spark erosion. Surfaces were treated by grinding the samples with 800 grit SiC paper and subsequently degreasing them with ethanol in an ultrasonic bath.

Testing was carried out at 1100°C in laboratory air for a maximum of 5200 h. In situ deformation was applied in the form of tensile creep strain. All samples were tested in single specimen creep rigs which were equipped with a sensitive strain measurement device. Reference samples which served to establish the oxidation behaviour in the stress-free condition were generally suspended around the creep specimens by means of Pt-wires. Heating up to test temperature was achieved in about 3 h using an automatic temperature control (6°C min^{-1}). Specimens were cooled in the furnace at the end of a test. Scale growth was measured by mass gain and by thickness measurements in the scanning electron microscope (SEM).

Table 1 Composition of MA 956 (in mass %)

Fe	Cr	Al	Ti	Y_2O_3
Bal.	20	4.5	0.5	0.5

RESULTS

The oxidation kinetics which are illustrated by the scale thickness time relation in Fig. 1 obeyed the common oxidation law:

$$S^n = kt \tag{1}$$

where S is the scale thickness, k the rate constant, t the exposure time and n the thickness exponent. The scale growth followed subparabolic kinetics for both non-stressed and stressed samples with $n \approx 3$ for the whole range of exposure. As was to be expected from the development of a rather compact scale the same kinetics were observed for the mass gain.

Microstructural studies revealed the formation of a continuous flat scale which was principally composed of Al_2O_3. Thin scales of up to about 8–10 µm, corresponding to about 1000 h exposure, exhibited a columnar grain structure with grains about 4–10 µm long and a width of 1.5–2.5 µm. For thicker scales variations in the grain shape appeared for the region beneath the layer of columnar grains. Apart from columnar grains small equiaxed grains with a diameter of about 0.5–1.5 µm were observed.

As illustrated in Fig. 2 the scale exhibited numerous compositional and physical defects which became more accentuated for longer exposure times, i.e. increasing scale thickness. According to the analysis given in Fig. 3, the compositional defects

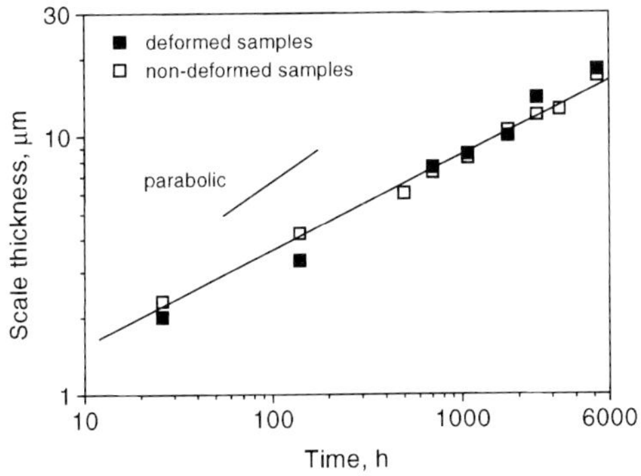

Fig. 1 Scale thickness versus time for MA 956 at 1100°C.

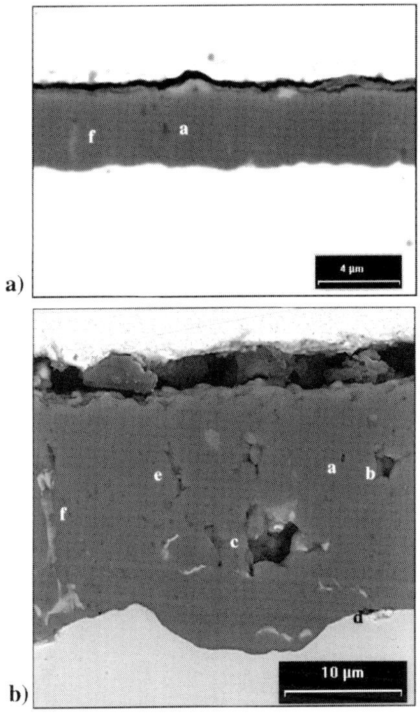

Fig. 2 Typical defects in the Al_2O_3-scale on MA 956, (a) 130h/1100°C, (b) 5200h/1100°C, a = micropore, b = midsize-pore, c = macropore, d = interfacial pore, e = alignment of defects and f = Y(Al)/Ti-oxides.

consisted of Y(Al)- and Ti-oxides (see also bright contrast in Fig. 2b) which corresponds with earlier observations.[1,2] The Y(Al)-oxides for which a composition of $Y_3Al_5O_{12}$ has been reported[1] appeared as thin figures, about 0.2–0.4 µm thick. In general, these figures were at one end in contact with the alloy-scale interface, in some

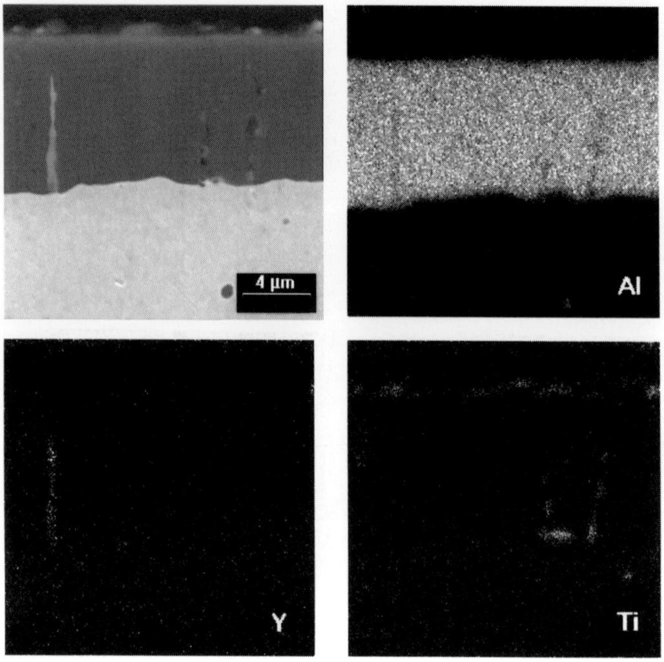

Fig. 3 Element distribution in the Al_2O_3-scale on MA 956, 1100h/1100°C.

cases almost traversing the entire scale (Fig. 3). The Ti-oxides appeared in the form of small inclusions up to about 2 µm in size but in general much smaller. Frequently they were found to be associated with pores by partially filling them (Fig. 3). As illustrated in Fig. 4 both Y(Al)- and Ti-oxides also appeared as nodules on top of the scale as also widely reported.[1–5] After longer exposure times some Fe- and Cr-oxides were also detected on the outer surface. Figure 4 illustrates that the largest nodules were Ti- and Fe-rich, whereas the medium size particles contained in particular Y.

The physical defects of the scale could be classified as macro-, midsize- and micropores according to their size. An overview of the time dependent evolution of their maximum diameter is presented in Fig. 5. It can be seen that macropores appeared solely after about 2000 h exposure corresponding to a scale thickness of about 10 µm. Moreover, they emerged only in the middle or lower half of the scale. For relatively thin scales, midsize-pores were also rarely observed whereas micropores could be detected for all scales independent on the exposure conditions. A few pores up to about 8 µm for long term exposure, Fig. 6, but generally smaller than 2 µm, were locally observed at the scale-alloy interface (Fig. 2 and 8). Differences existed for the pore shape. Micropores were always round shaped whereas midsize- and macropores often revealed irregular geometries. The distribution of physical defects in the scale was rather non-uniform, i.e. local concentrations existed but on the other hand large areas of the scale remained almost defect-free, apart from micropores which were found to be rather uniformly distributed.

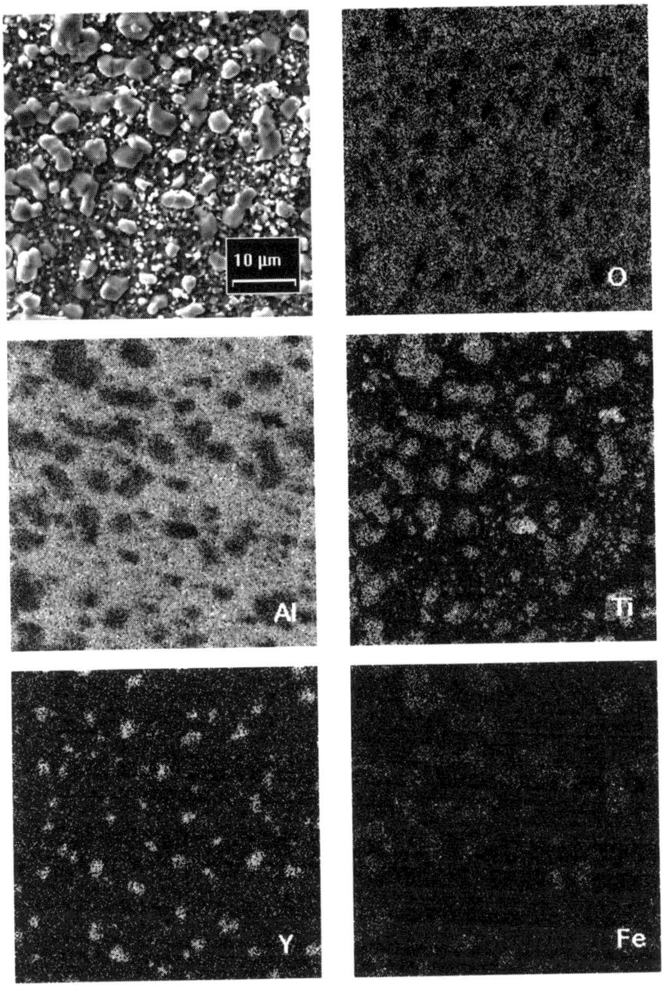

Fig. 4 Element distribution on the top of the Al_2O_3-scale on MA 956, 1100h/1100°C.

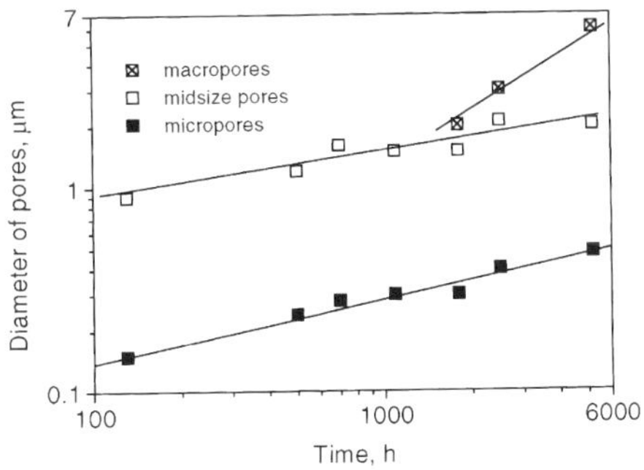

Fig. 5 Maximum size of pores versus time in the Al_2O_3-scale on MA 956 at 1100°C.

Fig. 6 Fracture surface of the Al_2O_3-scale on MA 956. Large defect at the scale-alloy interface 5200 h/1100°C.

Some details on the physical defects can be taken from Figs 7–13 for a sample which has been subjected to 5200 h exposure. The most obvious defects concern the macropores which might be even denoted as holes rather than pores. As illustrated in Fig. 7 the macropores were generally surrounded by relatively small grains, about 0.5–1.5 µm in size (for the grain shape see also Fig. 12). Frequently, Ti-oxide particles could be observed in combination with larger pores Fig. 8 (left side, bright particles). In addition, Fig. 8 (lower left corner) displays Ti-oxides at the scale–alloy interfaces which frequently were detected. A specific type of physical defect concerns chain-like pore arrangements which consisted of micro- and midsize-pores, Fig. 8 (right side). The length of these chains extended with exposure times according to Fig. 9.

Evidence was found for scale growth perturbations. Figure 10 illustrates that for instance at the end of a set of columnar grains scale growth occurred by the formation of equiaxed small grains. It can be seen that not all of the new grains were in direct contact with the preceding columnar grains (Note: the scale was growing inwards). In this way particularly midsize- pores were created which were frequently found at grain boundary triple points. Concentrations of midsize-pores together with micropores and clusters of small grains composed the macropores. Figure 11 demonstrates the initial stage of a macropore whereas in Fig. 12 a very strongly pronounced macropore can be seen. The latter seems to be generated at the lower part of a defect rich channel, which grew along a substrate grain boundary.

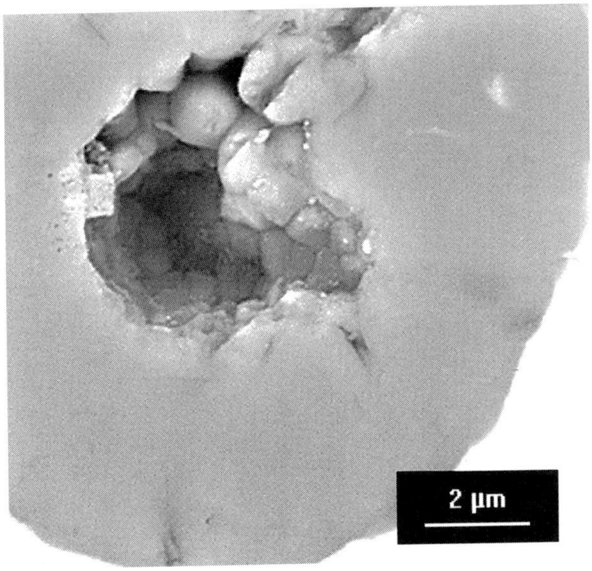

Fig. 7 Fine-grain microstructure around a macropore in the Al_2O_3-scale on MA 956, 1100°C/5200 h.

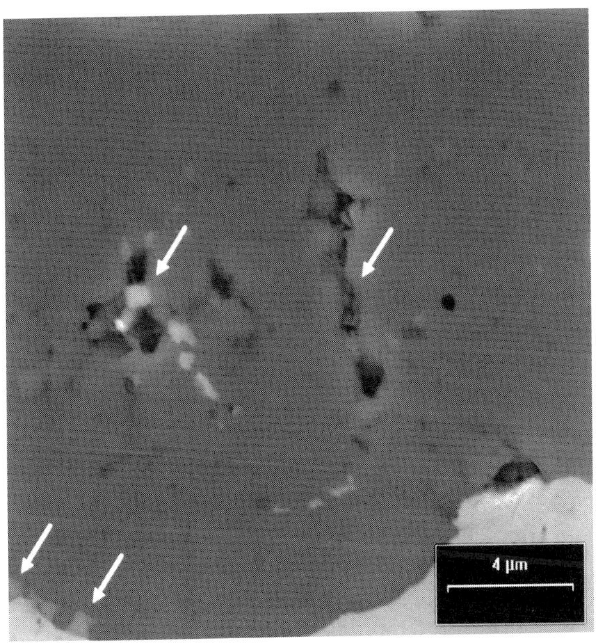

Fig. 8 Chain-like arranged midsize-pores in the Al_2O_3-scale on MA 956. The left part of the micrograph shows Ti-oxides linked to a macropore, respectively located at the scale-alloy interface (lower left corner), 5200 h/1100°C.

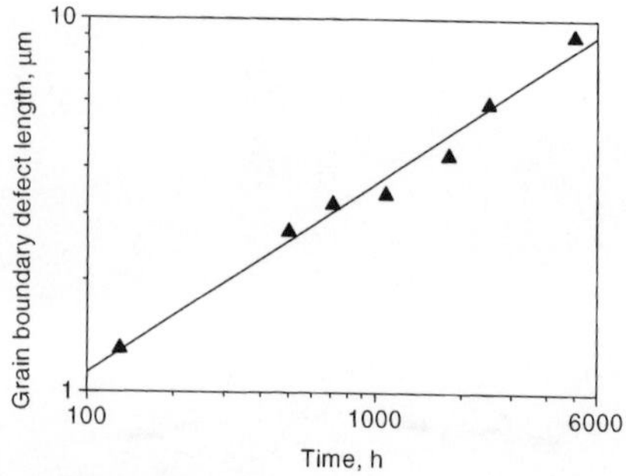

Fig. 9 Grain boundary defect length versus time in the Al_2O_3-scale on MA 956 at 1100°C.

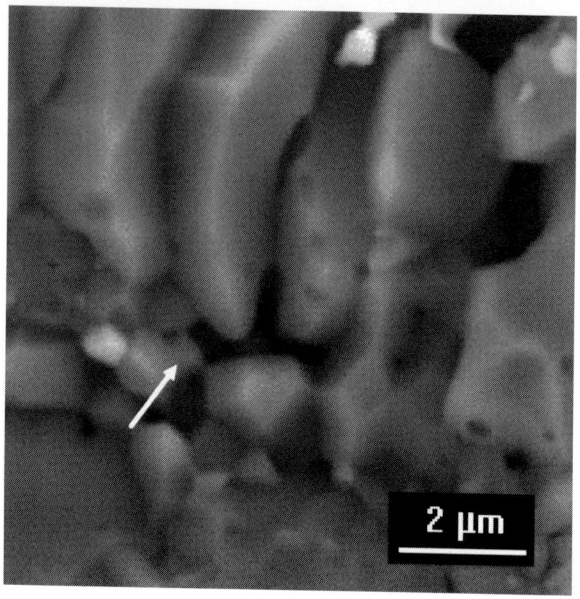

Fig. 10 Growth defects in the Al_2O_3-scale on MA 956, 1100°C/5200h.

As illustrated in Fig. 13(a), micropores typically appeared on scale grain boundaries and could be observed for all exposure times. A long chain of micro and midsize pores along columnar grain boundaries is shown in Fig 13(b), to be compared with that displayed in Fig. 8.

Mechanical deformation of the scale revealed typical equidistant cracking as the characteristic damage mechanism whereas scale spallation only took place very locally, if at all. Figure 14 shows that in the severely deformed necking area of a creep sample in spite of multiple cracking the scale adherence was still rather acceptable, Fig 14(a). From a larger magnification, Fig 14(b), it can be seen that strong deformation of the substrate surface took place at the bottom of a crack, but the neighbouring

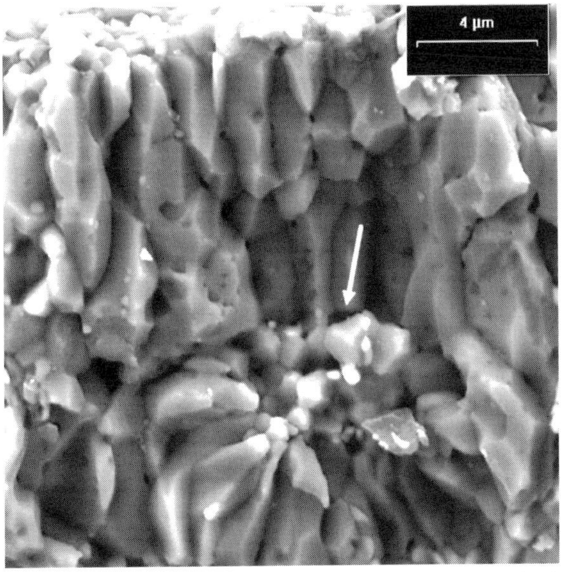

Fig. 11 Fracture surface of the Al_2O_3-scale on MA 956. Cluster of midsize-pores and fine grains, 5200 h/1100°C.

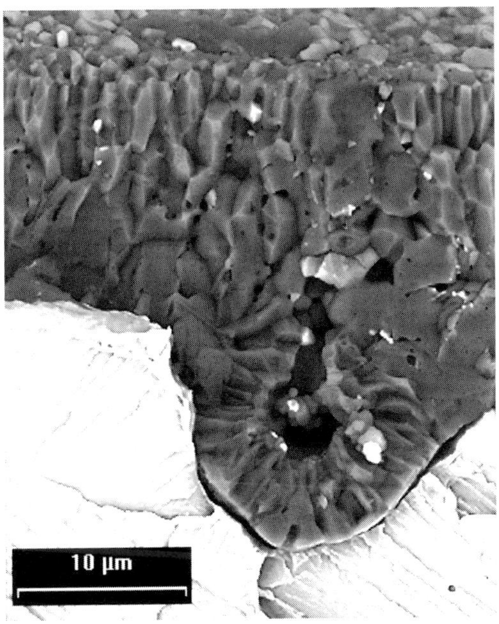

Fig. 12 Fracture surface of the Al_2O_3-scale on MA 956. Macropore and irregular scale growth 5200 h/1100°C.

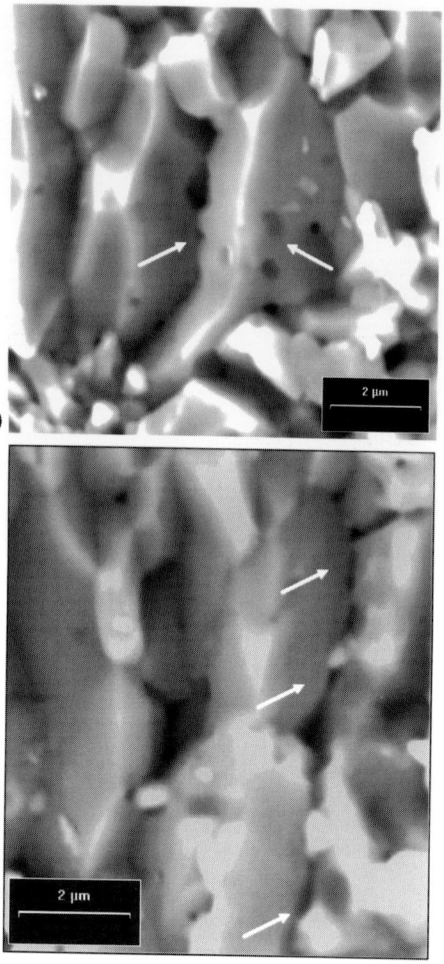

Fig. 13 Fracture surface of the Al_2O_3-scale on MA 956, showing the distribution of small defects, 5200 h/1100°C; (a) micropores on grain boundaries, (b) alignment of midsize- and micropores along grain boundries.

scale sections remained intact and no scale detachment occurred. Moreover, Fig 14(b) reveals the failure mode typically to be observed for relatively thin scales; single cracks propagated through the scale and no parallel cracks or marked crack deflection was observed. In contrast, for thicker scales exceeding about 8–10 µm scale disruption was characterised by the appearance of multiple cracking, Fig. 15. Typically a large number of embedded cracks appeared and in general a concentration of cracks adjacent to through scale cracks was observed. Figure 15(a) illustrates the case of respectively two and three adjacent cracks of which the right one appears to be a through scale crack. The outer parts of the cracks reflect a rather straight orientation which mirrors

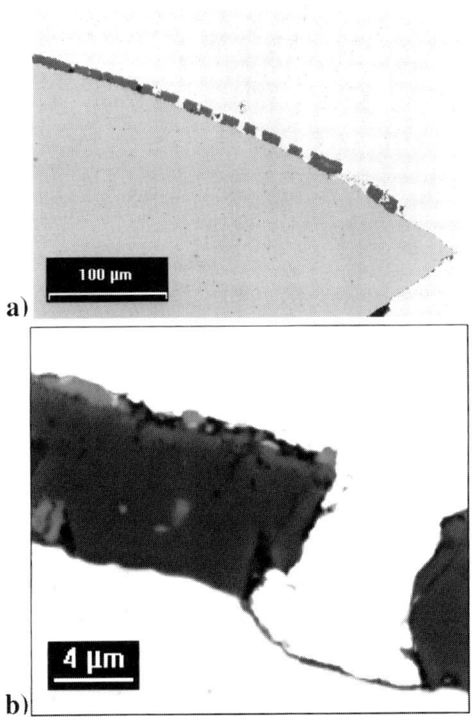

Fig. 14 Cracking of the Al_2O_3-scale in the necking region of a fractured creep sample of MA 956, 1000h/1100°C; (a) demonstration of scale adherence and (b) larger magnification showing a typical single crack.

intergranular fracture along the columnar grains while the irregular shape in the lower half of the scale should be according to the propagation through the area of a mixed grain shape. When through thickness cracks came in contact with the alloy-scale interface new oxides formed which were found to be composed of Al_2O_3. Additional tests on precracked scales have confirmed this restoration ability. Figs 15(a) and (b) allow the conclusion that failure was due to chain-like arranged pores. In contrast the competitive macropores seemed to present barriers for crack propagation as is demonstrated in Fig 15(b) which shows that a crack has by-passed a macropore when propagating through the scale. Figure 15(b) also reveals initiation of scale decohesion along the scale-alloy interface but no intensive scale spallation occurred.

The low critical strain to scale failure of 0.1–0.2% independent on scale thickness implies a first indication of a brittle scale behaviour. The latter was confirmed by microstructural studies which revealed brittle intergranular rupture. Such a material response allows the description of the scale properties in terms of fracture toughness. It has been proposed to treat scale rupture by inherent defects like the situation to be encountered in technical components as for instance welds containing a population of

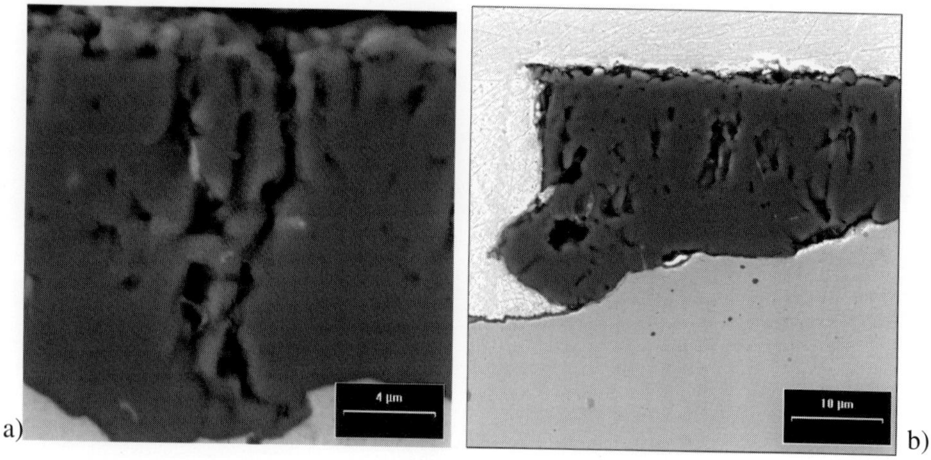

Fig. 15 Typical cracking features of the Al_2O_3-scale on deformed MA 956 after long term exposure, 5200h/1100°C; (a) multiple cracking and formation of new Al_2O_3 at the scale-alloy interface and (b) crack passing around a macropore.

flaws.[6–8] In this case, calculations of the critical stress intensity K_{IC} require to assess composite defects by combining the individual defects under the aspect of possible stress interactions.[9,10] Since oxide scales originally exhibit embedded defects, the following general equation holds:

$$K_I = \sigma_{ap}\sqrt{\pi \times a} \times M / \sqrt{Q} \tag{2}$$

where K_I is the stress intensity, σ_{ap} the applied stress, a the half of the crack depth, M the membrane stress correction factor and Q the defect shape parameter. While M characterises the eccentricity of the defect in combination with its size compared to the scale thickness, Q depends on the defect shape (aspect ratio) and the applied stress/yield stress ratio.

The stress intensity factor of the Al_2O_3 scale investigated has been determined on the basis of the two largest defects, namely the pore chains and the macropores. Figure 16 presents a schematic view of different defects as being used for the calculations. Chain-like arranged midsize and micropores are thought to exhibit overlapping stress fields. For the macropores short cracks are assumed to emanate from a hole.[11,12]

Based on these geometries and on the critical stress σ_c which was calculated from the measured critical strain ε_c by means of Hook's law, the average critical stress intensity factor for linear defects based on samples which suffered different exposure times yields:

$$K_{IC} = 0.78 \text{ MPa m}^{1/2}$$

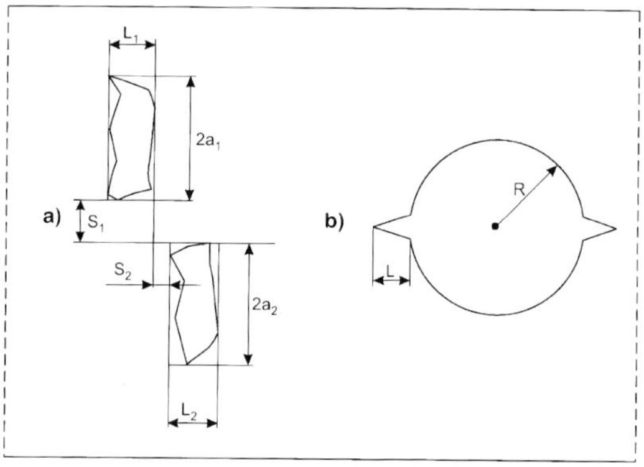

Fig. 16 Schematised defects in the scale; (a) aligned pores; for interaction, (i) $S_1 \leq 2a_1$ or $2a_2$ (whichever is greater) and S_2 as S_1, (ii) $2a_1 + 2a_2 + S_1 = 2a = l$ (2a and l form the sides of the composite defect) and (b) large single pore with circumferential cracks.

where $M = 1.13$, $1/(Q^{1/2}) = 0.66$ and the Young modulus $E_{1100°C} = 324800$ N/mm^2.[12] As the lowest value $K_{IC} = 0.38$ MPa m$^{1/2}$ has been calculated.

For the macropores one gets an average stress intensity factor of:

$$K_I = 0.72 \text{ MPa m}^{1/2}$$

This value is based on:[14]

$$K_I = F(L/R)\sigma\sqrt{\pi L} \tag{3}$$

L and R are given in Fig. 16. Measured R values are presented in Fig. 5. L was taken to be about 1 μm according to the grain size around the macropores assuming that cracks exist along the grain boundaries. The correction factor F (L/R) was taken from Ref.14.

DISCUSSION

The present investigations have shown that the well known excellent high temperature oxidation resistance of MA 956[1,2,4,5,15,16] holds also for exposure under superimposed creep. A compact scale of basically Al_2O_3 was observed for all testing conditions employed and scale restoration after cracking occurred by the formation of a new Al_2O_3-type oxide. The scale growth followed subparabolic kinetics which were identical with those reported earlier.[17] Parabolic kinetics which have also been found for the oxidation of MA 956[5,18–20] might be particularly related to higher temperatures. Subparabolic kinetics have been explained by the observation that at the beginning of exposure preferentially small equiaxed grains formed followed by columnar grains. Consequently the number of grain boundaries in the scale acting as diffusion paths was thought to be reduced and therefore oxidation could slow down.[21] However, in the present case, the opposite was observed, i.e. columnar grains dominated at the beginning of exposure, whereas at a later stage fine grains also appeared. The most realistic explanation for a subparabolic rate law is probably based on the stress state the scale.[30] Due to compressive stresses resulting from the volume increase during oxidation, the vacancy concentration will be reduced at the scale–alloy interface. This should impose a change in the random migration of the anion vacancies through the scale leading to a reduced inward flux of oxygen ions. Assuming that the compressive stresses at the scale–alloy interface increase with time,[30] the retardation effect will become enhanced and the oxidation rate will tend to be subparabolic. A further explanation might be that diffusion along the scale grain boundaries becomes gradually retarded due to continuous accumulation of the various species which were found to diffuse towards the top of the scale.

Evidence for inward scale growth typically to be observed when reactive elements as Y are available[1,4,17,22] could be obtained easily from the position of columnar grain areas in the scale. These appeared in the whole scale after short term testing while for long term exposure they were observed as a continuous band only in the outer part of the scale. Inward scale growth was also indicated by the observation that beneath the macropores scale intrusions penetrated into the substrate, i.e. enhanced oxygen inward diffusion along the macropore areas must have occurred.

Concerning the generation of the various defects in the Al_2O_3 scale compositional defects obviously related to an outward diffusion of cations and inward moving oxygen which was confirmed by the appearance of the oxides of Y(Al) and Ti on top of the scale. Oxide formation inside the scale by such a counterflux of cations and anions has been discussed in detail in the literature.[23,24]

With respect to scale failure, Y(Al)-oxides did not cause any detrimental effect. It could even be observed that when numerous cracks appeared, indicating severe scale deformation, the grain boundaries exhibiting Y(Al)-oxides did not suffer crack formation. Thus particularly good adhesion obviously existed along the Y(Al) oxide/Al_2O_3 interfaces.

Frequently, Ti-oxides were combined with relatively large pores. However, very probably, these pores were not induced by the Ti-oxides but they were already existing at an early stage and Ti was attracted to them and became oxidised. Ti-oxides may even have improved scale integrity for instance when pores which emerged at the scale-alloy interface were completely filled with Ti-oxides. Such a process of pore closure might contribute to the beneficial effect which was reported for small amounts of Ti on scale adherence for MA 956.[17]

Physical defects obviously resulted from scale growth perturbations particularly as far as midsize- and macropores are concerned. Especially the midsize pores were obviously generated at the alloy–oxide interface and became incorporated into the inward moving scale. Possible mechanisms have been proposed by various authors.[25,26] Crystallographic perturbations caused the formation of midsize pores together with areas of small grains. In combination with the micropores this finally resulted in a less compact scale morphology which could cause local initiation of macropores. Their final size was partially due to the pull out of small grains from such areas of bad cohesion which could happen during sample preparation. This means that the large pores to a certain extent reflect defects which originally did not exist as pores but as regions of loosely packed small grains. This feature, however, does not influence the calculation on scale integrity made on the defect size.

After long term exposure a few relatively large pores appeared, which seemed to enable the scale to progress locally rather rapidly (see Fig. 12). The scale around these large defects was characterised by fine columnar grains, which often tended to grow with a radial symmetry into the alloy. It can be assumed that these defects also originated from relatively large areas of a fine-grain noncompact microstructure. According to Fig. 12 such areas probably could be created along original alloy grain boundaries. On the other hand, the direct formation of macropores at the scale–alloy interface was also possible (Fig. 6). This can be explained by the fact that small pores will grow as a consequence of the oxide growth stresses in combination with interfacial defects.

Regarding the formation of micropores, it seems that these could also be generated by scale growth perturbations. Possible mechanisms of micropore formation based on the condensation of excess vacancies inside the scale have also been proposed.[18,26,27]

The question arises if oxide growth stresses additionally account for the formation of physical defects. An important feature concerns the fact that not only the newly formed inner parts of the scale revealed a gradual accentuation of physical defects with time but also in the outer part which initially appeared to be rather defect free became filled with mainly micropores. Therefore it seems unlikely that scale growth perturbations only were responsible for the physical defects and probably growth stresses were contributing in the formation of physical defects. In the present case typical indications of growth stresses as scale buckling or even decohesion[29,30] were not observed, which is consistent with earlier observations on MA 956.[15] On the other hand, due to an inward scale growth, internal stresses must be expected.[23,24,31,32] Relaxation of these stresses, for instance by diffusional creep, might have caused some microstructural decohesion in the form of creep damage as small pores and microcracks.

Scale failure was found to be attributed to intergranularly aligned defects. For thin scales only rows of micropores could be detected but their appearance did not come out very clearly. Possibly also the few midsize pores to be observed inside the scale or at the scale-alloy interface were acting as failure sources in thin scales and also crack initiation from the surface could have been possible. It seems surprising that the large macropores which emerged in thicker scales did not induce scale fracture, taking into account that the calculated stress intensity factor related to the macropores, in the frame of scatter, was comparable to that for the aligned defects. The essential reason for macropores to resist cracking seemed to be related to the microstructure. Since macropores were surrounded by relatively small equiaxed or irregularly formed columnar grains, crack propagation was likely to be retarded by crack deflection and crack branching.

No systematic correlation could be established between the critical strain and the scale thickness. Since thicker scales exhibited larger defects according to LEFM scale failure probability should have increased as it was confirmed for instance for Fe- and Ni-oxide scales although severe scatter was observed.[6,33] However, thickness independent failure strains for Al_2O_3 scales grown on Fecralloy with a maximum thickness of about 6.0 μm have been reported as well.[34] In this case the defect size was assumed to be rather constant. The missing correlation between the scale thickness and the critical strain for the present case is thought to result primarily from an experimental scatter based on the very low scale ductlity. A second interpretation can be related to the microstructure. It can be argued that the columnar grain morphology dominating for thin scales should be more prone to cracking than the fine grain one which was partially existing in thicker scales.

The calculated critical stress intensity factor of $K_{IC} = 0.78$ MPa $m^{1/2}$ which implies brittle fracture is comparable with the few literature values on Al_2O_3 scales. For an 18Cr/Al/Ce steel tested at 800°C 0.35–1 MPa $m^{1/2}$ was reported[35] while for the scale grown on Fecralloy at 1100°C but ruptured at *RT*, 0.38 MPa $m^{1/2}$ were found.[34] The associated critical strains were 0.08 and 0.07% respectively compared to 0.1–0.2% for the present work. A theoretical calculation based on the surface fracture energy γ[36] yields 1.37 MPa $m^{1/2}$. However, taking into account the small grain size of the scale and the relatively high test temperature of about 0.6 times the melting temperature, the question arises if creep might have been additionally involved in scale deformation. Based on theoretical calculations the actual deformation mechanism is provided by 'deformation maps'.[37] These indicate for a given material of a certain microstructure for a selected temperature–stress combination the dominating deformation mechanism together with the associated strain rate. Conversely for an imposed strain rate the stress value is supplied which is required for a certain deformation mode to become dominant. Using the present data, the Al_2O_3 deformation map indicates diffusional creep (Coble creep) as a feasible process[37] together with elastic deformation. It should be remembered that such a reflection on the contribution of creep has a purely theoretical background. The experimental conditions did not allow a decision as to whether creep actually took place.

SUMMARY

- The mechanical integrity of the Al_2O_3 scale on MA 956 has been investigated at 1100°C for oxidation times of up to 5000 h.
- Short term exposure led to a very compact, nearly defect free scale.
- With increasing exposure time pronounced chemical defects as Y(Al)- and Ti-oxides and physical faults in the form of pores of varying sizes emerged.
- Scale growth perturbations played the decisive role in the generation of pores. In addition, oxide growth stresses very probably supported the formation of micropores.
- Alignments of micro- and midsize pores became responsible for scale failure which occurred after 0.1–0.2% strain.

– Although brittle scale failure formed the essential damage mode, some contribution of diffusion creep might have occurred according to theoretical considerations based on the Al_2O_3 deformation map.

ACKNOWLEDGEMENT

The authors would like to acknowledge the technical help of Messrs. J. Casado, P. Frampton, K. Jackson, J. Mendes, K. Schuster from the I.A.M. Petten. Thanks are also to Mr. E. Bullock for a critical discussion of the manuscript. This research was carried out within the European Commission's Research and Development Programme.

REFERENCES

1. T. RAMANARAYANAN, R. AYER, R. PETKOVIC-LUTON and D. P. LETA: *Oxid. Met.*, 1988, **29**, 445.
2. W. J. QUADAKKERS, A. ELSCHNER, H. HOLZBRECHER, W. SPEIER and H. NICKEL: *Appl. Surf. Sci.*, 1991, **52**, 271.
3. E. SCHUMANN: *private communication.*
4. B. A. PINT: *Oxid. Met.*, 1996, **45**(1/2), 1.
5. M. J. BENNETT, H. ROMARY and J. B. PRICE: *Heat resistant materials I*, K. Natesan *et al.* eds, Metals Park, OH, ASM, 1991, 95.
6. P. HANCOCK and J. R. NICHOLLS: *Mater. Sci. Technol.*, 1988, **4**, 398.
7. P. HANCOCK and J. R. NICHOLLS: *Mater. High Temp.*, 1994, **12**(2–3), 209.
8. J. P. WILBER, J. R. NICHOLLS and M. J. BENNETT: *Microscopy of Oxidation 3,* The Institute of Materials, UK, 1997, 207–220.
9. Guidance on some methods for the derivation of acceptance levels for defects in fusion welded joints, British Standards Institution, PD 6493,1980.
10. The American Society of Mechanical Engineers ASME Boiler and Pressure Vessel Code, Section Xl, 1995.
11. H. HERTZBERG: *Deformation and Fracture Mechanics of Engineering Materials,* 2nd Edition, Wiley, 1983, 269.
12. R. W. RICE: *J. Mater. Sci.*, 1984, **19**, 895.
13. J. R. NICHOLLS, D. J. HALL and P. F. TORTORELLI: *Mater. High Temp.*, 1994, **12**(2–3), 149
14. O. L. BOWIEJ: *Math. Phys*, 1956, **35**, 60.
15. V. K. TOLPYGO and H. J. GRABKE: *Oxid. Met.*, 1994, **41**(5/6), 343.
16. J. CHAO, J.-L. GONZALEZ-CARRASCO, J. IBANEZ, M. L. ESCUDERO and G. GONZALEZ-DONCEL: *Met. Mater. Trans A*, 1996, **27A**, 3809.
17. W. J. QUADAKKERS, A. CZYRSKA-FILEMONOWICZ, T. MALKOW and H. NICKEL: *Heat Resistant Materials II*, Natesan *et al.* eds, ASM International, 1995, 1.
18. T. A. RAMANARAYANAN, M. RAGHAVAN and R. PETKOVIC-LUTON: *J. Electrochem. Soc.*, 1984, **131**, 923.

19. D. M. MACDONALD: *Frontiers of High Temperature Materials I* (Proc. Conf.), Inco Alloy International, New York, 1991, 101.
20. M. TURKER and T. A. HUGHES: *Oxid. Met.*, 1995, **44**(5/6), 505.
21. W. J. QUADAKKERS: *Werkst. Korros.*, 1990, **41**, 659.
22. B. A. PINT, J. R. MARTIN and L. W. HOBBS: *Oxid. Met.*, 1993, **39**,167.
23. F. H. STOTT and G. C. WOOD: *Mater. Sci. Technol.*, 1988, **4**,1072.
24. F. H. STOTT, A. ATKINSON: *Mater. High Temp.*, 1994, **12**(2–3), 196.
25. M. BOBETH, W. POMPE and M. ROCKSTROH *Microscopy of Oxidation* 2, S. B. Newcomb *et al.* eds, The Institute of Materials, 1993, 412–422.
26. B. A. PINT: *Oxid. Met.*, 1997, **48**(3/4), 303.
27. J. SMIALEK and R. GIBALA: *High Temperature Corrosion*, R.A. Rapp ed., NACE, Houston, TX, 1983, 274.
28. P. KOFSTAD: *Oxid. Met.*, 1985, **24**, 265.
29. A. G. EVANS, G. B. CRUMLEY and R. E. DEMARAY: *Oxid. Met.*, 1983, **20**(5/6), 193.
30. H.E. EVANS: *Int. Mater. Rev.*, 1995, **40**(1),1.
31. D. DELAUNAY, A. M. HUNTZ and P. LACOMBE: *Corros. Sci.*, 1980, **20**, 1109.
32. A. M. HUNTZ AND M. SCHUTZE: *M. Mater. High Temp.*, 1994, **12**,151.
33. M. M. NAGL, S. R. J. SAUNDERS, W. T. EVANS and D. J. HALL: *Corros. Sci.*, 1993, **35**, 965.
34. J. P. BANKS, D. D. GOHIL, H. E. EVANS, D. J. HALL and S. R. J. SAUNDERS: *Materials for Advanced Power Engineering 1994,* Liège, 1994, 1543.
35. M. SCHUTZE: *Doctoral Thesis*, RWTH Aachen, 1983.
36. J. ROBERTSON and M. I. MANNING: *Mater. Sci. Technol.*, 1990, **6**, 81.
37. H. J. FROST and M. F. ASHBY: *Deformation-Mechanism Maps,* Pergamon Press, Oxford, 1982.

Optimisation of MCrAlY Type Coatings for Single Crystal and Convential Cast Gas Turbine Blades

N. CZECH and W. STAMM
Siemens AG Power Generation, D–45466 Muelheim a.d. Ruhr, Germany

1. INTRODUCTION

In order to improve the efficiency and power output of stationary gas turbines, the turbine gas inlet temperature and the compressor ratio, respectively will increase in the future. As with all turbines the high pressure turbine blades are subjected to the highest thermal and mechanical stress. By using a combination of convection, impingement and film cooling the thermal-mechanical loadings of the high pressure turbine blades have been reduced. To improve the creep rupture strength conventional Ni-based alloys have been replaced by single-crystal materials. The high surface temperatures experienced promote corrosive and oxidation attack. To prevent the chemical attack of the base material, the high pressure turbine blades are coated with a MCrAlY overlay coating, with M being either Ni and/or Co. Recently, Siemens has developed a second generation of overlay coatings with a low content of Rhenium. This improves the oxidation and the thermal mechanical fatigue properties as shown in different publications.[1,2] The transition from conventionally cast blades to single crystal blades can lead to a change of the interdiffusion behavior between coating and base material. Ideal systems to study the difference in interdiffusion and mechanical properties are the conventional cast IN 738 LC alloy and the single crystal SC 16. The composition of both materials is given in table 1. The main differences between both alloys are that in SC 16 the grain boundary strengthening elements and Co have been eliminated. Due to the fact that blades for the stationary gas turbines require a high level of sulfidation resistance, chromium is an essential alloying element. By adding aluminum and titanium, both alloys can be precipitation strengthened by the intermetallic γ'-phase.

Table 1 The composition of the conventionally cast IN 738 LC and single crystal SC 16.

	C	Cr	Ni	Co	Mo	W	Ta	Nb	Al	Ti	B	Zr
IN 738 LC	0.1	16	Rem	8.5	1.7	2.6	1.7	0.9	3.4	3.4	0.01	0.1
SC 16	<0.01	16	Rem	-	3	-	3.5	-	3.5	3.5	<0.005	<0.008

2. EXPERIMENTAL

2.1 Oxidation Test

Static oxidation tests were carried out using cylindrical specimens with 10 mm diameter. The specimens were coated with a NiCoCrAlYRe-powder containing 12% Al and 3% Re using the VPS-method. A detailed description of the coating procedure is given in Czech *et al.*[2] The coating thickness amounts to about 230 μm. To achieve the chemical bonding and the desired microstructure the specimens have been heat treated. Finally the surface was shot peened to obtain a surface roughness of R_Z=15 μm or less. To characterise the oxidation behaviour static oxidation experiments were carried out in air under isothermal conditions (950°C) up to 20 000 hours. After six different exposure times the specimens were characterized microstructurally using light optical microscopy and scanning electron microscopy (SEM).

2.2 Thermal Cycle Fatigue Test

A discus shaped test specimen with a thin rim (2mm) and a thick core (12mm) was heated up in a quartz radiation furnace and cooled by compressed air. This kind of sample geometry was used because large temperature gradients in the specimens were experienced during heating or cooling, this lead to high stresses at the rim. Two quartz lamps were used for heating the specimen. The maximum power available was 72 kW. The test specimen was fixed on a special sample holder which could be transported in and out of the oven for heating and cooling. In the cooling process compressed air was directed to both sides of the sample. Due to the difference of the inner and outer temperature during this cycle, thermal stresses were introduced in the base material and the coating, which lead to cracks after a period of time. The circular stress at the rim was in the order of ± 400 MPa and the maximum strain was about 0.4% (tensile) and 0.8% (compression) at a maximum temperature of 1050°C. During heating compressive stress was produced at the rim. After about 26s the specimen was cooled down using compressed air for approximately 40s tensile stresses resulted at the rim. Further information about the test rig is given in Ref. 3.

3.0 RESULTS AND DISCUSSION

The microstructure of a MCrAlYRe-coating after 20 000 h at 950°C is given in Fig. 1. A homogeneous distribution of all phases below the β-phase depleted zone can be seen. The coating consists of four phases: γ-Ni, β-(NiCo)Al and two Cr/Re-rich phases, i.e. α-CrRe and σ-CrRe phase. A β-phase and γ′-depletion in the coating/base material interface is already observable after the precipitation heat treatment. The interdiffusion zone is desired to have a good bonding between coating and substrate. In Fig. 2 the interdiffusion behavior of the MCrAlYRe-coating on IN 738 LC and SC 16 is shown in comparison.

The interdiffusion zone can be divided into two parts, the β-phase depletion of the MCrAlY-coating and the γ′-depletion of the base material. As shown by the diagram

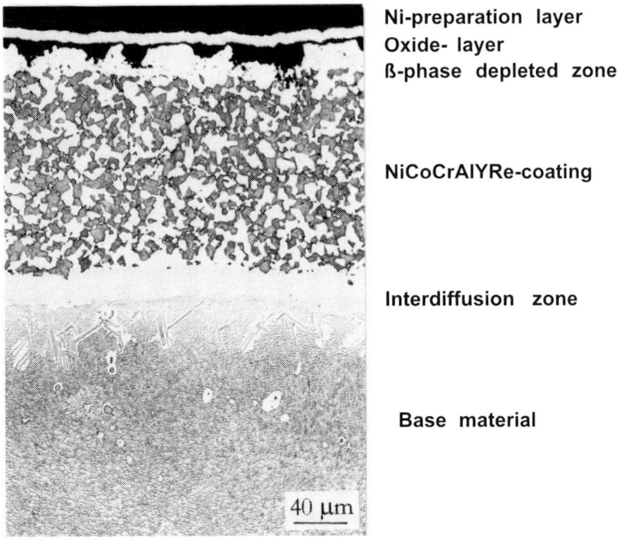

Fig. 1 The microstructure of a MCrAlYRe-coating after 20 000 h at 950°C.

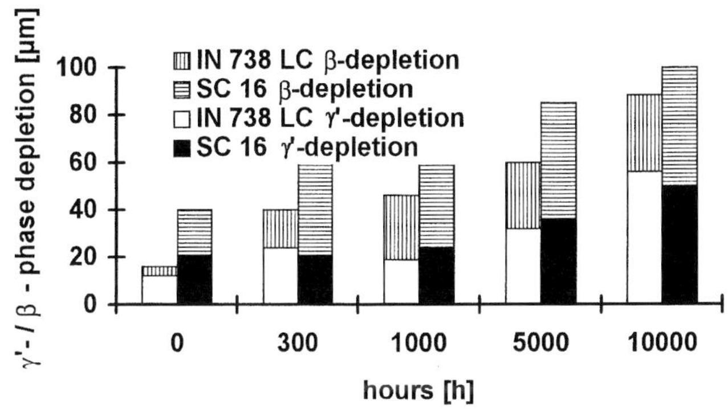

Fig. 2 γ' and β-phase depletion of IN 738 LC and SC16 after exposure at 950°C up to 10 000 h.

the base material attack on both systems is of the same order while the β-phase depletion of the coated SC 16 single crystal is slightly enhanced each time, this is caused by the different heat treatment procedure (SC 16: 1100 °C, 4h; 850 °C, 24h; IN 738 LC: 1120 °C, 2h; 845 °C, 24h) by which the β-phase depletion of the coating on the single crystal is increased in comparison to IN 738 LC. With increasing time both coated systems show the same rate of depletion. From this result it can be seen that the investigated NiCoCrAlYRe coating shows the same interdiffusion behavior on both alloys. Interdiffusion effects between coating and base material are very difficult to describe in a quantitative manner because the activities of the elements can be influenced by precipitations as in our case by α-CrRe and σ-CrRe. As shown by our results the diffusion kinetics in IN 738 LC, in comparison to SC 16, are not greatly influenced by grain boundaries. Instead, the kinetics for both systems are of the same order and determined by the same thermodynamic activities of the elements and the diffusion coefficients. This is understandable as the element concentration in both base materials are comparable.

For the uncoated SC 16 sample and for the coated SC 16 and IN 738 LC, thermal cycle fatigue investigations have been carried out and the number of cycles to the first crack initiation have been measured. Figure 3 shows the number of cycles for all systems at 1050°C. The base material SC 16 shows no cracks at 3000 cycles. For the coated base material some cracks, mainly in the coating, are observable after 2167 cycles. This shows that the coating has an influence on the thermal cycle fatigue behaviour of the blade material. As demonstrated by a microsection analysis the coating on SC16 shows a β-phase enrichment at the surface. This leads to brittleness at the surface and to an enhanced crack sensitivity. A comparable result also has been found by thermal mechanical fatigue (TMF) investigations on NiAl- and CoNiCrAlY-coated SRR99 single crystal alloys.[4] A reduction in TMF life in comparison to uncoated SRR99 by a factor of 2–5 is found. The pronounced rise of the crack number density during cycling in the CoNiCrAlY coating is related to the coating porosity in the as received condition, and to the generation of new pores during TMF cycling.

Whether a comparable theory can be applied to our coating is not clear because the development of pores is not clearly seen. For the coated IN 738 LC sample a cycle to crack number of 1875 is observable. The cracks are mainly in the coating with a crack length between 20 and 60 μm. One crack extends approximately 280 μm into the base material. Measurements on uncoated IN 738 LC were not acceptable because the grain boundaries go in a straight line from the rim to the middle of the discus shaped sample. This strange grain boundary structure reduced the cycle to crack number dramatically to around 50.

4. CONCLUSION

Re additions can lead to a significant improvement of long term performance of NiCoCrAlYRe coatings on IN 738 LC and SC 16 to thermomechanical, oxidation and interdiffusion properties as shown in some papers. In coatings with an optimised composition in respect to Re-, Cr-, Co-, and Al-content excellent oxidation behaviour and interdiffusion performance were obtained on SC 16 and IN 738 LC respectively.

In this paper it has been shown that after long term oxidation on the conventional cast IN 738 LC and the single crystal SC 16, the influence by interdiffusion on the life time reduction for both alloys is nearly the same. This result can be explained by the presence of nearly the same element concentration in both base materials. After 20 000h, as shown in Fig. 1, the β-phase and γ-depletion (interdiffusion zone) is remarkably small. Re can, however, have an indirect effect on long term oxidation and interdiffusion

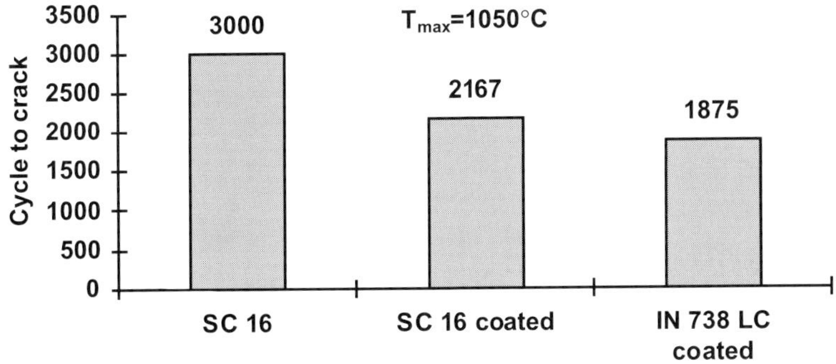

Fig. 3 Comparison of the cycle to crack number for SC 16, coated SC 16 in IN 738 LC.

performance of the coating; it tends to reduce the oxidation rate and decrease the extent of the interdiffusion zone in both alloys. Furthermore, it has been shown that a non-optimised coating can reduce the life time of coated systems. In our case the coating of the coated SC 16 sample shows an enriched β-phase distribution in the subsurface layer after the base material heat treatment. This leads to a decrease of the fatigue life time. For the future, more investigations especially on the Re influence on the diffusion behaviour, are necessary.

REFERENCES

1. N. CZECH, F. SCHMITZ and W. STAMM: *Surface and Coatings Technology*, 1994, **68/69**, 17–21.
2. N. CZECH, F. SCHMITZ and W. STAMM: *Materials and Manufacturing Processes*, 1995, **10**(5), 1021–1035.
3. N. CZECH and W. STAMM: *Surface and Coatings Technology*, 1996, **86/87**, 15–21.
4. J. BRESSERS, K. OSTOLAZA, M. ARANA: *in press*.

Characterisation of MCrAlY Coatings Sprayed by HVOF and Reactive Plasma Spraying

E. LUGSCHEIDER, C. HERBST and L. ZHAO
Materials Science Institute, Aachen University of Technology, D–52070 Aachen, Germany

INTRODUCTION

Vacuum plasma sprayed (VPS) is still one of the most frequently used thermal spray processes for deposition of high temperature oxidation resistant materials such as McrAlY alloys onto turbine parts.[1,2] The process is recognised for providing dense and oxide free coatings with superior performance under thermal and oxidative conditions.[3] It is however, also a comparably expensive process since equipment coats are in the range of two million US$. Furthermore, the chamber process requires time consuming evacuation and purging cycles which inhibit efficiency. Additionally, an appropriate process control by diagnostic means such as particle or substrate pyrometry is difficult.

The new generation of high-velocity oxygen-fuel (HVOF) spraying systems offer the potentials for processing MCrAlY materials in atmosphere. Producers of these systems claim gas velocities to reach over 2000 m/s.[4] Resulting high particle velocities allow ther spraying of particles in a plastic rather than in a melted state thus lowering oxidation and heat input into the substrate. Nevertheless, the high impact energy of the particles still allows a reasonable flattening and adherence even in a plastic state leading to dense, pore-free coatings. The residual stress states in HVOF sprayed coatings are favourable values when considered in terms of thick coatings. Within this project coatings with over 5 mm thick have been sprayed for target production. The investment coats for modern HVOF equipment are roughly a tenth in comparison with VPS.[5]

The main topic of this research work is to investigate the correlation between HVOF spraying parameters and the resulting coating properties, with particular focus on the oxygen content. For correlation analysis a statistical factorial plan was used to assess the effects of the parameters. Coatings were also subjected to cyclic oxidation tests to determine potentials for replacing VPS coatings.

On the other hand, today's McrAlY alloys show considerable creep under high temperature load.[6] Thus the intention is to reinforce McrAlY coatings by a fine distribution of hard phases such as nitrides. Reactive plasma spraying (RPS) is a new process,

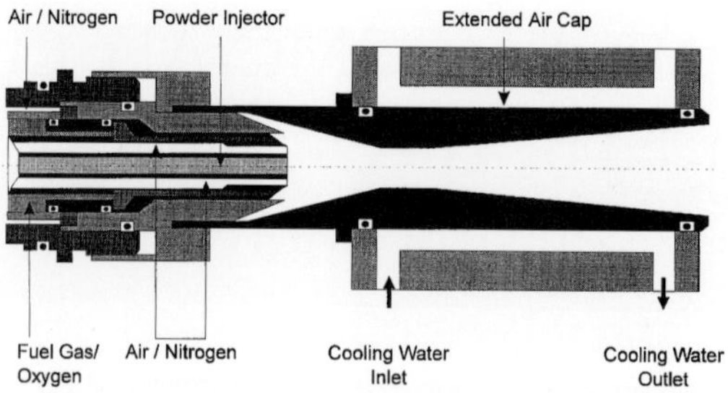

Fig. 1 Cross-section through HVOF system DiamondJet-Hybrid.

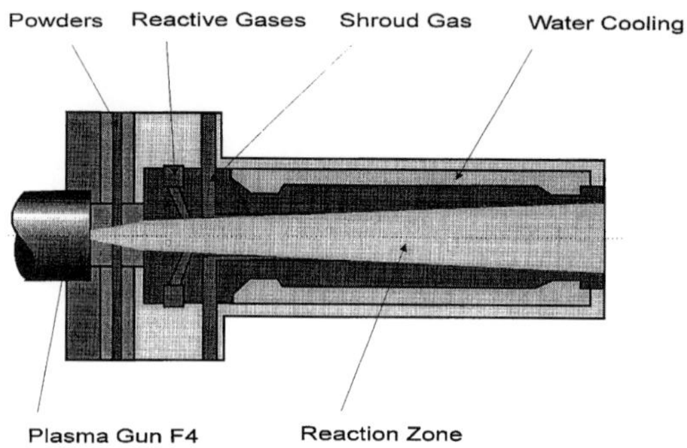

Fig. 2 Cross-section through custom made plasma reactor attached to F-4 plasma gun.

which has successfully demonstrated its potential for in-situ formation of hard phases in titanium alloys in COST 503. Therefore a second focus of this research work is the use of RPS in order to produce in-situ nitride reinforced McrAlY coatings.

EXPERIMENTAL

For HVOF spraying a commercial Sulzer-Metco Diamond Jet Hybrid system was used. The system can be driven with hydrogen, propane and propylene as fuel gases. Shroud and cooling gases for the burner can either be compressed air or nitrogen. Water cooling is only required for the nozzle extension. Powder is fed into the combustion area and accelerated through a convex–concave nozzle. Figure 1 shows a cross-section through the system.

RPS experiments were conducted in a custom-made reactor. The fundamental of the system was a tube-like reactor being attached to a commercial plasma gun (SM-F4), Fig. 2. The whole arrangement was installed into a commercial vacuum chamber. The reactor consists of a conically shaped water-cooled barrel. The reactive gas, nitrogen, uniformly enters the barrel behind the exit of the gun nozzle through a gas distributor. A little further downstream argon as a shroud gas is added to avoid extensive sticking of the spray material onto the barrel surface.

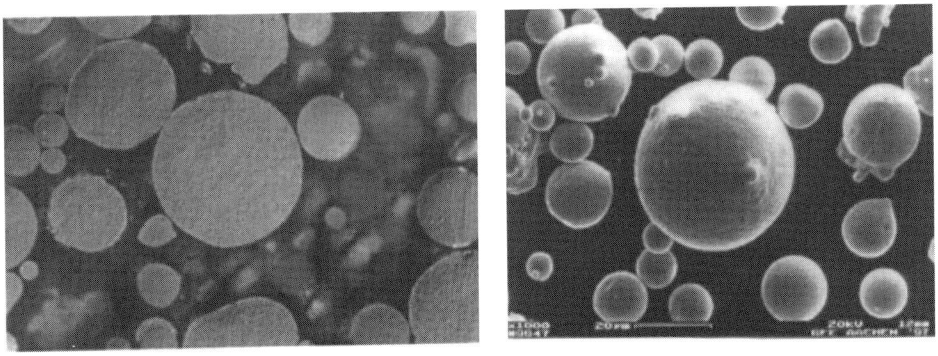

Fig. 3 Cross-section of initial powder (x1000) and REM photograph showing glibular shape.

Metallographical investigations were carried out using LM and REM. Phase constitution was evaluated via XRD and EDX. Microhardness was determined with a Vickers indenter. The oxide and nitrogen content of the coatings were evaluated using a O_2N_2-analyser (LECO TC136) after peeling them from the substrate, oxygen units are in wt% or ppm. Oxidation tests were carried out with coating pieces which were peeled form the substrate and dimensioned to a size of 10 mm x 10 mm. After spraying, coated sheets were heat treated in a vacuum furnace at 1100°C/4hrs. Cyclic oxidation tests were conducted in an atmosphere furnace at 1100°C in air. One cycle was for 8h followed by cooling in air. Weight gain was measured with a precision wage.

RESULTS

Powder

The powder used was a commercial NiCoCrAlY powder (Type PWA 1365); relevant data can be seen in Table 1. The powder exhibits a globular shape from gas atomisation as can be seen from Fig. 3. The microstructure of the powder depicts a dendritic structure, with a huge fraction of small particles visible. Measurements with a Sedigraph revealed a grain fraction with 10% below 10 μm. 8% of particles were larger than 38 μm, the rest in the range –38 μm, +10 μm. XRD measurements showed a (Ni,Co)Al phase. In the literature this phase is being described as the b-phase. A Cr-rich CoNi phase, often described as the g-phase, could not be clearly detected. However, due to the rapid solidification after atomisation, particles were not expected to form equilibrium phases so that a shift of elements is supposed to occur.

Table 1 Chemical composition of NiCoCrAlY powder (-38 μm, =10 μm, O-content 374 ppm).

	Ni	Co	Cr	AL	Y
wt%	Bal.	21.7–23.0	17.0–17.2	12.5–12.6	0.45–0.65

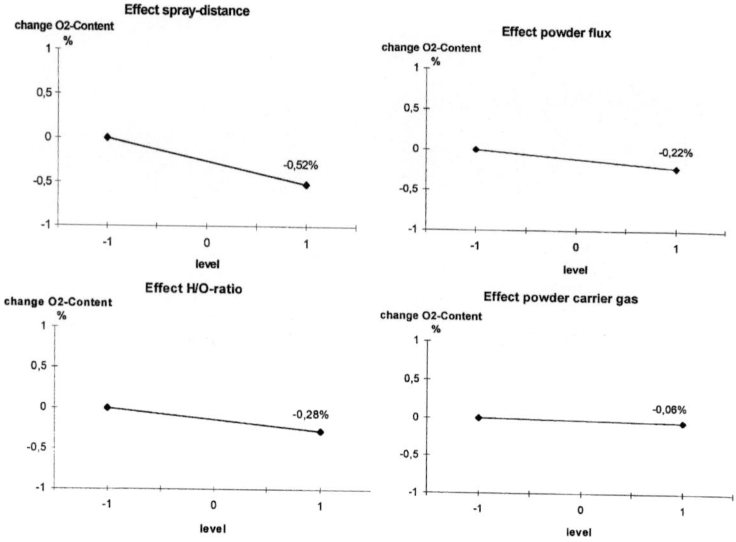

Fig. 4 Single effects of spray parameters on oxygen content of HVOF-sprayed NiCoCrAlY.

HVOF

Effects of HVOF parameters on coating oxygen content

The effects of spray-distance. fuel-gas composition, carrier and powder gas flow on the oxygen content of the coatings can be concluded from Fig. 4 The most significant parameter is spray-distance. Increasing from a base-line of 250 mm to beyond 300 mm reduces the oxygen content from 1.5% to 1%. The reason for that is that the hot combustion gases increase oxidation on substrates near the plasma gun by the entrapment of oxygen into the jet and blowing it onto the hot surface. By increasing the distance of the heat load on the substrate is reduced suppressing such severe oxidation. This can be measured by thermocouples attached to a bore hole in the back of the substrate. By increasing the stand off distance the temperature drops from 260°C to below 150°C.

Also increasing the powder feed rate has a reducing effect on the oxide content. It can be explained by two reasons: (i) Increasing the amount of spray material at a constant combustion power results in a lower specific heat input for a single spray particle such reducing melting and oxidation tendency during flattening. Metallographical investigations revealed that the coatings with the lowest oxygen content mainly consist of unmelted particles, which do not show any oxidation. In contrast particles, which have been completely melted clearly depict oxide phases, Fig. 5. (ii) A higher feed rate requires a fewer number of passes thus lowering oxidation by the combustion gas stream. However, it has to be considered that thicker coating layers required to build up a certain thickness will result in a higher residual stress. As a result traverse speed has to be increased as well.

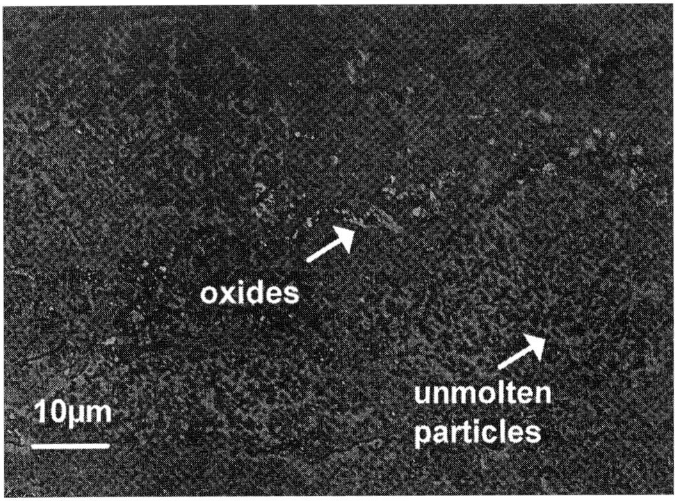

Fig. 5 LM micrograph of as-sprayed structure of HVOF NiCoCrAlY.

High amounts of powder in the carrier gas do not cool down the combustion gas much, but result in merely plastification of the particles. The parameter has a negligible influence on O-content and as such can be freely adjusted for optimum powder transport.

A high H/O ratio is beneficial for low oxide content in the coatings. With the tightening up of oxygen for combustion, the flame-jet cools down and particles do not melt completely. However, decreasing the flame enthalpy by such a means also decreases the gas velocity, observable by disappearing mach knots in the flame. A compromise has to be found. The best results have been obtained for a H/O ration of 1:0.23.

A combined increase of both spray-distance and powder feed rate does not lead to a reduction of oxygen content by just adding both single figures. The absolute benefit is less (roughly minus 0.31%) since a reduction of passes by higher powder feed rates is not that efficient in terms of oxidation reduction at higher spray distances. In contrast, a combined increase of powder feed rate and H/O ratio boosts oxide reduction. Both single effects reinforce each other to reduce the degree of melting of the particles and such oxidation during flattening.

Also it is beneficial for the oxide content to keep substrate temperatures down and to use nitrogen instead of air for the shroud/cooling gas. A NiCoCrAlY coating produced by HVOF under best conditions reduced oxygen contents of 0.49 wt%. Porosity in such a coating can not be detected by image analysis systems. A further reduction of oxygen can be achieved by using a powder with a reduced fraction of fine particles to prevent complete melting. Additional experiments need to be conducted to prove this. Coatings in the as-sprayed and heat treated conditions can be seen in Figs 6 and 7 respectively. The as-sprayed coatings showed g¢- and b- phases in an unclear matrix phase, whereas after heat treatment a dominant b-phase is observed in a matrix phase.

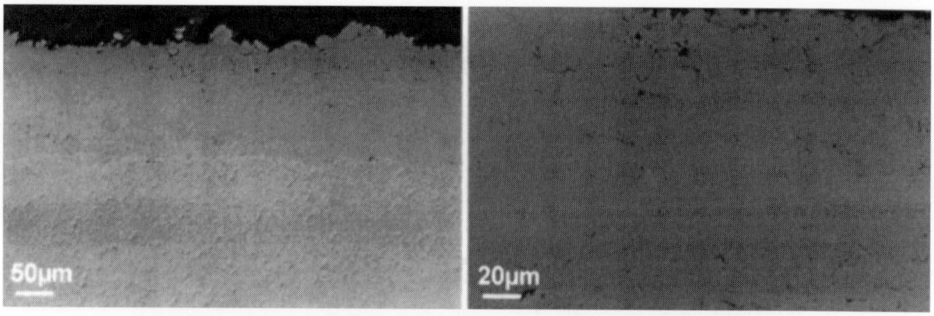

Fig. 6 LM micrographs of optimised HVOF sprayed NiCoCrAlY caotings in as-sprayed condition.

Fig. 7 LM micrograph of optimised HVOF sprayed NiCoCrAlY after 1100°C/4hrs/vacuum.

Oxidation behaviour

Isothermal oxidation of a HVOF sprayed NiCoCrAlY coating at 1100°C/10hrs/air results in the formation of a thin oxide layer, a few microns thick, which protects the coating from further degradation, as can be seen in Fig. 8.

Cyclic oxidation tests revealed that HVOF coatings can compete very well with VPS ones. Figure 9 shows the results of weight gain over three cycles. The VPS coatings kept at a constant level whereas the HVOF coatings show a gentle incline, but at a lower level. Metallographic investigations on aged coatings revealed for both coating types the formation of oxide layers.

Reactive Plasma Spraying

In these initial HVOF trials it was identified that the electrical power input needed to be reduced towards those employed in VPS to prevent powder sticking and overheat-

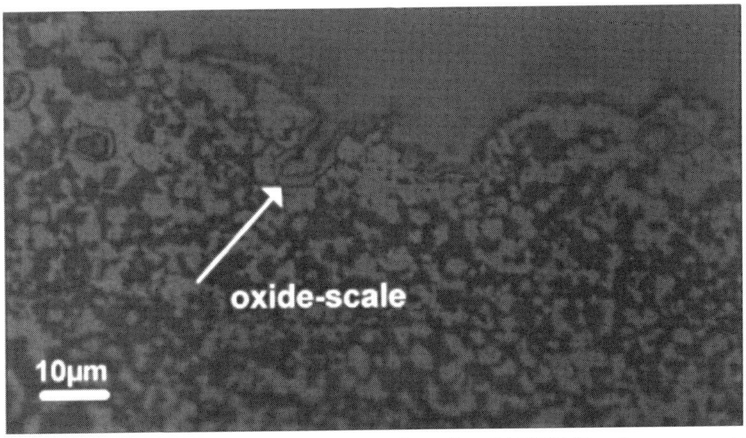

Fig. 8 LM micrograph of a NiCoCrAlY coating after 1100°C/4hrs/vacuum + 1100°C/10hrs/air showing the formation of an oxide layer.

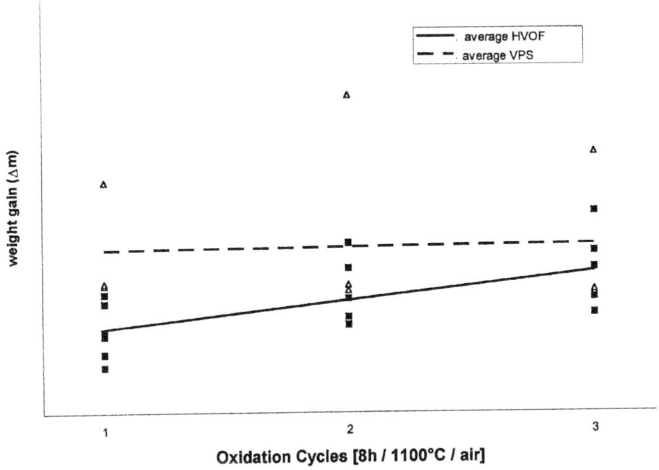

Fig. 9 Cycling oxidation behaviour of VPS (Δ) and HVOF (■) coatings.

ing of the system. Furthermore higher power levels presumably will lead to a higher acceleration of the spray particles thus reducing dwell times in the reactor. Further parameters having been taken into consideration were flow rates of reactive gas, plasma gas and chamber gas and the preheating temperature of the substrate. The parameter ranges used are given in Table 2.

Figure 10 shows the results of nitrogen analysis measurements. It is obvious that even under favourable conditions, as used for *Reactive 2*, it is possible to raise the nitrogen content towards the level seen in relatively cold sprayed HVOF coatings. This indicates the effect that the substrate temperature has during spraying for nitro-

Table 2 Parameter range used in experiments.

Power level:	31–34 kW	Reactive gas flow:	5–8 SLPM
Plasmagas Ar/N_2:	35/5–8 SLPM	Shroud gas flow:	25 SLPM

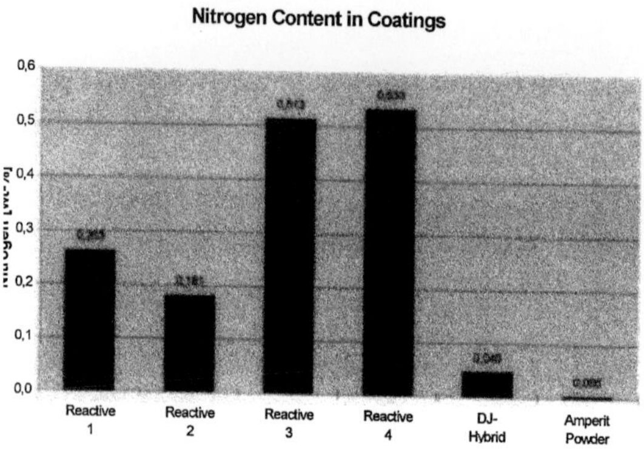

	R1	R2	R3	R4
Chamber Gas N2/Ar	**Ar**	**N2**	**Ar**	**N2**
Reactor N2 On/Off	**On**	**Off**	**On**	**On**
Preheating Yes/No	**No**	**No**	**Yes**	**No**

Fig. 10 Nitrogen-content in RPS sprayed NiCoCrAlY coatings depending on parameters.

gen absorption. From column, *Reactive 1*, the influence of the reactor itself can be deduced. The reactor increases the nitrogen content in the coating by twice as much as simply a nitrogen-rich atmosphere in the vessel. Thus, a combination of both parameters becomes even more beneficial for increasing the nitrogen content as shown in *Reactive 4*. As mentioned above, the substrate temperature plays a major role in on-substrate reactions. Preheating the substrate to 800°C prior to spraying, as done with *Reactive 3*, lead to a significantly higher nitrogen level. It has been doubled compared with a cold substrate. Thus, with these parameters it was possible to increase the nitrogen content in the coating by a factor of 100 compared with the powder, and it was still a factor of 10 higher when compared with the HVOF sprayed MCrAlY.

Figure 11 shows a section of the microstructure after spraying. It depicts big round particles and insufficiently particles that show the original dendritic powder microstructure. This means that these particles have not reached the liquid state. In contrast lots of small round particles are visible that show a single phase microstructure. These droplets have presumably been solidified in-flight and embedded into the coating in the solid state. This leads to the conclusion that the powder particle size distribution is a very sensitive parameter with reactive plasma spraying and has to be selected very carefully. In view of the tendency of particles sticking in the barrel, they must not be too fine, but must be kept below an upper level for a high surface-to-volume ration, which is essential for a high reaction degree.

After annealing at 1100°C for 4 hours in vacuum the microstructure of reactively sprayed coatings showed a typical two phase MCrAlY microstructure consisting of a matrix phase and homogeneous b-phase detected by XRD.

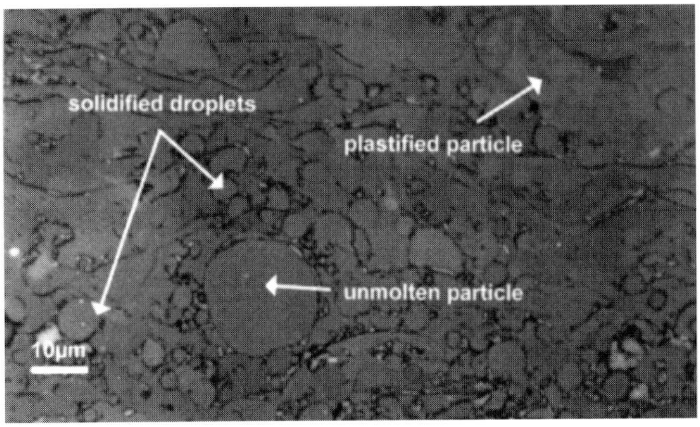

Fig. 11 Influence of particle size on solidification behaviour of NiCoCrAlY with RPS.

However, reactive sprayed coatings did release an in-situ formation of nitride phases only when reactive elements such as chromium have been added to the initial powder. This has been confirmed by WDX investigations. It could be shown, for instance, that adding fine Cr particles increased the nitrogen content up to 13 at.% in Cr-rich phases, Fig. 12. The reason for this is that Cr when distributed in Ni- or Co-crystals are not reactive enough. Therefore, reactive plasma spraying requires a different powder design. Reactive elements must be subjected to the reactive media, thus atomised powders must be replaced by mechanical mixtures or preferably by agglomerated powders.

SUMMARY AND OUTLOOK

Influences of HVOF parameters on coating microstructure and oxygen content were investigated by statistical means. Significant parameters are spray-distance, H/O ratio, powder feed rate, powder size fraction and substrate temperature. By improving the parameter sets oxygen content in the coatings could be reduced to less than 0.49 wt%. Further improvement is likely by adapting the powder size fraction. Oxidation tests show that HVOF sprayed NiCoCrAlY coatings form protective oxide scales. VPS and HVOF coatings show very similar oxidation behaviour under cycling conditions. Further tests will be conducted.

Reactive plasma spraying (RPS) is an appropriate tool to increase reactive gas content in coatings aimed at hard phase formation. The nitrogen content could be increased by a factor of 100 from the initial powder. However, the formation of nitrides could not be observed. Tests with mechanical mixture powders containing free reactive elements instead of pure atomised powders show higher reactivity during spraying. Thus further tests with agglomerated and mechanical mixture powders have to be conducted in order to promote hard phase formation.

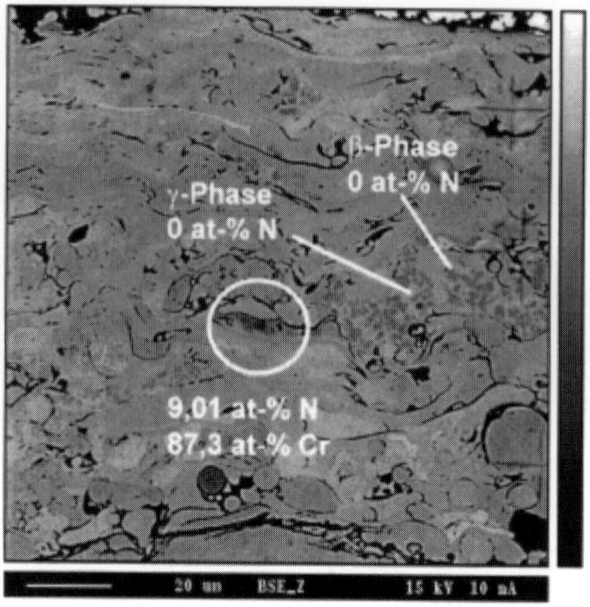

Fig. 12 REM micrograph of RPS sprayed NiCoCrAlY + 5 wt% Cr showing reactivity free Cr.

REFERENCES

1. R. YAMASAKI, J. TAKEUCHI, A. NAKAHIRA, m. SAITOH and Y. ITOH: 'Characteristics of Pack Aluminized VPS Sprayed MCrAlY Coatings', *Proceedings of ITSC '95*, Kobe, May 1995, 863–868.
2. P. J. Meyer: 'Conomics of MCrAlY Coatings', *Proceedings of ITSC '95*, Kobe, May 1995, 217–222.
3. L. B. TEMPLES, M. F. GRUNUNGER and C. H. LONDRY: 'Influences of Oxygen Content on McrAlYs', *Proceedings of the 1993 NTSC*, Anaheim, USA, 359–363.
4. M. C. NESTLER, S. HARTMANN and H.-M. HÖHLE: 'DiamondJet- eine wirtschaftliche Lösung für hochnbeanspruchte Schichten', Tagungsunterlagen zum 3, *Kollquium Hochgeschwindigkeitsflammspritzen*, H. Kreye et al. Eds, 10–11 November 1994, 93–104.
5. G. IRONS and V. ZANCHUK: 'Comparison of MCrALY Coatings Sprayed by HVOF and Low Pressure Processes', *Proceedings of the 1993 NTSC*, Anaheim, USA, 191–197.
6. A. BENNETT: 'Properties of Thermal Barrier Coatings', *Materials Science and Technology*, 1986, **2**(3), 257–261.

EB-PVD Zirconia Thermal Barrier Coatings for Experiments in Space

E. LUGSCHEIDER*, G. DOEPPER, H. G. MAYER†, A. SEIDEL¶ and W. DREIER§

**Materials Science Institute, Aachen University of Technology, 52056 Aachen, Germany*

†Forschungsinstitut für Kerntechnik und Energiewandlung e.V., 70569 Stuttgart, Germany

¶Daimler-Benz Aerospace, Dornier GmbH, 88039 Friedrichshafen, Germany

§DARA GmbH, 53227 Bonn, Germany

ABSTRACT

Ceramic thermal barrier coatings based on partial stabilised zirconia (psz) are commonly known for applications in energy converting machines like gas turbines and, therefore, such coatings have been developed to meet these high demands. However, it is appropriate to use psz-coatings for other applications, as will be presented in this paper.

The paper deals with zirconia coatings on W26Re substrates, which are used as covers for small crucibles. In these crucibles metallic molten material, e.g. Cu, Ni and Zr, will be investigated concerning its behaviour under microgravity. Therefore, the following demands are required: electrical isolation during the entire experiments, thermal shock and thermal cycling resistance up to the highest temperatures as well as a low light reflection coefficent. As a ceramic coating material for metallic W26Re substrates, partial stabilised zirconia was chosen because of its distinct properties. As a deposition process, which results in a coating with a very high potential concerning thermal cycling resistance, the electron beam-physical vapour deposition (EB-PVD) process was used. Finally, the coated substrates were examined concerning the special demands listed above, to ensure their practical value under microgravity. This paper describes the special application and the developed coating process as well as the results of the testing procedures.

INTRODUCTION

Experiments carried out in space under microgravity conditions put high demands on the experimental setups. On one hand, high costs have to be considered because the cost of one mission is estimated at between £120M and £190M. On the other hand, a space mission has high risk because the time for experiments is restricted and all experimental setups are forced to work, since usually a repair of components is impossible. Therefore, much efforts have to be done to realise such experiments successfully.

This paper presents the development of zirconia coatings for crucible lids used in the space experiments which were part of the 83rd Space Shuttle Mission. The whole equipment was installed in the NASA Microgravity Science Laboratory (MSL) and launched first as MSL-1 on 4 April 1997. This mission was aborted for safety reasons after four days because one out of three fuel cells for power generation didn't work. Later the restart mission MSL-1 R was realised on 1 July 1997, within a very short time of three month and was at least successful. The mission lasted sixteen days. Part of the experiments on board were done in the frame of TEMPUS, which is a German acronym meaning containerless processing of molten materials under microgravity conditions. 22 experiments were carried out at least, with 6 German and 4 American investigation teams in charge. Samples of 7 mm to 8 mm diameter were processed, made up of pure metals and alloys such as Zr, Ni, Zr–Al–Cu–Ni, Pd–Cu–Si. The aim was undercooling of samples while the material properties were measured, mainly by optical measurement methods. Undercooling defines the process when a molten material is cooled down below its 'normal' solidification temperature still staying in the liquid state. With undercooling, sudden solidifaction occurs when a specific temperature is reached, which is depending on the material. This process requires an absolute clean environment so that a contact between the crucible walls and the molten samples should be avoided. Otherwise nucleation takes place starting locally at the contact zone which, on the other hand, can be used for triggered nucleation.

The complete experimental equipment was installed in the space lab including the processing chamber where 18 sample holders with one sample each were arranged in a circular way. Every holder could be placed in the investigating position for the measurements to be carried out while the samples were fused and fixed in the desired position with rf-modules. Two types of holders were used, one so called a 'wire cage holder' and another named a 'cup holder'. Whereas wire cage holders enabled a very good view on the samples and were used for samples which showed nearly no evaporation while being liquid, the cup holders just had horizontal slits to avoid contamination of the investigation chamber by evaporants. Figure 1 shows the wire cage holder.

EXPERIMENTAL DETAILS

Both types of holders had metallic lids on the top which had to be coated with a material showing constant thermal stablity and electrically insulating properties up to 1500°C. In the same time the coating had to be chemically inert against the molten samples and had to work as a thermal barrier due to thermal shocks and thermal cycling in case of a contact between the sample and lid. In addition, the coating material had to show an appropriate mechanical and electrical behaviour considering the handling during mounting the lids as well as the influence of the rf-field during the experiments. Therfore, partially yttria stabilised zirconia was chosen as the coating material due to its special properties. The lids were coated in an EB-PVD equipment because EB-PVD zirconia coatings show outstanding strain tolerance due to their columnar micro stucture.[1] The EB-PVD equipment had a second electron beam gun as a substrate heater which is a good method for temperature control with varying substrate volumes.[2] In the present case, the lids were of a ring-type shape and they were quite

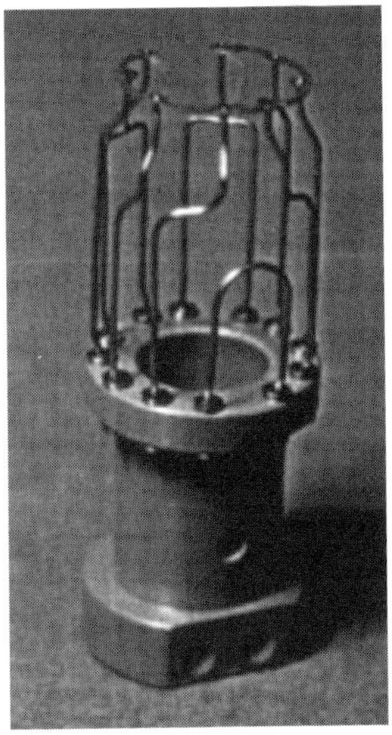

Fig. 1 Sample holder type 'wire cage'.

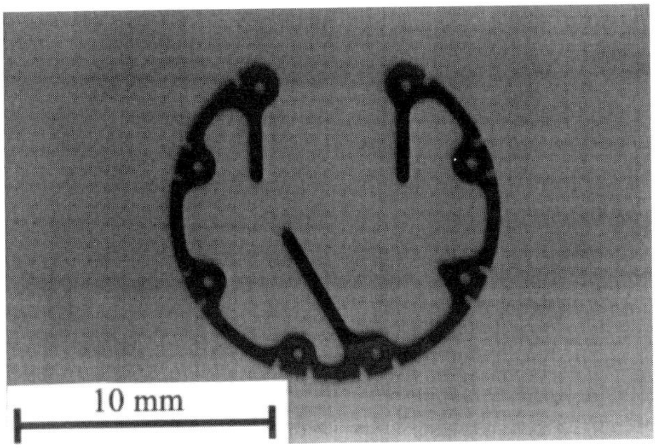

Fig. 2 Lid.

small with a thickness of 0.3 mm and a diameter of about 12 mm, as can be seen in Fig. 2.

The base material was W26Re which shows a poor oxidation behaviour especially at high temperatures. When zirconia is deposited with EB-PVD typical temperatures (in the range of 950°C) and by adding oxygen as reactive gas, oxidation of the base material occured when the reactive gas was added directly from the beginning of the deposition process. Therefore, the following variations of deposition process were carried out, to avoid a direct oxidation of the W26Re during deposition and to realise best adhesive coatings. The stochiometry of the coatings was varied with a first non-

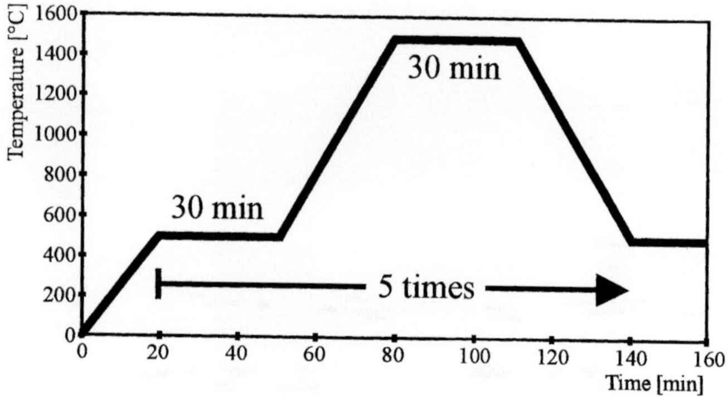

Fig. 3 Thermal cycle mode.

stochiometric layer of 1.5 μm or 3.5 μm thickness, whereas the rest of the coating was stochiometric. In addition, the whole coating thickness was varied between 5 and 10 μm.

The coated lids were tested by different methods: there were bending tests, until plastic deformation of the base material occured, as well as thermal cycling tests and thermal shock tests. The thermal cycling was done between 500°C and 1500°C for five times. The temperature was kept constant at each level for 30 minutes This testing procedure is shown in Fig. 3. The thermal shock tests were carried out by dropping molten copper, steel and zirconium of a realistic size on coated samples while these were put on a big steel plate. Therefore, a temperature gradient between melting temperature and room temperature was created throughout the cross section of the coated lids to model realistic conditions. Smaller thermal shocks were initiated when trigger needles were EB-welded into some of the lids. After treating the lids in that way the electric resistance was measured at room temperature.

RESULTS AND DISCUSSION

Best results – considering adhesion and the mentioned testing procedures – were obtained with a non-stochiometric base layer of 1.5 μm thickness and a stochiometric coating on top of it, leading to a complete coating thickness of at least 7.5 μm and no bond coat was necessary. A typical microstructure of such coatings can be seen in Fig. 4. The lids were homogeniously coated and the electric resistance as coated was 200 MΩ.

When lids were bended elastically the coatings showed elastical behaviour too. Coating spallation just occured when the base material was plastically deformed, as can be seen in Fig. 5. After thermal cycling none of the coated lids failed since only the grain boundaries could be observed to be a bit more distinctive. The electric resistance was still as coated. Figure 6 shows the surface of the coating typically seen after thermal cycling.

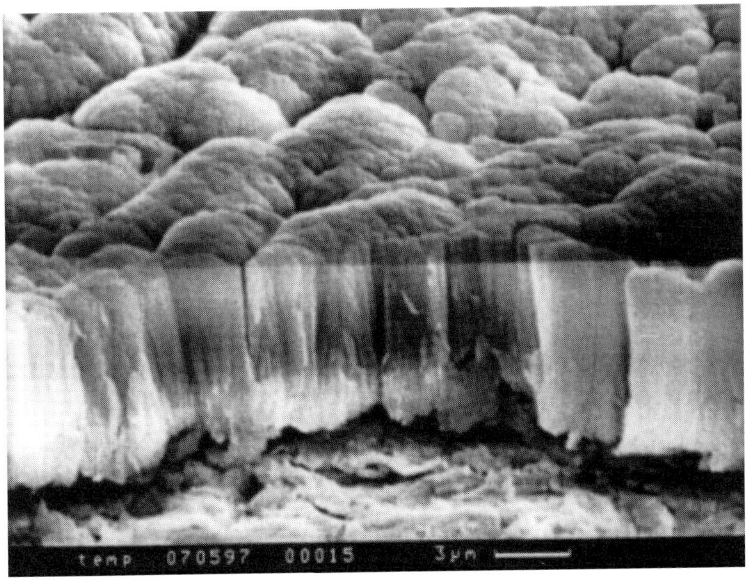

Fig. 4 Typical microstructure of EB-PVD zirconia coating.

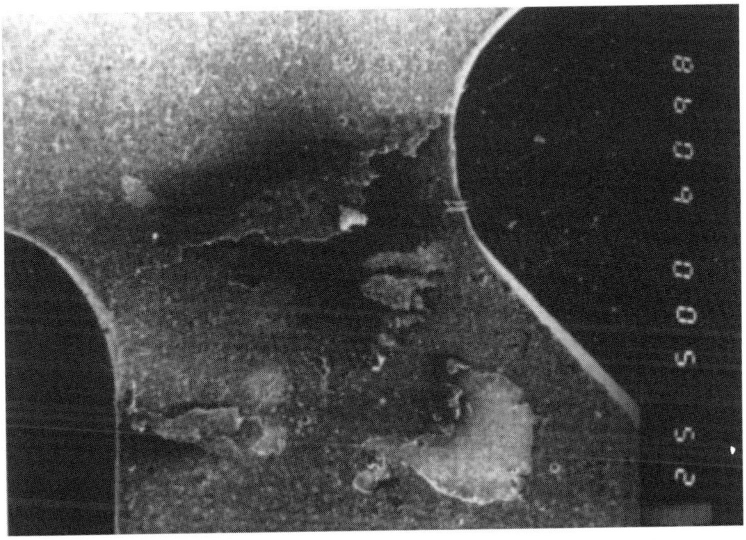

Fig. 5 Coating spallation after bending to plastic defrmation.

Thermal shock tests carried out with the molten metals showed different results. The copper drop could easily be removed from the coating, although the electric resistance dropped down to 40 MΩ. Figure 7 shows a copper drop on a coated lid as an example for this testing method. Steel drops could easily be removed too. The grain boundaries were seen more distinct (see Fig. 8) but the electric resistance still was as

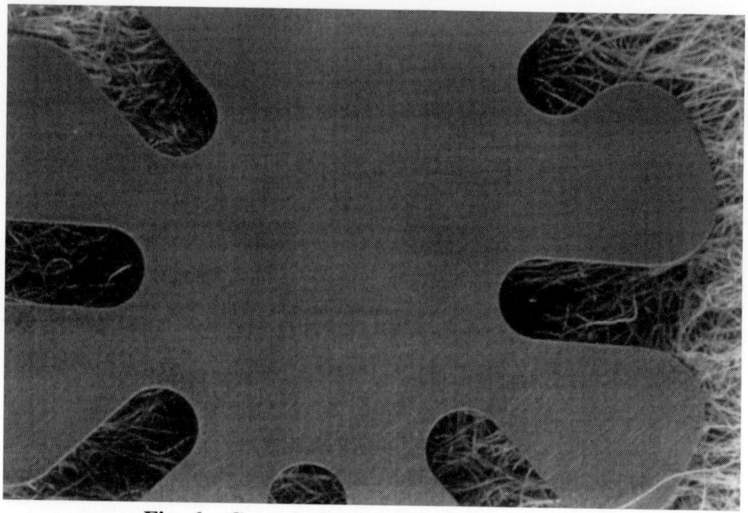

Fig. 6 Coated sample after thermal cycling.

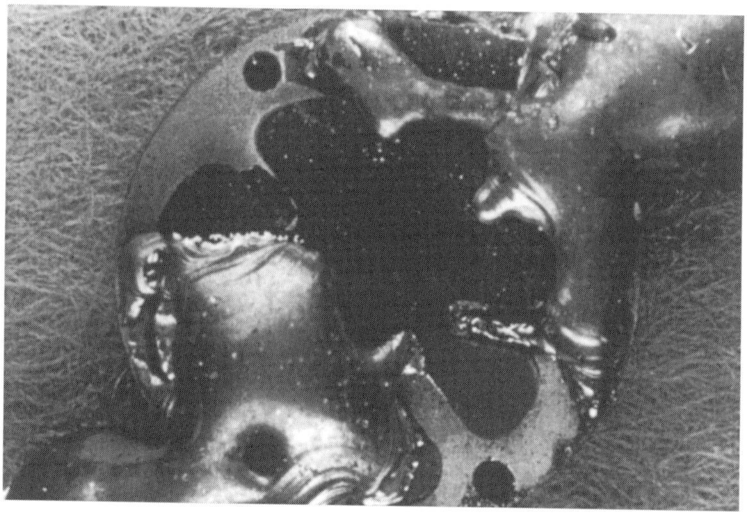

Fig. 7 Coated lid with copper drops.

coated. On top of the coatings residues of steel were left, as can be seen in Fig. 8 too. When zirconium was dropped onto the zirconia coatings they couldn't be removed without damaging the coatings. Therefore, electric resistance was measured between Zr-residues and the undamaged part of the coating showing values of 1 MΩ, which was defined as the minimum. When EB-welding of the trigger needles in coated lids was carried out the coating still remained intact. Near the welded areas the electric resistance just dropped down to 100 MΩ.

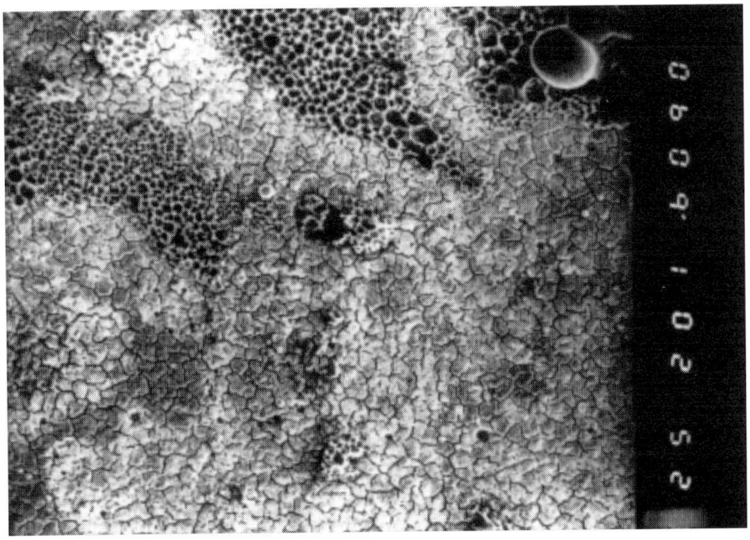

Fig. 8 Surface of the coating after removing the steel drop.

In summary, the coated lids passed all the testing procedures successfully and a real good adhesion could be observed between coating and base material without using any bond coat.

CONCLUSIONS

Components for experiments in space have to fulfill special demands. An outstanding high quality of the experimental setup is desired due to high costs and high risk of these space laboratory tests. In TEMPUS molten metals and alloys were investigated under microgravity conditions, requiring special investigation equipments. These were sample holders with lids to be coated for electrical isolation as well as themal and chemical stability. Because the lids were made up of W26Re with poor oxidation behaviour a special coating process for EB-PVD zirconia coatings had to be developed. The coated lids passed all testing procedures; which were bending tests, thermal cycling tests and thermal shock tests. A very good strain tolerance of the EB-PVD zirconia coatings up to temperatures of 1500°C could be observed due to their columnar microstructure. The surface was dull due to avoid influencing the optical measurement methods. At least, the whole mission MSL-l was successful because all experiments were carried out and no coating failed under realistic conditions too.

ACKNOWLEDGEMENT

This work was done on behalf of DARA GmbH, the German aerospace agency, under code number 50 WP 9598. The publication is done with permission of the customer. The authors are responsible for the contents of this publication

REFERENCES

1. E. LUGSCHEIDER, C. BARIMANI, M. MOELLER and G. DOEPPER: 'Aufbau und Wirkprinzip keramischer Wärmedämmschichten', *Werkstoffe für die Enerietechnik,* H. W. Grunling ed., Symposium 3, Tagungsband zur Werkstoffwoche '96, DGM-Informationsgesellschaft-Verlag, Frankfurt, 1997, ISBN 3–88355–231–3, 177–183.
2. E. LUGSCHEIDER, C. BARIMANI and G. DOEPPER: 'Ceramic thermal barrier coatings deposited with the EB-PVD-technique', International Conference on Plasma Surface Engineering (PSE), 09.09-13.09.1996, Garmisch Partenkirchen, *Surface and Coatings Technolgy*, 1997, to be published.

Possibilities and Limits of Temperature Control in EB-PVD Equipment Used for the Deposition of Zirconia Thermal Barrier Coatings

E. LUGSCHEIDER, C. BARIMANI and G. DOEPPER
Materials Science Institute, Aachen University of Technology, 52056 Aachen, Germany

ABSTRACT

The deposition of ceramic thermal barrier coatings, e.g. coatings based on partially stabilised zirconia, is of great interest for applications in energy converting machines like gas turbines. In this context the EB-PVD technique shows a really high potential because of its benefits in adapting the coatings microstructure to special demands. To achieve the desired columnar microstructure with respect to the optimised thermal cycling resistance, it is important to practice exact control of the substrates temperature.

This paper deals with the possibilities and limits of temperature control in special EB-PVD equipment used at the Materials Science Institute in Aachen. In this case a second electron beam gun was used as a substrate heater which results in a heating source inside the substrates. This direct heating method is seen to be contrary to indirect ones, e.g. radiant heaters. Therefore, accurate temperature control is more difficult to attain.

The pros and cons of the different methods for temperature measurement will be discussed. As a conclusion, the optimum method of temperature control using an electron beam as a substrate heater will be described. Because of the bad reproducability of temperature measurements by thermocouples, the decision was made to use an FEM program to obtain more accurate temperature values of the substrates surfaces, and correlating them to the energy which is converted into heat inside the substrates. Confirmation of the FEM calculation was made by temperature measurements inside the samples. However, additional investigations are necessary and in progress.

INTRODUCTION

The use of ceramic thermal barrier coatings based on partially stabilised zirconia is of great interest to the gas turbine industry. With the deposition of such coatings on turbine blades the thermal stresses can be significantly reduced. Especially EB-PVD coatings which show an outstanding strain tolerance, and therefore, a very good thermal cycling behaviour due to their columnar structure, which can be seen in Fig. 1.[1] To realise such a columnar microstructure the substrate temperature plays an important role with EB-PVD deposition. Because the ionisation of vapour particles is less than 1% without additional ionisation equipment, energy can only be adopted into the de-

Fig. 1 Microstructure of thermal barrier coating when EB-heated.

posits by heating the substrates.[2] This can be done in different ways, as there are indirect heating methods, e.g. commonly used radiant heaters, or direct heating methods such as the electron bombardment of substrates using a second electron beam (EB)-gun The last method was used in the present investigation.

EXPERIMENTAL DETAILS AND MEASURING RESULTS

Heating of substrates can be done prior to or during the deposition process. Whereas in industrial EB-PVD plants the substrates are heated prior to deposition usually with radiant heaters. The substrate temperature during deposition is kept constant by a high deposition rate; the heat radiation of the melting pool and by electrons which are reflected on the surface of the melting pool. In contrast to these plants, heating in laboratory equipments was used during deposition to uncouple the evaporation parameters from the substrate temperature. In the EB-PVD equipment used for the present investigation a second EB-gun was installed as substrate heater. The experimental setup is shown in Fig. 2.

Especially when using an electron beam for heating the substrates, temperature measurement becomes more difficult. If, for example, a thermocouple is used as an in contact temperature measuring method it is also a target for the electrons. This results in two heating sources, one inside the substrate and one inside the thermocouple. This leads to wrong temperature values. Therefore, the thermocouples had to be arranged on the back of the samples, i.e. on the opposite side to the heating electrons and the vapour particles reaching the substrate.

Besides thermocouples, the only other appropriate measuring method considering the desired temperature range is a pyrometer. Pyrometers are quite expensive, com-

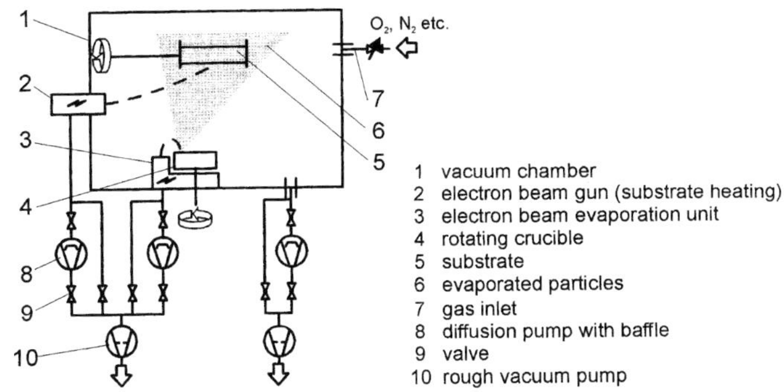

Fig. 2 Scheme of EB-PVD equipment.

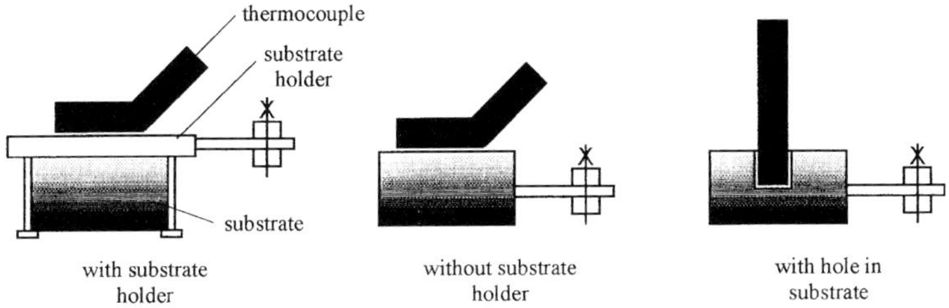

Fig. 3 Arrangement of thermocouple.

plex to set-up and use, but on the other hand they are a non-contact measuring system. Incorrect temperature results can also be observed when changes in the emission coefficient are occuring, e.g. a ceramic top layer is deposited on a metallic surface, and are not being considered.

Comparing both measuring methods, thermocouple and pyrometer, the thermocouple is the more desirable system because it is cheaper and simpler, although the substrate temperature can only be measured when the substrate is in a stationary coating arrangement. If the substrates are rotated as is usually the case for turbine blades, the temperature is not measurable with a thermocouple. As an alternative, an FEM-calculation was carried out to consider the situation in the EB-PVD plant, so as to overcome the boundaries and to learn more about the temperature distribution inside a EB-heated substrate. To check the reliability the calculated values, were compared with actual thermocouples measurements. As can be seen in Fig. 3, different arrangements of thermocouples were investigated and compared to each other.

Besides temperature measurement, EB-heating has the advantage of being able to measure the current of the heating electrons flowing through the substrate. This allows an easy determination of the actual heating power. Figure 4 shows the dependency between the temperature and the current of the heating electrons, for the different thermocouple arrangements, (see Fig. 3). It was found that the temperature value is

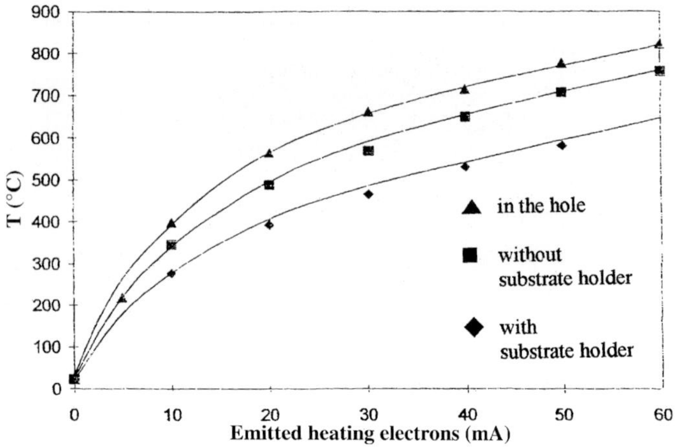

Fig. 4 Substrate temperature vs. current of heating electrons.

strongly dependent on the thermocouple arrangement. Although a specially flatened thermocouple was used to optimise the contact between the thermocouple and the sample, the difference in temperature values between a measurement on the substrate holder and inside a hole in the substrate was up to 26% which is shown in Fig. 4. This means, that the arrangement of a thermocouple is critical when considering the temperature data.

FEM-CALCULATION AND CALCULATED TEMPERATURE DISTRIBUTION

On the basis of the measured heating power input and the resulting substrate temperature the desired heating parameters can be determined, independent of a rotating or non-rotating substrate to be coated. Therefore, the FEM-calculation was carried out considering several boundary conditions; they are;

- Heat exchange by only heat conduction inside the substrate and heat radiation between substrate and all parts inside the process chamber, but no heat transfer by convection because of the high vacuum environment.
- Electron beam effects a target leading to backscattered electrons and secondary electrons as well as creating an area of energy conversion inside the substrate. X-rays and thermal electrons were neglected due to their low energy level.[3]
- As an additional boundary condition the temperature of the process chamber walls were found to be less than 200°C and could therefore be neglected because their influence on the substrate temperature is to small. In addition to this, the heat conduction through the substrate holder was neglected too.

The energy distribution of the electrons inside the substrate was calculated with a Monte-Carlo-Simulation. The results of the calculations are shown in Fig. 5, where

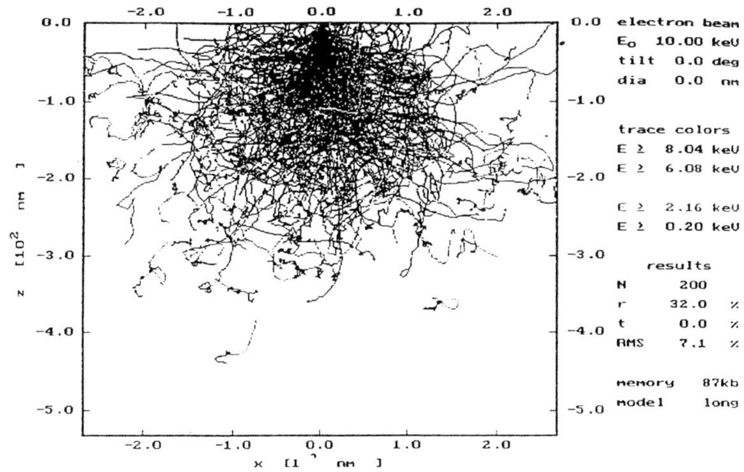

Fig. 5 Monte-Carlo-Simulation of electron distribution inside a substrate.

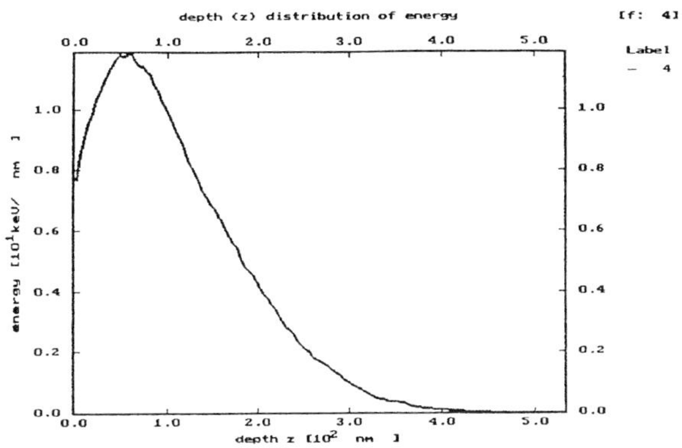

Fig. 6 Depth energy distribution due to heating electrons.

the tracks of 10 000 electrons were simulated with a spot-like penetration of the electron beam with an acceleration potentiel of 10 keV. The base material was an IN 738 LC which is a commercially available super alloy. With a rate of 33.6% backscattered electrons and 3.40% secondary electrons the electron distribution lead to an energy distribution which is two-dimensionally shown in Fig. 6 and a penetration depth of 0.5 μm.

On the basis of these data the temperature values were calculated with a commercially available FEM-program using a modular structure to include a flexibility that considers changing conditions, e.g. the change in substrate volume. Before the calculation could be carried out, a standard substrate had to be modelled. Figure 7 shows the model of the substrate with elements of the same surface dimensions but different

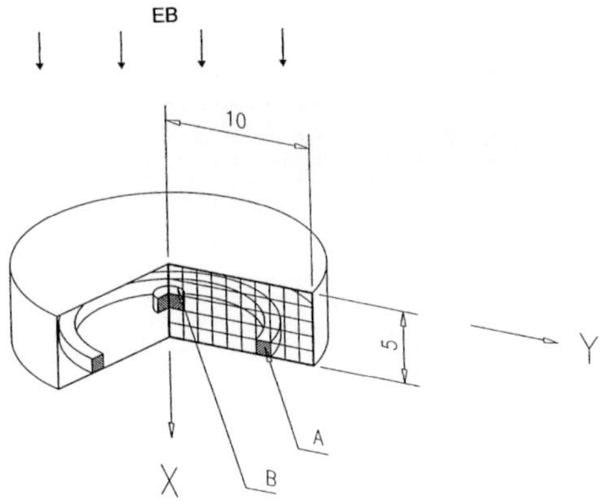

Fig. 7 FEM-model of the standardised substrate.

thicknesses. The whole substrate was made up in a rotational mode, rotating the cross section about a central axis. This cross section was divided into several elements of different thicknesses, see Fig. 8. This was so the energy input by the electron beam into a small area underneath the substrate surface could be considered.

The micoroscopic depth temperature distribution in a thin area of 5 μm-thickness underneath the substrate surface is polynomic, which is coincident with data taken from the litertature.[2,4] This independent of energy input. The macroscopic depth temperature distribution through the whole of the substrate thickness is polynomic as well. With raising the heating power the temperature is increasingly of a polynomic nature.

The two-dimensional temperature profile for realistic conditions and the quasi-stationary state after a heating time of 30 minutes with a heating power of 21 W is shown in Fig. 9. As can be seen, the temperature difference between the front and the back of the sample on the middle axis is just 20 K. Therefore, unrealistic temperature values are more likely to be due to a bad measurement arrangement than to a polynomic temperature distribution inside the substrate.

A second result is that the time until the temperature reaches 98% of its maximum value is dependent upon the effective heating power put into the substrate. Increasing the power input means that the time to reach the desired substrate temperature is drastically reduced. The resulting curves are shown in Fig. 10.

COMPARISON OF CALCULATED AND MEASURED DATA

Figure 11 shows the measured and calculated temperature values with the corresponding power input values are marked. Whereas the measured values represent the measurement arrangement inside the samples (see Fig. 3), the calculated values show the temperature on the substrate surface. As can be seen in Fig. 11, there is a difference between both types of data which is getting larger with increasing input heating power.

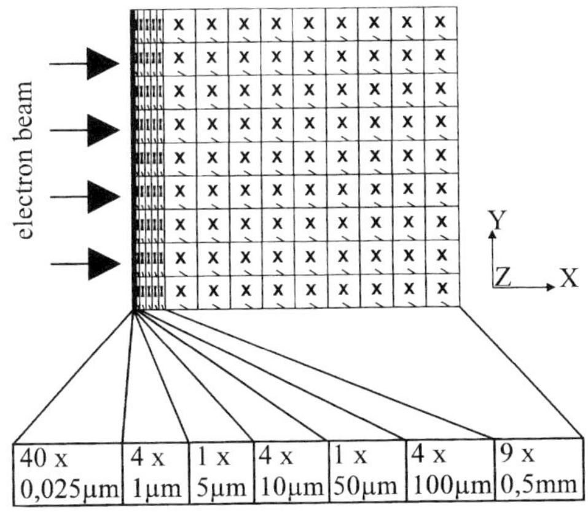

Fig. 8 Element grid of cross section of the standardised substrate.

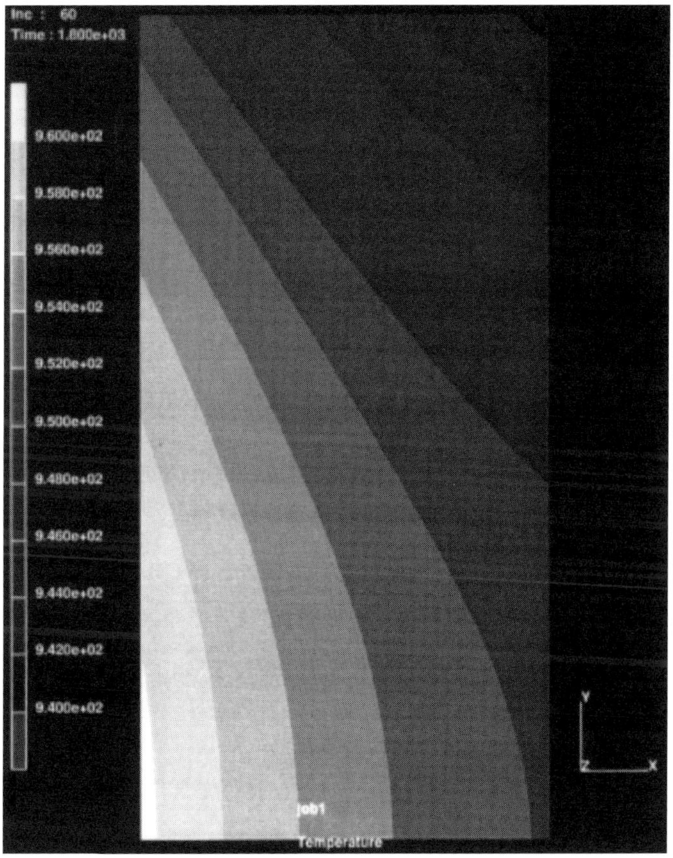

Fig. 9 Temperature profile inside substrate under quasi-stationary conditions.

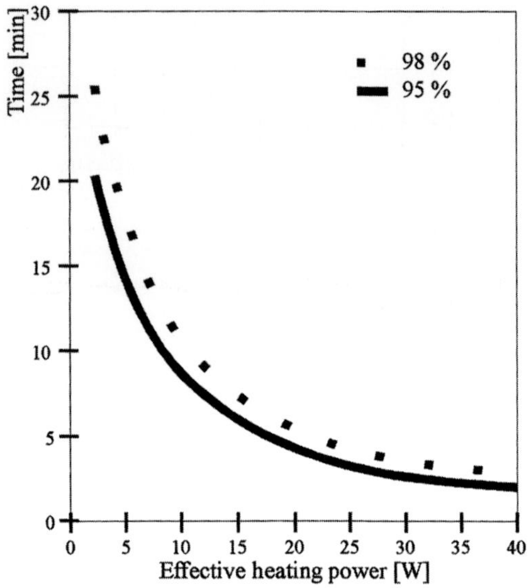

Fig. 10 Time until temperature reaches 95 % or 98 % of its maximum value.

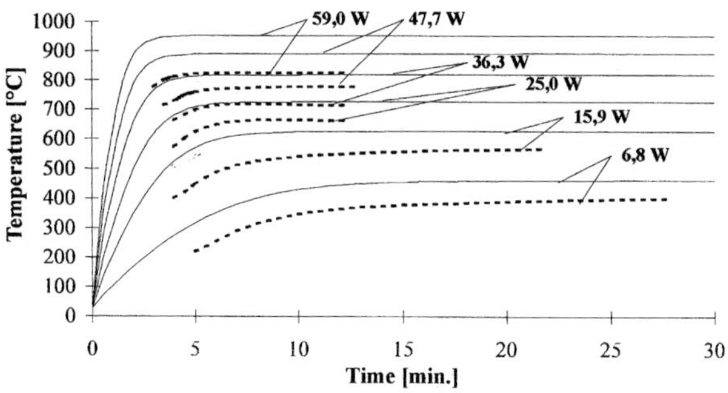

Fig. 11 Substrate temperature measured and calculated with different EB-power.

It is believed that the calculated data values are more likely to be correct, because the contact between the thermocouple and the substrate although, in a hole, was still bad.

CALCULATION OF SUBSTRATE TEMPERATURE WHILE DEPOSITING A THERMAL BARRIER COATING

The temperature changes during deposition of a zirconia thermal barrier coating were also calculated. The zirconia deposit was simulated as a differential change of material inside the substrate in steps of 50 μm because it was not possible to model the

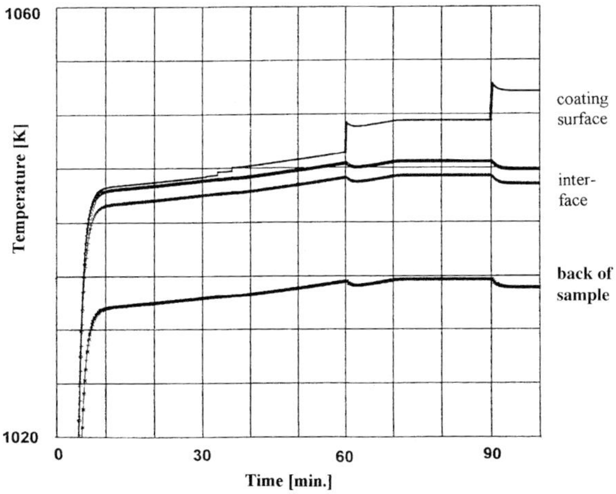

Fig. 12 Temperature development during deposition of TBC Deutscher Verlag fir.

coating as a real overlay material. This was also followed by modelling the temperature changes when a coating of 50 μm was 'deposited'. The changes in temperature are shown in Fig. 12. As can be seen, the main temperature change occurs on the coating surface and is increasing, whereas the interface temperature and the temperature on the back of the sample remained at the same level. The zirconia coating is therefore already reacting as a thermal barrier coating during the deposition process.

CONCLUSIONS

Thermal barrier coatings (TBC) are of great interest to the gas turbine industry. To optimise reproducability and microstructure of EB-PVD TBC substrate temperature control is very important. In the EB-PVD equipment used for TBC deposition substrates are directly heated using a second electron beam gun. By measuring the electron current flowing through the substrate, heating power can be determined and related to the resulting microstructure.

Depending on the arrangement of the thermocouple to the substrate a measuring error of up to 26% can be observed. An alternative to thermocouple temperature measurements is the FEM-calculation of the temperature distribution inside the substrate and the temperature development with time. This was done considering the important boundary conditions in the EB-PVD plant.

The microscopic and the macroscopic temperature distributions were found to be coincident with data taken from literature showing a polynomic nature. The temperature difference betweeen the front and back of the substrate is less than 30 K and cannot made be responsible for measurement deviations. Calculated temperature changes follow the same course as the measured ones, but there is still a difference between them. It is believed the calculated values are likely to be the more accurate. The time for reaching a desired substrate temperature is reduced with increasing heating input power. The temperature distribution inside the substrate during the deposition process was also calculated. Zirconia is working as a TBC while deposition process is going on. Based on these results temperature control is more reliable when using an EB-gun as a direct substrate heating method.

ACKNOWLEDGEMENTS

The authors appreciate Dr rer. nat. P. Karduck, Gemeinschaftslabor für Elektronenmikroskopie der RWTH Aachen, for his support and realising the Monte-Carlo-Simulations of the electron distributions. Also, the authors say thank you to Dipl.-Ing. T. Mahnken for his scientific support.

REFERENCES

1 E. LUGSCHELDER, C. BARIMANI, M. MOELLER and G. DOEPPER: 'Aufbau und Wirkprinzip keramischer Wärmedämmschichten', *Werkstoffe für die Enerietechnik,* H. W. Grunling ed., Symposium 3, Tagungsband zur Werkstoffwoche '96, DGM-Informationsgesellschaft-Verlag, Frankfurt, 1997, ISBN 3-88355-231-3, 177–183.

2. E. LUGSCHEIDER, C. BARIMANI and G. DOEPPER: 'Ceramic thermal barrier coatings deposited with the EB-PVD-technique', *Surface and Coatings Technolgy,*1997, to be published.

3. S. SCHILLER, U. HEISIG and S. PANZER: 'Elektronenstrahltechnologie', Wissenschafiliche Verlagsgesellschaft mbH, Stuttgart, 1977.

4. D KUPER: 'Strahltechnik 2', *DVS-Berichte 4*, SchweiBtechnik, Dusseldorf, 1986, 217–244.

Some Effects of Metallic Substrate Composition on Degradation of Thermal Barrier Coatings

I. G. WRIGHT, B. A. PINT, W. Y. LEE*, K. B. ALEXANDER AND K. PRÜSSNER

Oak Ridge National Laboratory, Oak Ridge, Tennessee, USA

**Stevens Institute of Technology, Hoboken, New Jersey, USA*

ABSTRACT

Comparisons have been made in laboratory isothermal and cyclic oxidation tests of the degradation of oxide scales grown on single crystal superalloy substrates and bond coating alloys intended for use in thermal barrier coatings systems. The influence of desulphurisation of the superalloy and bond coating, of reactive element addition to the bond coating alloy, and of oxidation temperature on the spallation behavior of the alumina scales formed was assessed from oxidation kinetics and from SEM observations of the microstructure and composition of the oxide scales. Desulphurisation of a nickel-base superalloy (in the absence of a Y addition) resulted in an increase in the lifetime of a state-of-the-art thermal barrier coating applied to it, compared to a Y-free, non-desulphurised version of the alloy. The lifetime of the same ceramic coating applied without a bond coating to a model alloy (NiAl doped with Zr but not desulphurised) that formed an 'ideal' alumina scale was found to be, in the same test, at least four times longer than on the Y-doped superalloy plus state-of-the-art bond coating combination. Some explanations are offered of the factors controlling the degradation of such coatings.

BACKGROUND

As part of the Materials and Manufacturing segment of the U.S. Department of Energy's Advanced Turbine System (ATS) program,[1] efforts are being made to facilitate the scale-up of single crystal casting processes for industrial turbine-sized vanes and blades,[2,3] and to develop improved thermal barrier coating systems for these components.[4,5] An integral part of the casting program is the development of a production-scale process for desulphurising the alloys to avoid the need for small alloying additions of yttrium, the levels of which are difficult to control in the single crystal casting process without using alumina cores and face coats on the molds.

Studies by several groups, including Ikeda,[6] Funkenbusch, Smeggil and Borstein,[7–9] Luthra,[10] Smialek *et al.*,[11–16] Schaeffer *et al.*,[17] and Meier et al.,[18] have shown that lowering the tramp sulphur content of alumina scale-forming alloys to sub-ppm (atomic) levels can dramatically improve the resistance to scale loss by spallation. For those studies, the desulphurisation was typically achieved by hydrogen annealing of relatively thin specimens at temperatures of 1200°C or greater, under conditions where the alloy surface remained scale-free so that any sulphur diffusing to that surface could be removed by reaction with hydrogen. Smialek[13] suggested that, for optimum improvement in resistance to spallation, the total sulphur content of a component should be reduced to less than the level required to produce one monolayer upon segregation; for specimens of the thickness used in this study, this corresponded to approximately 0.4 ppma.[14,16]

Comparisons of the influence of sulphur and reactive element additions led to the suggestion that the role of reactive elements such as yttrium is to prevent sulphur segregation during oxidation.[7] Identification of yttrium sulphides in the alloy substrate[19] was used to justify an argument for sulphur gettering by yttrium, whereas others have proposed that the role of yttrium is to modify the surface energy of the relevant interfaces, thereby inhibiting the segregation of sulphur to those locations.[20] Other observations suggest that reactive element additions exert the further effect of slowing the rate of growth of alumina scales as a result of incorporation of the reactive element ions into the oxide, and their segregation to the oxide grain boundaries.[20–22]

The focus of the study reported here is on routes to developing a practically 'ideal' protective alumina scale on alloys used as bond coatings on nickel-base superalloys, as a contribution to maximising the life of the thermal barrier coatings systems (TBCs) that will be an essential part of their application in advanced turbines. This paper details the oxidation behaviour in which one of the Ni-based single crystal superalloys, René N5 (General Electric Aircraft Engines) from the ATS single crystal casting program conducted by PCC Airfoils, Inc.,[2] is used as the substrate. In addition, cast (not single crystal) β-NiAl doped with an optimum amount of Zr, which has been found to form an essentially 'ideal' alumina scale[21–23] was used as a experimental bond coat alloy.

EXPERIMENTAL PROCEDURES

A yttrium-doped single crystal superalloy, René N5, was used as a basis for comparison of desulphurised, Y-free versions of the same alloy (denoted as N5X). Desulphurisation was performed by PCC Airfoils Inc.[2] using a proprietary in-melt technique, a post-solidification hydrogen treatment, and a combination of both. Alloy N5A is nominally René N5 without Y and not desulphurised; alloy N5AH was hydrogen-desulphurised; alloys N5B and N5C were melt-desulphurised; and alloy N5BH was melt-desulphurised and then hydrogen annealed. A Zr-doped NiAl alloy was produced by vacuum arc melting. The chemical compositions of the alloys are shown in Table I. The sulphur levels of the alloys subjected to desulphurisation were measured by the glow discharge mass spectroscopy technique (GDMS). At present there are uncertainties when using GDMS to measure very low levels of sulphur because of the lack of a sulphur standard in the appropriate range. Hence, the absolute values of the sulphur contents shown in Table I are questionable. Nevertheless, in the absence of more definitive sulphur measurements, the values reported from GDMS analyses are

Table I Chemical compositions of alloys studied (atomic percent).

Element	Nominal René N5	René N5	N5A	N5B	N5C	N5AH	N5BH	NiAlZr
Al	13.79	13.87	13.79	13.41	13.49	13.33	13.31	49.65
Cr	8.08	7.79	8.09	8.02	8.14	8.16	8.13	ND
Co	7.64	7.28	7.43	7.48	7.49	7.49	7.58	–
W	1.63	1.61	1.38	1.65	1.68	1.66	1.65	–
Ta	2.16	2.11	1.98	2.17	2.14	2.16	2.16	–
Mo	0.94	0.90	0.77	0.89	0.88	0.89	0.93	–
Re	0.97	1.02	0.92	0.98	0.99	0.98	0.98	–
Ti	0.03	0.01	ND	0.01	0.01	0.01	0.01	ND
C	0.25	0.25	0.25	0.25	0.03	ND	ND	0.02
Hf (ppma)	538	537	398	539	506	540	505	–
Y (ppma)	17–67	34	ND	ND	ND	ND	ND	ND
Zr (ppma)	0.00	33	<65	66	66	66	66	400
Si	0.00	0.15	0.11	0.19	0.09	0.06	0.11	0.21
Fe	0.00	0.08	0.35	0.09	0.09	0.09	0.09	0.02
Mn	0.00	ND	ND	ND	ND	ND	ND	–
B (ppma)	0.00	277	274	278	278	223	167	–
S (ppma)	0.00	7.5	5.5	1.5	0.8	1.3	0.6	47
P (ppma)	0.00	39	57	ND	78	39	58	–
O (ppma)	0.00	52	96	49	68	53	2166	20
N (ppma)	0.00	ND	25	22	26	ND	17	–
Ni	64.47	64.82	64.81	64.74	64.88	65.02	64.68	50.04

ND: not detected

used in the paper. Overall, the compositions of René N5 and the Y-free derivatives were close to the nominal composition with the exceptions that the Y content of René N5 was half that expected, and the Ti contents of all the alloys were low. Note, also, that the alloys subjected to hydrogen annealing had been not only desuphurised, but also essentially decarburised.

Specimens of approximate dimensions 20 x 10 x 1 mm were cut by electrode discharge machining from specially cast single crystal bars, so that the large surfaces of the specimens corresponded to the same crystal orientations as the surfaces of an airfoil. The specimen surfaces were polished through 600 grit SiC papers prior to oxidation; specimens intended for examination of the scales by electron microscopy were polished through 0.3 μm alumina prior to oxidation (the difference in surface finish was found to make a small difference in the initial weight gain, but no difference in the shape of the oxidation curve). All specimens were measured to 0.01 mm, weighed to 0.01 mg, and then ultrasonically degreased in acetone and methanol before oxidation. Specimens were suspended in the oxidation furnaces from Pt–Rh wires. Appropriate corrections for evaporative loss of Pt and Rh were made in evaluating the oxidation kinetics.

The cyclic oxidation test cycle involved heating from near ambient temperature to test temperature, holding at temperature for 1 h or 100 h, before cooling. Tests with a 1 h dwell time at temperature were conducted in an automatic rig in which individually suspended specimens were introduced directly into a hot furnace into which oxygen was slowly flowing. At the end of each exposure period, the specimens were withdrawn into individual ceramic housings for a 10 min. cooling period. This cycle was interrupted for weighing every 20 or 50 cycles. In a second cyclic oxidation test, individual specimens were placed in lidded, preannealed alumina crucibles and exposed in air in an electric box furnace. The specimens were loaded into the cold furnace, which was then heated and held at temperature for 100 h, and then allowed to cool, after which the specimens were removed and weighed. By weighing each crucible with and without the specimen, the specimen weight change and the weight change due to the total uptake of oxygen were obtained. Isothermal oxidation kinetics also were measured for direct comparison. These exposures were carried out in dried oxygen in a Cahn model 1000 microbalance, which weighed to 0.01 mg. The runs were started by lowering specimens into a heated furnace; at the end of the exposure period, they were quickly removed from the furnace.

The test temperatures ranged from 1050 to 1200°C; the projected maximum temperature experienced by the bond coating-ceramic interface of TBC-bearing airfoils in advanced land-based gas turbines is at the low end of this range, whereas airfoils in aircraft gas turbines may experience peak metal temperatures at the higher end of this range.

After oxidation the specimens were examined in a low-power stereo optical microscope. Selected specimens were mounted, cross sectioned and metallographically polished for examination by conventional light microscopy. Both oxidised surfaces and polished cross sections were characterised using a scanning electron microscope (SEM) equipped with a field-emission gun (FEG) and energy-dispersive X-ray detector (SEM/EDS). Concentration profiles across mounted and polished cross sections were measured using wavelength dispersive spectroscopy in an electron probe microanalyser (EPMA).

EXPERIMENTAL RESULTS

Isothermal Oxidation

René N5

A parabolic plot of the isothermal oxidation kinetics for René N5 in oxygen at 1100, 1150, and 1200°C, shown in Fig. 1(a), indicates that the overall oxidation behaviour of this alloy approximated a parabolic rate law over the 100 h exposures reported, even though the values of the rate law exponent (n) calculated from the data ranged from 1.9 to 3.6. These values are shown in Table II, which also lists derived parabolic rate constants (k_p). The scale thicknesses calculated from the microbalance data (as-

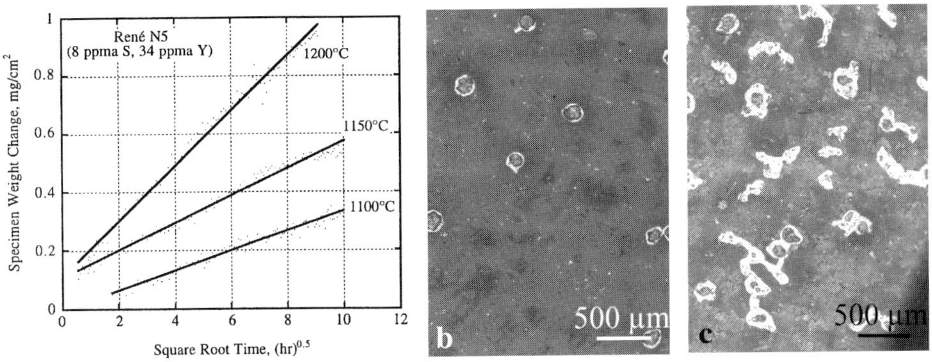

Fig. 1 Isothermal oxidation kinetics (a) parabolic plot of René 5 and topographies of scale formed at (b) 1150°C and (c) 1200°C.

Table II Parabolic rate constants (k_p) for alloys studied (10^{-13} g^{2}cm^{-4}s^{-1}).

Alloy	1100° C		1150° C		1200° C	
	n	k_p	n	k_p	n	k_p
René 5	1.9	3.2	3.6	6.0	2.7	26.6
N5A	2.0	13.2	–	–	2.0	60.8
N5B	2.2	4.5	2.9	10.6	2.2	24.2
N5C	2.8–1.9	6.0–8.1	3.6	9.6	2.4–2.8	36.7–14.8
N5AH	2.4	3.6	2.3	13.3	2.9	32.1
N5BH	1.9	2.4–6.4	3.0	10.5	2.1	44.9
N5 mean		6.3		11.0		35.6
NiAlZr	2.2	4.7	2.2	17.9	2.1	26.8

suming the scale to be all Al_2O_3) were 0.9, 1.4, and 4.1 μm at 1100, 1150, and 1200°C, respectively, which corresponded well with measurements from cross sections.

René N5 formed a grey-appearing, adherent oxide scale which, with increasing time and temperature, developed a blue cast which was associated with partial spallation of an outer layer of Ni, Al-based spinel. The surfaces also contained small (approximately 150–200 μm in diameter) blue-appearing areas of localised spallation which increased in number with time and temperature. These areas are shown in low-magnification topographical views of the scale after 100 h at 1150°C and 1200°C in Figs 1(b) and 1(c), respectively, and were associated with 2 to 10 μm-sized Ta-rich oxide phases in the alloy surface. These blue-appearing areas were not present in the initial scale, but developed with time. The blue appearance was associated with the initial, partial spallation of the spinel layer in the vicinity of Ta-rich particles. Figure 2(a) indicates that such Ta-rich particles that intersected the alloy surface were undercut by the growing alumina scale, and effectively isolated from the alloy. At some point, the scale surrounding them fractured and spalled, typically at or near the metal-oxide interface, Fig. 2(c). For the isothermally-exposed specimens, this event probably occurred on cooling from the test temperature at the end of the test since the bare alloy surface was visible in plan views, Fig. 2(b). On specimens exposed in the cyclic tests these areas were overgrown with voluminous scales of mixed oxides.

Typical images of the intact oxide scales on René N5 are shown in plan view in Figs 3(a) and (b) and in cross section in Figs 3(c–f). Small (approximately 0.1 μm-sized) Ta precipitates were found in the grain boundaries of the adherent alumina scale, and as small particles in the outer scale, as is evident in cross sections after 100 h at 1150°C and 1200°C, Figs 3(c) and 3(d). Tantalum-rich particles incorporated into the alumina scale would be expected to migrate to the outer surface of the scale;[20,24] since the number of such particles was greatly increased at 1200°C compared to 1150°C, it appears that they may have resulted from outward diffusion rather than being part of the initially-formed transient scale. The bulk of the relatively thick, outer layer of presumed transient scale was a Ni,Al spinel. Fracture cross sections, Figs 3(e) and 3(f) showed the structure of the main alumina scale to be mostly columnar, but the structure was not as consistently columnar as, for instance, on an ideally-doped β-NiAl alloy.[22]

Y-Free and desulphurised alloys

The isothermal oxidation kinetics of the Y-free variants of René N5 with different degrees of desulphurisation are shown as parabolic plots in Fig. 4, and the calculated rate law exponents and derived parabolic rate constants are shown in Table II. In general, the isothermal oxidation behavior of the N5 alloys was little different from that of René N5. The rate constants in Table II suggest that the oxidation rate of René N5 was slightly slower than the Y-free alloys at all three temperatures, but any such trend was small.

The scale formed on alloy N5A spalled extensively on cooling from the test temperatures. The appearance of the scales of the other N5 alloys was similar to that of

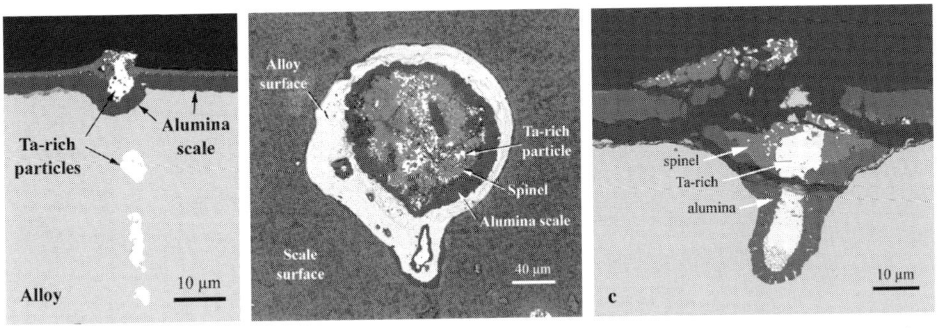

Fig. 2 Scale morphologies associated with iocal spallation on René N5 after isothermal oxidation at 1150°C for 100 h showing (a) cross section showing incorporation of Ta-rich alloy phase into the scale; (b) topography of an area where spallation has occured to the metal surface, and (c) a cross section of an area similar to (b) after 100 h at 1200°C.

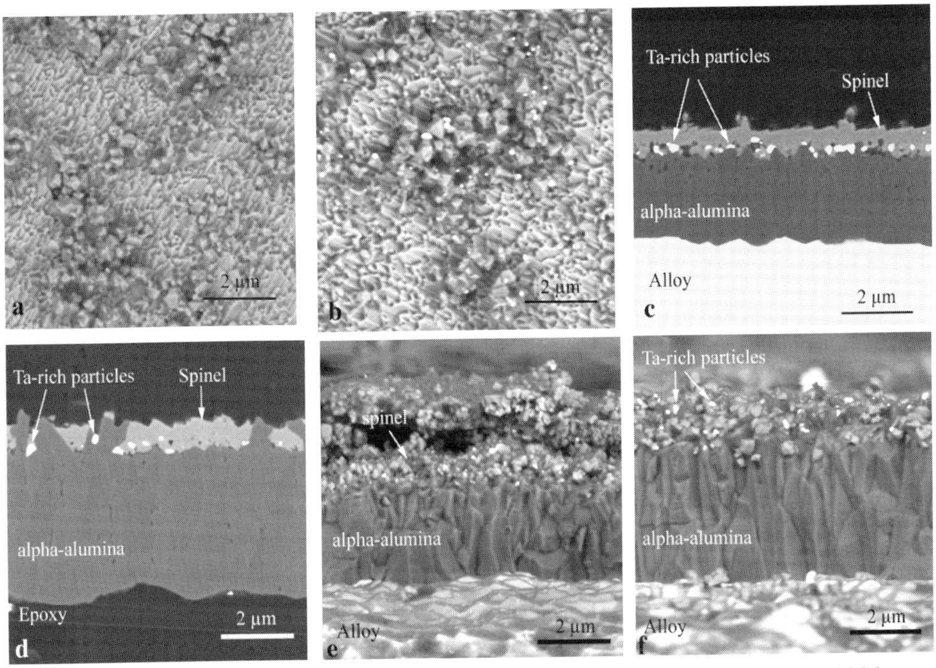

Fig. 3 Morphology of scale formed on René N5 after isothermal oxidation for 100 h at 1150°C and 1200°C: (a) and (b) plane views; (c) and (d) SEM cross sections; and (e) and (f) fracture sections of the scales.

those formed on René N5: they were essentially grey with a few blue spots, and developed an increasingly blue cast and more numerous blue spots with increasing temperature. The surfaces of the scales formed on N5AH after 100 h at 1150°C, and on N5BH after 100 h at 1200°C are shown in Figs 5(a) and b, and 6(a) and (b), respectively. On N5AH, the circular (blue) areas did not exhibit spallation to the alloy sur-

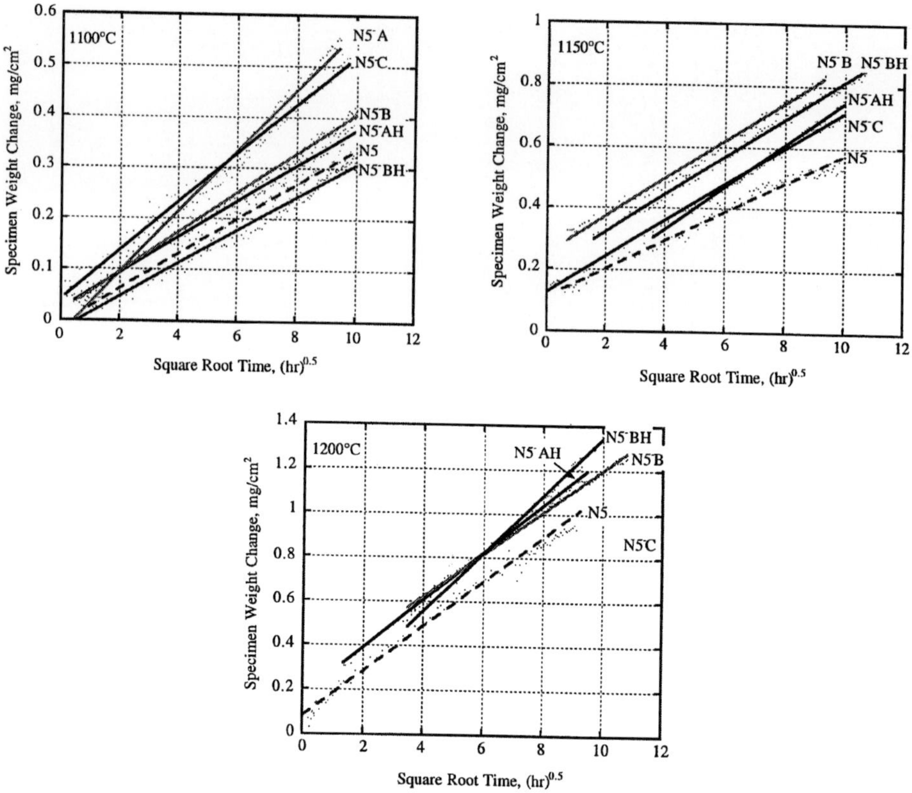

Fig. 4 Isothermal oxidation kinetics of desulphurised N5 alloys, parabolic plots.

Fig. 5 Morphology of scale formed on N5AH after isothermal oxidation for 100 h at 1150°C: (a) and (b) plan views; (c) SEM of fracture section of the scale.

face, but the scale formed over these areas contained faceted crystals of NiO, Fig. 5(b), and bright-appearing Ta-rich particles were evident on the surface of the alumina scale. Figure 6(a) indicates spallation to the alloy at some of the circular areas and along the edge of the specimen of N5BH oxidised for 100 h at 1200°C. The surface of

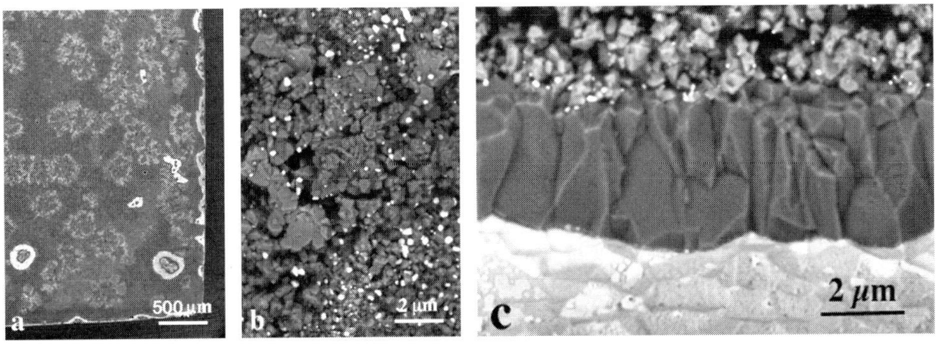

Fig. 6 Morphology of scale formed on N5BH after isothermal oxidation for 100 h at 1200°C: (a) and (b) plan views; (c) SEM of fracture section of the scale.

the intact scale exhibited large crystals of Ni, Al spinel (shown in Fig. 6(b)). Bright-appearing Ta-rich particles also were visible on the surface of this scale.

Fracture cross sections of these scales, shown in Figs 5(c) and 6(c), indicated that the main alumina scale on both alloys had a columnar structure that was as well developed as that on René N5, despite there being no Y dopant present. This structure is indicative of doping of the scales with a reactive element; in these cases the dopant is presumably Hf (Table I) although the transmission electron microscopy of these scales required to confirm this assumption has not yet been completed. In comparison with René N5, the alumina layer formed on N5BH after 100 h at 1200°C was thinner, and the spinel layer was thicker. The thicknesses of the scales corresponded well with those calculated from the overall weight gains (assuming all the scale thickness was alumina).

Cyclic Oxidation

Rend N5 and desulphurised variants

Cyclic oxidation kinetics (1 h cycles) of the alloys at 1100 and 1150°C are shown in Figs 7(a) and (b). While an improvement in resistance to scale spallation is obvious from the data for 1100°C when the sulphur level was reduced from 5–6 ppma (N5A) to approximately 0.8 ppma (N5C) in the absence of Y-doping, there was not much discrimination among the kinetics of alloys with lower sulphur levels, nor between René N5 and the lower sulphur alloys at this temperature. It was assumed that the maximum in the specimen weight gain curve could be used to indicate the time at which major spallation of the scale started, and the isothermal kinetic data were used to estimate the oxide thickness at that time. For N5A, the estimated maximum scale thickness attained before spallation initiated was relatively small, approximately 0.9 µm. The scale thicknesses on the other alloys which did not exhibit major spallation ranged up to approximately 2.2 µm.

At 1150°C, alloys N5B and N5C (1.5 ppma S and 0.8 ppma S, respectively) suffered weight losses due to scale spallation at earlier times than René NS and alloys N5AH (1.3 ppma S) and N5BH (0.6 ppma S). Cyclic oxidation results from other

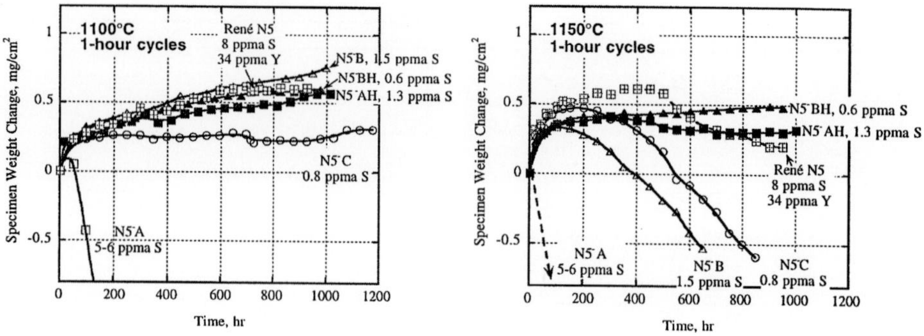

Fig. 7 Cyclic oxidation kinetics of desulphurised N5 alloys, 1 hour cycles.

work in this study indicated a consistent improvement in scale spallation behaviour for alloys N5AH, N5BH, and N5C, but that the order of improvement was N5BH>N5AH>N5C, although chemical analysis indicated that the sulphur content of N5C was lower than for N5AH. The estimated scale thicknesses at which major spallation started at 1150°C was 1.0 to 1 .2 μm for N5B and C, and 1.6 μm for René N5. Alloys N5AH and BH attained estimated scale thicknesses of0.8 to 1.2 μm without apparent spallation. Figs 8(a) and (b) present a comparison of the cyclic and isothermal kinetics for René N5, N5A and N5C at 1100 and 1150°C, respectively. Assuming that the difference between the isothermal and cyclic kinetic curves is a measure of scale loss by spallation in the cyclic test, the degree of spallation appears to decrease with decreasing S level, and to increase with increasing temperature. The spallation observed from René N5 apparently was a reflection of the fraction of the oxide surface affected by Ta-rich particles.

The kinetics from the test involving 100 h cycles at 1100°C are shown in Fig. 9. These data also indicate a trend to decreased scale spallation with decreased alloy sulphur level. Compared to the 1-hour cyclic test at 1100°C, this test provided more discrimination among the alloys, with results similar to the 1-hour cycle test at 1150°C. The fact that the N5BH alloy (0.6 ppma S) exhibited early spallation was unexpected; the estimated scale thickness at which this started was 1.9 μm. The scale thicknesses corresponding to the onset of spallation on N5B and N5C were 1.3, and 2.2 μm, respectively, whereas René N5 and N5AH attained estimated thicknesses of 2.5 to 2.7 μm without obvious spallation. The experimental procedures used in the 100 h cyclic test allowed a comparison to be made of the weight change of the specimen with the overall weight change due to oxygen pickup. As shown in Fig. 10, such comparisons for René N5 and N5AH indicate some degree of scale spallation from René N5, but very little spallation from N5AH. The observed difference in spallation behaviour was that on René N5 the oxide cap over the Ta-rich particles spalled together with surrounding oxide scale, whereas on N5AH there was more fracturing of the spinel part of the scale but less loss.

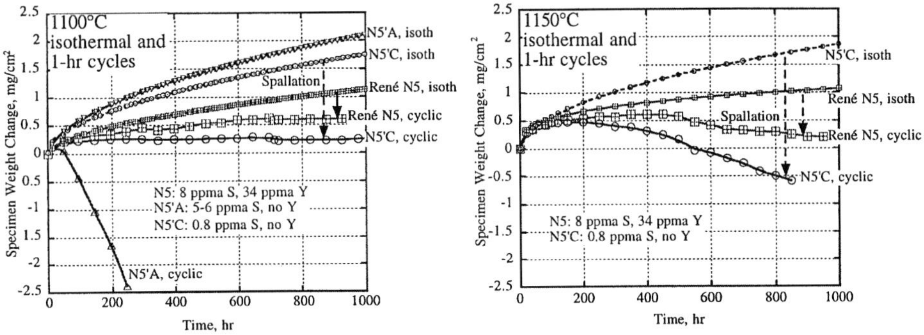

Fig. 8 Isothermal and cyclic (1 hour cycles) oxidation kinetics for René N5 and Y-free and desulphurised variants in oxygen at (a) 1100°C and (b) 1150°C.

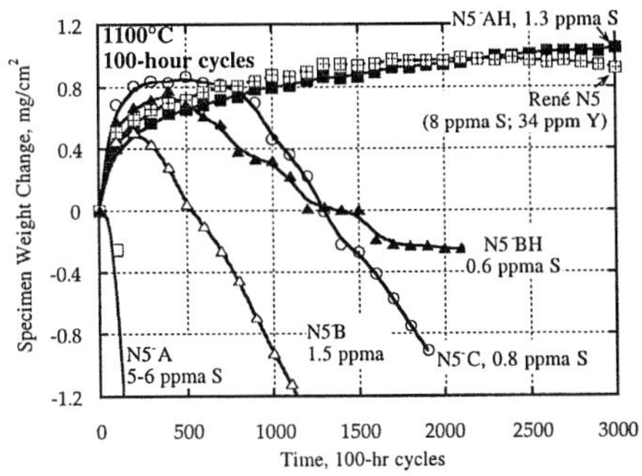

Fig. 9 Cyclic oxidation kinetics (100 hour cycles) at 1100°C in air.

Low-sulphur bond coating

A laboratory chemical vapor deposition (CVD) rig was modified to produce low-sulphur NiAl bond coatings on the desulphurised N5C (0.8 ppma S) substrate.[25] EPMA concentration profiles through the approximately 25 μm coating thickness (Fig. 11(a)) indicated relatively uniform levels of Ni and Al (approximately $Ni_{1.3}Al$), as well as Co (approximately 9 wt %) and Cr (approximately 3 wt %). There were also indications that heavy elements such as Ta, W, and Re had diffused into the aluminide coating, to levels of less than 1 wt %. In backscattered electron images the coating grain boundaries appeared much lighter than the adjacent grains, implying that these heavy element dopants were most likely segregated at the metal grain boundaries. Incorporation of sulphur into the coating was examined by GDMS, using sputtering on a flat surface of a coated specimen. Despite problems in interpreting the concentration profiles due to variations in sputtering rate within the sampled area, profiles for the major elements were consistent with the EPMA data. The sulphur level of the coating shown in Fig. 11(b) was approximately 11 ppmw at the beginning of sputtering, falling into the 1 to 2 ppmw range after 10 min. of sputtering, and then rising to approximately 7.5 ppmw

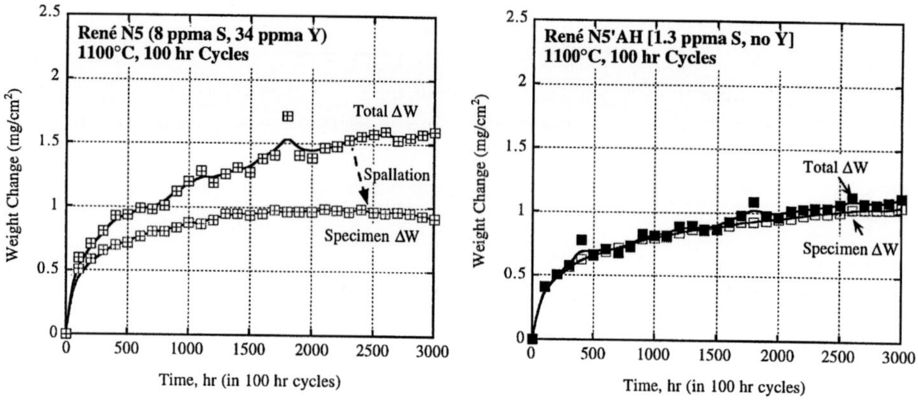

Fig. 10 Cyclic oxidation kinetics (100 hour cycles) at 1100°C in air showing the difference between the specimen weight change and that due to the total oxygen uptake for (a) René N5 and (b) a Y-free variant (N5AH) desulphurised to 1.3 ppma.

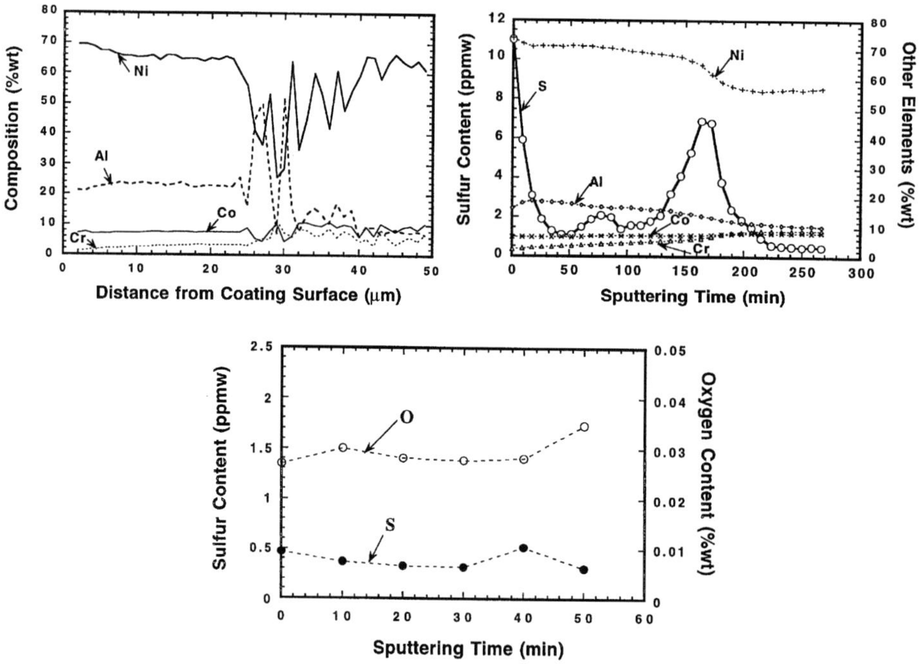

Fig. 11 Concentration profiles through the as-coated alloy by (a) EPMA, and (b) GDMS; and through the uncoated substrate by GDMS (c).

after 150 to 170 min. of sputtering. The concentration profiles for sulphur and oxygen made by the same technique in the uncoated N5C substrate remained relatively flat, with the indicated sulphur level in the range 0.3 to 0.5 ppmw (0.6 to 0.9 ppma), Fig. 11(c).

The sulphur peak on the coating surface was found to result from the coating process (from Viton 'O'-rings). This outer S-rich layer was removed before oxidation exposures by mechanically polishing the coating surface.[25] The inner sulphur peak appeared to be located in the same region as the refractory element-rich 'diffusion region' between the coating and the substrate. As shown in Fig. 12(a), the scale spallation behaviour of such a coating up to approximately 500 hours of thermal cycling in the 1 h cycle test at 1100°C was essentially the same as for the uncoated N5C substrate, and was significantly improved compared to the undoped, non-desulphurised alloy (N5A). When scale spallation occurred from the coating after 500 cycles, the loss of scale was from the vicinities of the surface grain boundaries of the coating; an adherent alumina scale continued to form over the grain bodies of this coating. Elements such as Ta, Co, Cr, and W from the superalloy substrate were found to be enriched in the coating grain boundaries, and led to the formation of less protective, less adherent oxides over these regions. Figures 12(b) and (c) illustrate the surface topography of the coating before and after oxidation, and show the nature of the scale spallation. The formation of grain boundaries in these coatings appeared to be associated with impurities such as oxide particles present on the substrate surface prior to coating.

TBCs on N5 substrates

Figure 13(a) shows the oxidation kinetics from a 1 hour cycle test at 1150°C for René N5, N5A and N5C coated (on all sides) with a state-of-the-art platinum–aluminide bond coating, and on one side with a top coat of yttria-stabilised zirconia (YSZ) deposited by electron beam physical vapor deposition (EB-PVD). The criterion used to assess failure of the TBCs was the loss of at least 20 percent of the ceramic layer on the flat surfaces of the specimens. The loss of small chips of ceramic from the edges of the specimens was discounted, since the specimen shape and dimensions were not optimised to accommodate a TBC. The ceramic layer on N5A failed after 300 cycles; on N5C failure occurred after 450–500 cycles; whereas the TBC on René N5 failed after 650 cycles. On this basis, the life of the coating on the non-desulphurised. Y-free substrate (N5A) was approximately one half that on René N5, whereas the life of that on N5C (~0.8 ppma S) was approximately 75 percent that on the Y-containing alloy.

TBCs on an 'ideal' alumina-forming bond coating

The same EB-PVD YSZ coating was directly applied to one side of specimens of the model bond coat alloy β-NiAl–Zr to demonstrate the potential spallation-free lifetime possible when the bond coating forms an 'ideal' alumina scale.[23] Previous studies[21,22] have shown that this alloy forms an α-Al_2O_3 scale that grows almost entirely by oxygen transport, at a rate up to 2 to 4 times slower than that formed on undoped NiAl (or other undoped alumina-forming alloys), and which shows only edge spallation at temperatures in the range studied here for times in excess of 2500 hours (oxide thickness greater than 9 μm). As shown in Fig 13(b), the lifetime of the ceramic layer on this substrate was more than 2500 cycles, at least four times that of the René N5–PtAl bond coat system in the same test. Surprisingly, a YSZ coating deposited by plasma-

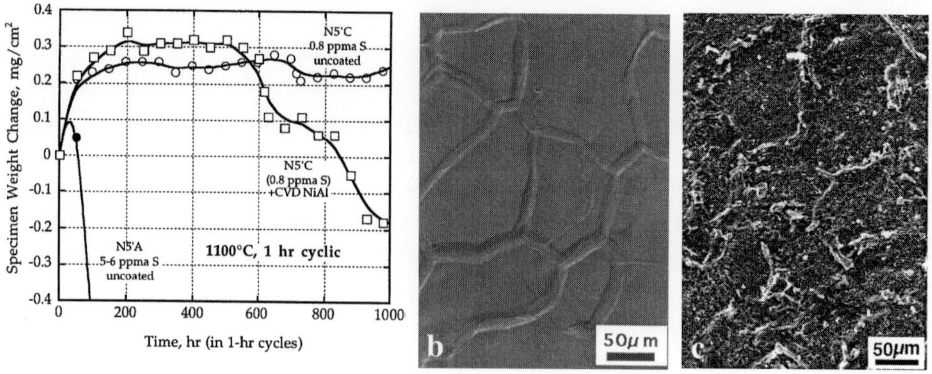

Fig. 12 Cyclic oxidation behaviour (1 hour cycles, 1100°C) of Y-free, desulphurised (0.7 ppma S) René N5 with and without a low-S NiAl coating (a) kinetics, (b) coating topography before exposure (c) coating topography after exposure.

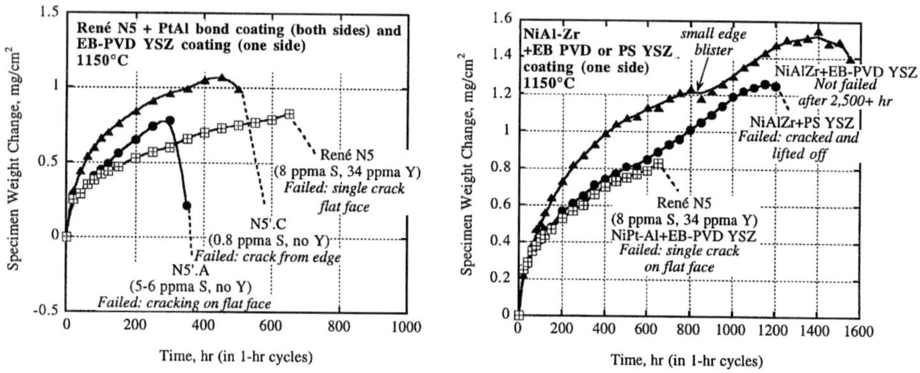

Fig. 13 Cyclic oxidation kinetics (1 hour cycles, 1150°C) of substrates with TBCs (a) René N5 and Y-free and desulphurised variants (b) René N5 compared to β-NiAlZr.

spraying directly onto the polished surface (600 grit finish) of the β-NiAl–Zr alloy was found to spall only after 1200 cycles, which was almost double the lifetime of the René N5–PtAl–EB-PVD YSZ system in the same test.[23]

DISCUSSION AND CONCLUSIONS

Desulphurisation of Y-free René N5 using a production-scale process was found to be capable of achieving excellent resistance to scale spallation, at least under the conditions of the tests used in this study. Overall, there appeared to be little obvious difference in the performance of René N5, and a Y-free version desulphurised by post-solidification hydrogen annealing to an indicated sulphur level of 1.3 ppma (<0.05 ppma C), or by a combination of in-melt processing and post-solidification hydrogen annealing to a sulphur level of 0.6 ppma (<0.05 ppma C). In 1 hr cycle tests at 1100°C, desulphurisation to the order of 1.5 ppma appeared sufficient whereas, in 100 hr cycle tests at 1100°C and 1 hr cycle tests at 1150°C, improved scale spallation resistance was not experienced until lower sulphur levels were achieved. Given the uncertainty of the absolute value of the final sulphur level of the desulphurised alloys, it appeared that the one of the best-performing alloys (N5BH, 0.6 ppma S) probably was desulphurised to a sulphur level near that found to be optimum by Smialek,[13] and which corresponded to a surface coverage of sulphur upon segregation of less than one monolayer. The other best-performing alloy (N5AH) had apparently been desulphurised to only 1.3 ppma.

The composition of René N5 includes small additions of Hf, Zr, and Ti (Zr is not a specified addition, but probably is present as an impurity with Hf) in addition to Y, all of which are elements that have been considered to exert a 'reactive element effect' during high-temperature oxidation. If indeed a major manifestation of the reactive element effect is to prevent or inhibit the segregation of tramp sulphur at the substrate-oxide interface, it is of interest to enquire why, in the absence of Y, Hf or Zr did not exert any beneficial effect in alloy N5A. There is significantly more Hf present in the alloy than there is Y, and Hf as well as Y were found to have segregated to the grain boundaries (and presumably the scale-substrate interface) of the protective oxide formed on René N5,[23] although Ta was typically the dominant segregant. One factor may be that, while both Y and Hf form extremely stable oxides (with Y_2O_3 being the more stable – having a more negative free energy of formation – than HfO_2), Y also forms very stable sulphides (Y_2S_3, and YS), whereas the stability of Hf sulphide (HfS_2) is significantly lower, of the same order as the most stable sulfides of Ti and Zr.[26] Also, while both Y and Hf form very stable carbides and nitrides, in both cases the Hf compounds (HfC and HfN) have a significantly more negative free energy of formation than the Y compounds (YC_2 and YN). As pointed out by Sigler,[26] these differences suggest that Hf most likely would react with carbon, nitrogen and possible oxygen before completely reacting with sulphur. In contrast, while Y will react strongly with oxygen and sulphur, it is likely to react preferentially with sulphur because of the much greater differential in free energy of formation between Y_2S_3 and Al_2S_3 than between Y_2O_3 and Al_2O_3.

The two Y-free, desulphurised alloys that exhibited the most improved resistance to scale spallation (N5AH and BH) also had been decarburised as a result of the hydrogen annealing part of the desulphurisation process (see Table I). Although analysis is not presently available, it is possible that decarburisation influenced the improvement in resistance to scale spallation by, for instance, allowing the Hf content of the alloys to play a more active role (similar Hf contents in NiAl produce slow-growing, very adherent alumina scales). Further, the less detrimental role of Ta-rich particles on local scale spallation from these two alloys may also have been a result of the removal of carbon although, for instance, there was no obvious change in the size or distribution of Ta-rich particles in the alloys.

The thermal cycle tests which used a 1 h dwell time at temperature were very similar to those used as standard screening tests in the gas turbine industry, intended as a simplified simulation of the most severe part of the duty cycle experienced by aircraft engines. In fact, the duty cycle of advanced power generation turbines will probably consist of a relatively fast ramp up from cold to maximum power, followed by long periods at essentially maximum power, punctuated by short periods of operation at partial load. The 1 h cyclic tests will not provide a representative simulation of this operation. Also, the maximum metal temperature experienced by the bond coating-ceramic interface is not expected to exceed approximately 1050ºC, significantly lower than the peak temperatures of 1150 to 1175ºC typically used in aircraft gas turbine simulation tests. From this perspective, comparison of the overall trends from the 1 h and 100 h cyclic tests at 1100°C reported here is pertinent. The observation that lower sulphur levels were required to resist scale spallation in the 100 h cycle test indicates that this was, in fact, a more severe test than the 1 h cycle test at the same temperature. Although the 1 h cycle test results in more frequent exposure to stresses from differential thermal expansion between the alloy and the scale, the incremental increase in scale thickness between individual cycles is relatively small so that, possibly, the increase in stress is easier to accommodate without scale buckling or cracking than with the thicker scale formed after a 100 h cycle. The longer cycle time may also allow time for the growth of interfacial voids, which may contribute to the increased spallation.

The initial results shown here for NiAl coatings suggest that desulphurisation of coatings is probably achievable in the relatively near-term. The growth of less protective scale associated with oxidation of elements diffusing from the superalloy substrate along the grain boundaries of such coatings remains a problem. However, the observation that grain boundaries in these coatings appear to grow from impurities in the original substrate surface suggests that the number of grain boundaries could be reduced by relatively straightforward modification of the coating process, which could lead to a further improvement in resistance to scale spallation. Nevertheless, the added benefit of reactive element doping to give reduced scale growth rates remains a desirable goal, since any alumina scale eventually will be prone to spallation when it attains some critical thickness at which the stresses from continued growth and the imposed stresses from thermal cycling and thermal expansion mismatch with the substrate and external YSZ layer cannot be accommodated without buckling or cracking.

The state-of-the-art Pt aluminide bond coating used in the TBC applied to specimens of René N5 typically forms an alumina scale that is very resistant to spallation. In the 1150°C/l h cycle test, the observed increase in life with decreasing sulphur level in the substrate of the TBC based on this bond coating suggests that sulphur from the substrate can decrease the resistance to spallation of the scale formed on the Pt aluminide. This is surprising since, as detailed elsewhere,[27] undoped Ni,Pt–Al (50 at% Al) alloys with sulphur levels of approximately 20 ppma formed alumina scales which were extremely resistant to spallation in this test at scale thicknesses greater than 4 μm. Also, there appeared to be significantly less Ta in the oxide scales formed on the state-of-the-art Ni,Pt aluminide bond coating applied to René N5.[28]

The effectiveness of the optimally-doped β-NiAl–Zr alloy (compared to the undoped Pt aluminide layer on the Y-doped superalloy) in retaining an external YSZ layer in the thermal cycling tests resulted from four main factors. The first two factors were the reduced rate of growth of the oxide and the lack of void formation at the substrate-oxide interface, both of which are manifestations of the reactive element doping in this alloy. The third and fourth factors resulted from the fact that a monolithic alloy was used as the 'bond coating'. Hence, there were no unwanted elements able to transport to the oxidising surface and become incorporated into the scale, and there was always sufficient aluminium available for protective scale formation since there was no loss by interdiffusion with a substrate. These advantages are obviously difficult to realise in a practical coating. A further difficulty is in controlling the dopant level and its uniform distribution in a practical coating process, especially since over- and under-doping may be detrimental to performance.[22]

ACKNOWLEDGMENTS

This research was supported in part by the U.S. Department of Energy Office of Industrial Technologies as part of the Advanced Turbine Systems Program, and in part by the Division of Materials Sciences under contract DE–ACOS–960R22464 with Lockheed Martin Energy Research Corporation. We would like to thank Dr B. Nagaraj of the General Electric Aircraft Engines Group for supplying samples of René N5 and the EB-PVD TBCs; Dr W. Brindley of NASA-Lewis Research Center for applying the plasma sprayed ceramic coating; and Dr C. T. Kortovich of PCC Airfoils, Inc., who supplied the desulphurised alloys. We would also like to thank colleagues at ORNL and in the aerospace alloy casting and coating supply industry for their useful comments and discussion throughout the course of this work.

REFERENCES

1. 'Materials/Manufacturing Plan for Advanced Turbine Systems Program', Office of Industrial Technologies, Rpt. No. *DOEIOR-2007,* April 1994.
2. C. S. KORTOVICH, R. M. GARLOCK and C. R. HAYES: 'Turbine Airfoil Manufacturing Technology,' Paper No. 97-GT-428, presented at the *42nd ASME Gas Turbine and Aeroengine Congress*, Exposition, and Users Forum, Orlando, Florida, June 1997.
3 B. A. MUELLER, A. R. PRICE, K. S. MURPHY and G. B. BELL: 'Land-Based Turbine Casting Initiative', paper No. 97-GT-430, presented at the *42nd ASME Gas Turbine and Aeroengine Congress*, Exposition, and Users Forum, Orlando, Florida, June 1997.
4. N. S. BORNSTEIN and J. MARCIN: 'Pratt & Whitney Thermal Barrier Coatings', *Proc. Advanced Turbine Systems Annual Program Review Meeting*, C. T. Alsup, C. M. Zeh, and S. Blazewicz, eds., DOE/METC-96/1023, Vol. 1 (DE96000561), 1996, 182–193.
5. J. GOEDJEN, S. SABOL and G. WAGNER: 'Westinghouse ATS Thermal Bar-

rier Coatings Development', *Proc. Advanced Turbine Systems Materials Workshop*, Charleston, SC., 13–14 Feb. 1996.

6 Y. IKEDA, K. NII and K. YOSHIHARA: 'High-temperature oxidation and surface segregation of sulphur', *Proc. JIMIS-3: High-Temperature Corrosion Transactions of the Japan Inst. of Metals*, Supplement 24, 1983, 207–214.

7. A. W. FUNKENBUSCH, J. G. SMEGGIL and N. S. BORNSTEIN, *Met. Trans.*, 1985, **16A**, 1164.

8. J. G. SMEGGIL, A. W. FUNKENBUSCH and N. S. BORNSTEIN, *High-Temp. Sci.*, 1985, **20**, 163.

9. J. C. SMEGGIL, A. W. FUNKENBUSCH and N. S. BORNSTEIN: 'A relationship between indigenous impurity elements and protective oxide scale adherence characteristics', *Met. Trans.*, 1986, **17A**, 923.

10. K. L. LUTHRA and C. L. BRIANT: 'Adhesion of oxide films on MCrAlY alloys', *Fundamental Aspects of High-Temperature Corrosion-II*, D. A. Shores and G. R. Yurek eds, The Electrochemical Soc., Pennington, NJ, 1986, 187–199.

11. J. L. SMIALEK: 'Adherent Al_2O_3 scales produced on undoped NiCrAl alloys', *Oxidation of Metals and Associated Mass Transport*, M. A. Dayanada, S. J. Rothman, and W. E. King, eds, The Metallurgical Society of AIME, 1986, 207–313.

12. J. L. SMIALEK: 'The effect of sulphur and zirconium co-doping on the oxidation of NiCrAl', *High Temperature Materials Chemistry IV*, Z. A. Munir, D. Cubicciotti and H. Tagawa, eds, Proc. Vol. 88–5, Electrochemical Society Inc., Pennington, NJ, 1988, 241–7.

13. J. L. SMIALEK: 'Effect of sulphur removal on Al_2O_3 scale adhesion,' *Met. Trans.*, 1991, **22A**, 739–752.

14. J. L. SMIALEK, D. T. JAYNE, J. C. SCHAEFFER and W. H. MURPHY: 'Effects of hydrogen annealing, sulphur segregation, and diffusion on the cyclic oxidation resistance of superalloys; a review', *Thin Solid Films*, 1994, **253**, 285–292.

15. J. L. SMIALEK and B. K. TUBBS, 'Effect of sulphur removal on scale adhesion to PWA 1480', *Met. Trans.*, 1995, **26A,** 427–435.

16. J. L. SMIALEK: 'Oxidation resistance and critical sulphur content of single crystal superalloys', *Paper No. 96-GT-519*, presented at the 41st ASME Gas Turbine and Aeroengine Congress and Exhibition, Birmingham, England, June 1996.

17. J. C. SCHAEFFER, W. H. MURPHY and J. L. SMIALEK: 'The effect of surface condition and sulphur on the environmental resistance of airfoils', *Oxid. Met.*, 1995, **43**(1/2), 1–23.

18. G. H. MEIER, F. S. PETTIT and J. L. SMIALEK: 'The effects of reactive element additions and sulphur removal on the adherence of alumina to Ni- and Febase alloys', *Materials and Corrosion*, 1995, **46**, 232–240.

19. E. SCHUMANN, J. C. YANG and M. J. GRAHAM: 'Direct observation of the interaction of yttrium and sulphur in oxidised NiAl', *Scripta Met.*, 1996, **34**(9), 1365–1370.

20. B. A. PINT: 'Experimental observations in support of the dynamic segregation

theory to explain the reactive element effect', *Oxid. Met.*, 1996, **45**, 1–37.

21. B. A. PINT, J. R. MARTIN and L. W. HOBBS: '$_{18}$O/SIMS characterisation of the growth mechanism of doped and undoped α-$A1_2O_3$', *Oxid. Met.*, 1993, **39**, 167–195.
22. B. A. PINT: 'The oxidation behavior of oxide-dispersed β-NiAl: I. Short-term performance at 1200°C', *Oxid. Met.*, 1998, **49**, 531–60.
23. B. A. PINT, I. G. WRIGHT, W. Y. LEE, Y. ZHANG, K. PRÜSSNER and K. B. ALEXANDER: 'Substrate and bond coat compositions: factors affecting alumina scale adhesion', *Mat. Sci. and Eng.*, 1998, **A245**, 201–11.
24. B. A. PINT, A. J. GARRATT-REED and L. W. HOBBS: 'A possible role of the oxygen potential gradient in enhancing diffusion of foreign ions on α-$A1_2O_3$ grain boundaries', *J. Am. Ceram. Soc.*, 1998, **81**, 305–14.
25. W. Y. LEE, Y. ZHANG, I. G. WRIGHT, B. A. PINT and P. K. LIAW: 'Effects of Sulphur Impurity on the Scale Adhesion Behavior of a Desulphurised Ni-Based Superalloy Aluminised by Chemical Vapor Deposition', *Met. Trans. A*, 1998, **29**, 833.
26. D. R. SIGLER: 'Aluminum oxide adherence on Fe-Cr-Al alloys modified with group IIIB, IVB, VB and VIB elements', *Oxid. Met.*, 1989, **32**, 337–55.
27. E. C. DICKEY, B. A. PINT, K. B. ALEXANDER and I. G. WRIGHT: 'The effect of platinum on the growth and adhesion of α-$A1_2O_3$ scales', *this proceedings*,
28. K. B. ALEXANDER and B. A. PINT, Oak Ridge National Laboratory, *unpublished work*, 1997.

The Effect of Platinum on the Growth and Adhesion of α-Al_2O_3 Scales

E. C. DICKEY, B. A. PINT†, K. B. ALEXANDER†
and I. G. WRIGHT†
*Department of Chemical and Materials Engineering,
University of Kentucky, Lexington, KY 40506–0046*
*†Metals and Ceramics Division, Oak Ridge National Laboratory,
P.O. Box 2008, Oak Ridge, TN 37831–6156*

ABSTRACT

Cast PtAl and NiPtAl alloys were used as model materials to study the effect of Pt on the oxidation behaviour of aluminide bond coats. The addition of Pt improves α-Al_2O_3 scale adhesion in cyclic testing at 1150°C and 1200°C. However, in comparison with aluminides which contain reactive elements (e.g. Zr and Hf), the beneficial effect of Pt was not as great. SEM and TEM analysis of the oxides indicated little effect of the Pt on scale microstructure. Using TEM/EDS analysis, Pt was not detected as a segregant to the alumina grain boundaries grown on PtAl, whereas Zr and Hf was routinely detected. Also, no sulphur was detected at the PtAl/Al_2O_3 interface. While Pt is somewhat beneficial in cast aluminides, Pt in an aluminide bond coat on a single crystal Ni-base superalloy substrate also may provide additional benefits to the overall oxidation behaviour.

INTRODUCTION

Thermal barrier coatings (TBCs) have played an important role in increasing the efficiency of turbine engines, because they allow the operating temperatures to be increased substantially while protecting the underlying metallic turbine blade. Ceramic TBCs are applied to the hot section of the turbine engine by plasma spraying (PS) or electron-beam physical vapour deposition (EB-PVD).[1–3] Yttria-stabilised zirconia (YSZ) has emerged as the preferred coating material because of its toughness, low thermal conductivity and relatively high thermal expansion coefficient for a ceramic material. One limitation of YSZ is that it has a very high oxygen diffusivity and therefore provides almost no protection from oxidation for the underlying alloy, necessitating an intermediate oxidation-resistant bond coat. This underlying bond coat must not only provide a barrier to oxidation, but must also be resistant to thermo-mechanical fatigue.

In most cases, the bond coat is designed to form an external, protective α-Al_2O_3 scale during oxidation. Particularly for EB-PVD TBCs,[2–3] the interface between the bond coat and the Al_2O_3 scale is a common site of mechanical failure resulting in loss of the YSZ top coat. To prevent such spallation, it is imperative to control the bond coat chemistry so that well-adhered oxide scales result. The two most common classes of bond coats are MCrAlY (M=Ni,Co) and diffusion aluminides. For MCrAlY coatings, Y is added to improve scale adhesion. The effect of reactive elements (REs), like Y, have been extensively studied (see for example reviews in Refs 4–9). For the case of aluminides, the addition of Pt is known to improve scale adhesion,[10–15] but there have been no fundamental mechanistic studies of the role of Pt reported in the literature.

This paper presents initial results on a study of the role of Pt in improving scale adhesion in Pt aluminide bond coats. Model Pt-containing alloys with and without RE additions were cast and both isothermal and cyclic oxidation studies were performed on the various bond-coat-type compositions to assess the oxidation behaviour. Complementary microstructural and chemical studies were carried out using scanning electron microscopy (SEM), energy-dispersive X-ray spectroscopy (EDS) and scanning transmission electron microscopy (STEM) to gain a better understanding of the role of Pt in improving oxidation resistance.

EXPERIMENTAL PROCEDURE

The chemical composition of the substrates investigated in this study are shown in Table I. The 2.3at.% (10wt.%) Pt level in the NiPtAl alloys was chosen to approximate the level in a Pt aluminide bond coat. The platinum and nickel aluminides were inductively melted and cast in a copper mold, and were then annealed for 4h at 1350°C.[16] Generally, test specimens were disks of 12–16 mm diameter and 1–2 mm thick.

Thermal cycling exposures were conducted in two ways: (1) in an automated cyclic furnace at 1150°C and 1200°C with dry O_2 flowing through a reaction tube and samples inserted into the hot furnace with 1 h cycle times (time at temperature) and 10 min. cooling periods between cycles, and (2) in individual pre-annealed alumina crucibles at 1200°C in laboratory air for 100 h cycles. With the alumina crucibles, the samples were also inserted into a hot furnace but heat-up time was somewhat slower. The crucibles had lids to collect spalled oxide and thus allow the measurement of the total amount of reaction. Isothermal (not cycled) kinetics were measured in dry, flowing O_2 using a Cahn Instruments, Model 1000 microbalance.

After oxidation, microstructural and chemical investigations were performed using a Philips XL30 field-emission SEM with a light-element EDS. Selected samples were Cu-plated and cross sectioned either for metallographic polishing and SEM analysis or thinned for higher spatial resolution analysis, performed on a Philips 200 field-emission STEM equipped with light-element EDS.

RESULTS AND DISCUSSION

Cyclic oxidation studies were carried out on NiAl, NiPtAl (10wt.%Pt) and PtAl to assess the effect of Pt on the oxidation behaviour of aluminides, Fig. 1. In 1 h cycles at both 1150°C and 1200°C, the addition of Pt significantly improves the spallation resistance as compared to undoped β-NiAl, where weight losses reflect continuous spallation of the alumina scale. However, for longer times in the same tests, spallation is

Table I Chemical composition of the substrate alloys in atomic percent. Compositions were determined by inductively coupled plasma analysis; S, C, O and N by combustion analysis.

Alloys	Undoped NiAl	NiAl +Zr	NiAl +Hf	PtAl	PtAl +Zr	NiPtAl	NiPtAl +Hf	NiPtAl +Hf/Si	René N5
Ni	48.67	50.04	48.83	<0.01	0.02	45.75	45.19	45.02	64.85
Al	51.22	49.65	51.05	49.33	51.58	51.83	52.23	52.25	13.88
Pt				50.32	48.07	2.33	2.36	2.30	
Fe	0.01	0.02	0.03	<0.01		<0.01	<0.01	<0.01	0.08
Cr	<0.01	<0.01	<0.01	0.02		0.01	0.01	<0.01	7.79
Si	0.06	0.21	<0.01	0.04	<0.01	<0.01	<0.01	0.19	0.15
Other	0.004 B	<0.01 Y	<0.01 Y	0.04 Cu		0.002 Ce	0.002 Ce <0.001 Y	0.01 Cu 0.004 B <0.001 Y	7.28 Co 2.11 Ta 1.61 W 1.02 Re 0.90 Mo
Dopant	<0.01 Hf <0.001 Y <0.01 Zr	0.04 Zr	0.05 Hf 0.01 Zr	<0.01 Hf 0.001 Y <0.01 Zr	0.04 Zr	<0.01 Hf <0.001 Y <0.01 Zr	0.05 Hf <0.01 Zr	0.06 Hf <0.01 Zr	0.05 Hf 0.003 Y 0.003 Zr
C	0.04	0.02	0.04	<0.01			0.04	0.04	0.25
O	0.004	0.02	0.003	0.252	0.027	0.072	0.102	0.128	0.005
N	<0.001	n.d.	<0.001	0.005	<0.001	<0.001	0.002	0.003	<0.001
S (ppma)	<4	47	<4	<10	<12	10	<4	<4	7

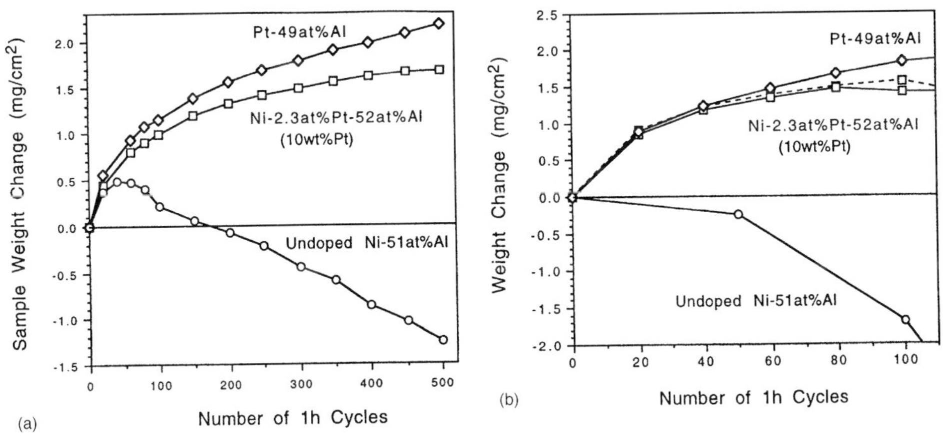

Fig. 1 Weight change of various uncoated aluminides cycled (1 h cycle time) from room temperature to (a) 1150°C and (b) 1200°C.[16] The addition of Pt improved scale adhesion compared to undoped β-NiAl

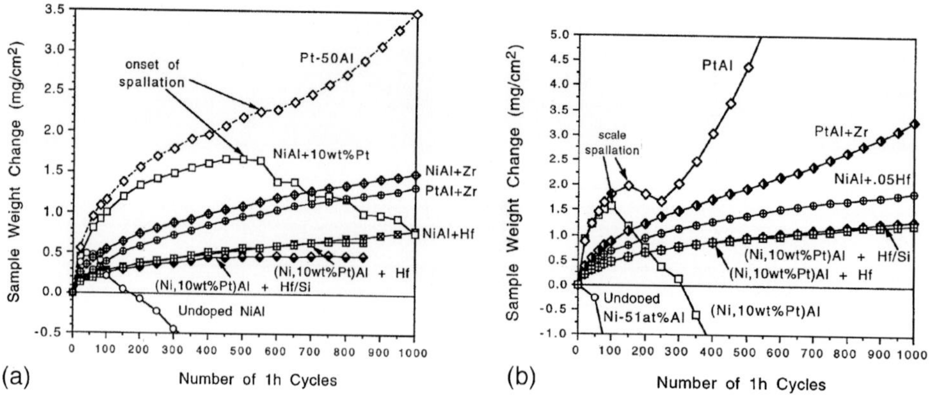

Fig. 2 Weight change of various uncoated aluminides cycled (1 h cycle time) from room temperature tp (a) 1150°C and (b) 1200°C.[16] The addition of reactive elements such as Zr and Hf further improve scale adhesion compared to aPt addition alone. The high weight gains for PtAl are due to internal cracking and oxidation.

observed for both Pt-containing alloys, Fig. 2. The onset of spallation was after approximately 500 1 h cycles at 1150°C and 100–150 cycles at 1200°C.

Comparing the effect of Pt on scale adhesion to that of RE additions such as Zr and Hf clearly indicates that the beneficial effect of RE-doping on scale adhesion is much stronger than that of Pt, Fig. 2. On comparing the performance of NiAl+Hf to NiPtAl+Hf and NiPtAl+Hf/Si, it is evident that Pt is not necessary in achieving good scale adhesion. This is similar to the results found for Pt and Hf additions to CoCrAl alloys.[14] Also, the addition of Si with Hf did not provide any additional benefit to the oxidation behaviour of cast NiPtAl. An important aspect of the specimen weight gain data in Figs 2(a) and (b) is that the absolute weight gains for the various samples are not easily compared. At both temperatures, initial weight gains for both PtAl and NiPtAl are higher than for Zr- and Hf-doped aluminides. This does not reflect better scale adhesion on the Pt-containing aluminides but rather a higher scale growth rate without RE-doping.[8,9,17,18] As illustrated in Fig. 3, the addition of Zr reduces the parabolic growth rate by a factor of approximately 3, while a Hf addition reduces the rate by a full order of magnitude. (An explanation for the additional benefit of Hf has not yet been determined.) Thus, the higher initial weight gains for the undoped Pt-containing alloys is the result of an adherent but faster growing scale.

Another point to note in Fig. 2 is the high weight gains for PtAl – the final weight gain after 1000, 1 h cycles at 1200°C was 20.8 mg.cm^{-2}. Again, this is not an indication of good performance. The high weight gains are the result of internal cracking and oxidation of the alloy, Fig. 4(a), which masks any weight loss due to spallation of the external alumina scale. Some alloy cracking may initiate due to casting defects, however, the PtAl+Zr alloy also had casting defects but did not show similar cracking

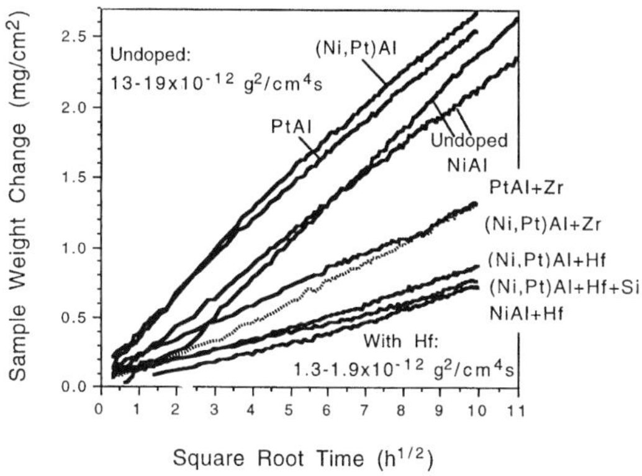

Fig. 3 Isothermal weight gains at 1200°C plotted versus the square root of time to show the parabolic reaction kinetics.[16] The addition of Hf reduces the scale growth rate by a factor of 10, while a Zr addition is slightly less effective. This reduction in rate is attributed to an inhibition of Al diffusion in the Zr- or Hf-doped a-Al_2O_3 scale.

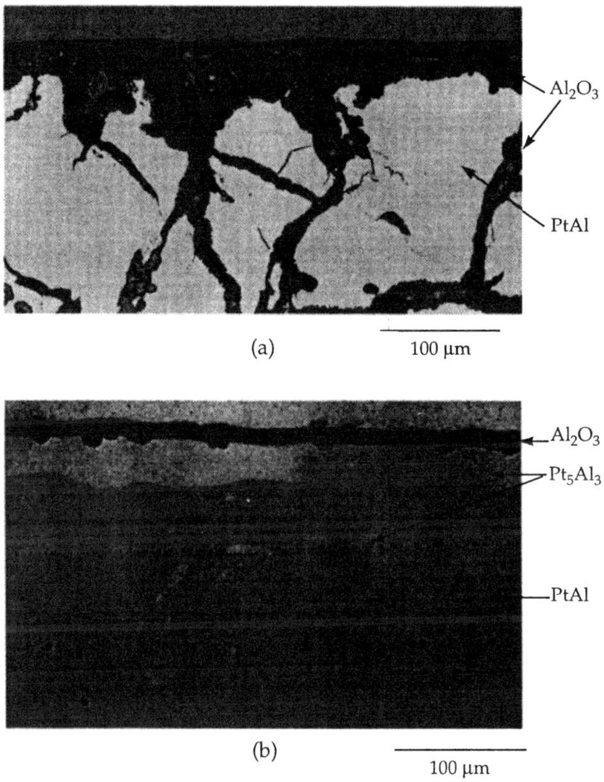

Fig. 4 (a) SEM back scattered electron image of metallographically-polished cross-sections of PtAl cyclically oxidised at 1200°C for 1000 1 h cycles, (b) light micrograph of PtAl+Zr after the same test. Without the reactive element (Zr) addition, massive cracking and internal oxidation occurs. A deletion layer is observed in the PtAl+Zr near the scale.

(a)

(b)

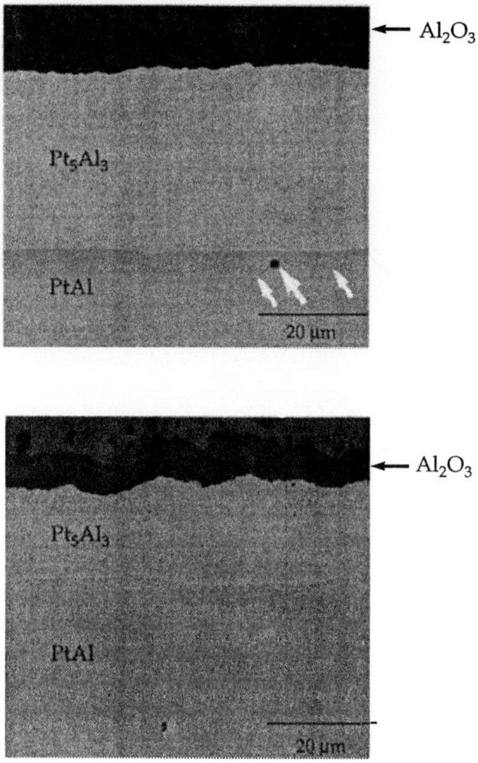

Fig. 5 Cross sectional SEM back-scattered images of (a) PtAl and (b) PtAl+Zr after isothermal oxidation at 1200°C for 100 h. Spallation of the Al2O3 scale is apparent in the undoped PtAl, while scale is well adhered in the Zr-doped specimen. The slower scale growth rate with Zr doping also results in a much thinner Pt5Al3 layer. The arrows in (a) mark a void with cracks near it.

and internal oxidation after cycling, Fig. 4(b). From the weight gain curves for PtAl+Zr, it is evident that there was virtually no spallation from this alloy. In the light micrograph, there appears to be a second phase forming near the surface, an approximately 40 µm thick layer formed after 1000, 1 h cycles at 1200°C, Fig. 4(b). This depleted Pt-Al phase has not been conclusively identified by XRD, but EDX analysis indicated Pt_5Al_3. The formation of a depletion layer is not surprising as PtAl is a line compound.[19] This outer layer appears cracked and brittle but this cracking may have occurred during specimen preparation. Because of the severe cracking on the cycled PtAl, it is difficult to compare the depleted areas in that material. Instead, the two Pt aluminides were compared after a 100 h isothermal exposure at 1200°C, Fig. 5 (the same samples used for kinetic data in Fig. 3). Using SEM back-scattered electron imaging (sensitive to substrate atomic number) of the metallographic cross-section, the Al depletion layer near the surface is clearly observed. Some spalling of the scale

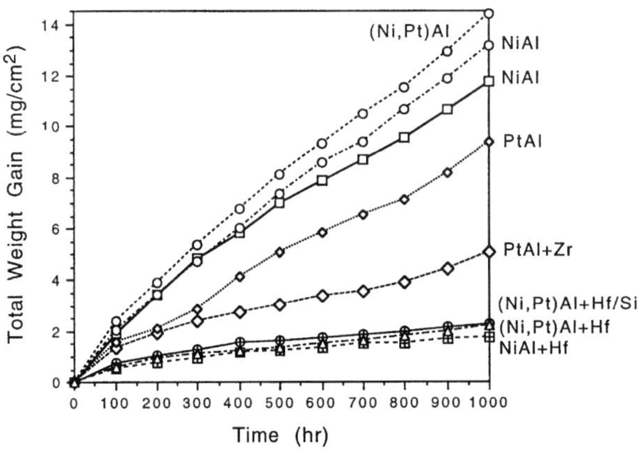

Fig. 6 Total weight gain (sample weight gain + spalled oxide) during 100 h cycles at 1200°C. The nearly straight line behaviour of undoped NiAl represents almost complete spallation during cooling after each cycle. The addition of Pt had only a marginal effect on adhesion in this longer term cyclic test. However, the addition of Hf or Zr continues to provide a benefit throughout the test with almost no spallation observed.

was observed on PtAl, Fig. 5(a), but the most important feature is the much thicker depletion zone on PtAl (~32 μm) compared to PtAl+Zr (~17 μm). This is a direct result of the slower oxidatiop rate due to Zr doping (Fig. 3) which means a slower Al consumption rate and, therefore, less Al depletion. If this Pt_5Al_3 layer is more brittle than PtAl, then the thicker Pt_5Al_3 layer on the undoped PtAl alloy may contribute to the internal cracking and detrimental behaviour observed during thermal cycling. Once the alloy cracks, faster oxidation (forming in the cracks) and more depletion will occur. Thus, Zr-doping not only produced a thinner scale (subject to lower strains on cooling[20]), but the thinner depletion layer in the alloy may also help in preventing the formation of internal cracks. The polished cross-sections also reveal a pore in the PtAl alloy, Fig. 5(a), with cracks emanating from it – this type of defect may nucleate alloy cracks. Also, the scale and metal-scale interface on PtAl±Zr, Figure 5(b), appear more convoluted than on PtAl. This is somewhat surprising because the convoluted scale is more adherent. However, the convolutions do not appear to be prevalent in the sample cycled for 1000h, Fig. 4(b).

The effect of cycle time on scale adhesion was also investigated in this study by performing complementary 100 h cycles for 1000 h at 1200°C, Fig. 6. In this figure, total weight gains are shown to illustrate the total amount of attack. Surprisingly, with the longer cycles, the benefit of Pt-doping was almost nonexistent. The near-linear weight gains reflect almost complete spallation from the undoped alloys. For PtAl, there was less external spallation but increased weight gains due to internal attack. A RE addition again overshadowed the Pt effect in this test. One possible explanation for the reduced effect of Pt in longer cycles is that Pt doping is unable to prevent the growth of interfacial voids to the same degree as RE-doping. Large interfacial voids

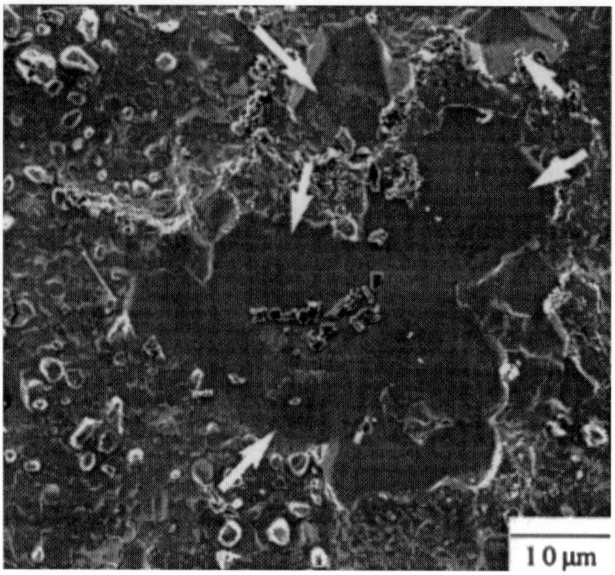

Fig. 7 SEM secondary electron image of the NiPtAl (2.3at.%, 10wt%Pt) substrate exposed by spallation of the alumina scale after 10, 100 h cycles at 1200°C. The arrows mark a variety of large voids.

were observed in the spalled areas (exposed substrate) of undoped NiPtAl after 10, 100h cycles at 1200°C, Fig. 7. One explanation of the role of Pt is that it reduces the growth rate of interfacial voids but to a lesser degree than RE-doping. A smaller fraction of interfacial voids would improve scale adhesion, particularly in short tests, Fig. 1. However, at longer times, the voids would be of sufficient size to cause scale spallation. The longer cycle time (100 h) allows enough time for the voids to grow and results in spallation after each cycle. Thus the benefit of the Pt addition is lost when the cycle time is increased. A reduction in the interfacial void growth rate could involve an interaction between Pt and the interfacial segregation of indigenous S (which is thought to accelerate the growth of interfacial voids[21,22]). However, a Pt–S interaction has not yet been investigated.

From the cyclic and isothermal oxidation results, it is clear that the Pt effect is significantly different than the RE effect. One additional effect of RE-doping is a change in the α-Al_2O_3 microstructure.[8,9,18,23] A comparison of the scales formed at 1200°C reveals that the addition of Pt did not produce a fully columnar grain structure or inhibit the formation of alumina whiskers, Fig. 8. This is in agreement with the assumption that the columnar structure is the result of a change in the growth mechanism to predominant O diffusion inward.[8,9,23] Platinum did not reduce the scale growth rate, Fig. 3, thus it is not likely that it changed the growth mechanism. The whiskers on the scales formed on undoped Pt aluminides are a further indication that the outward diffusion of Al is contributing to the scale growth rate, Fig. 8.

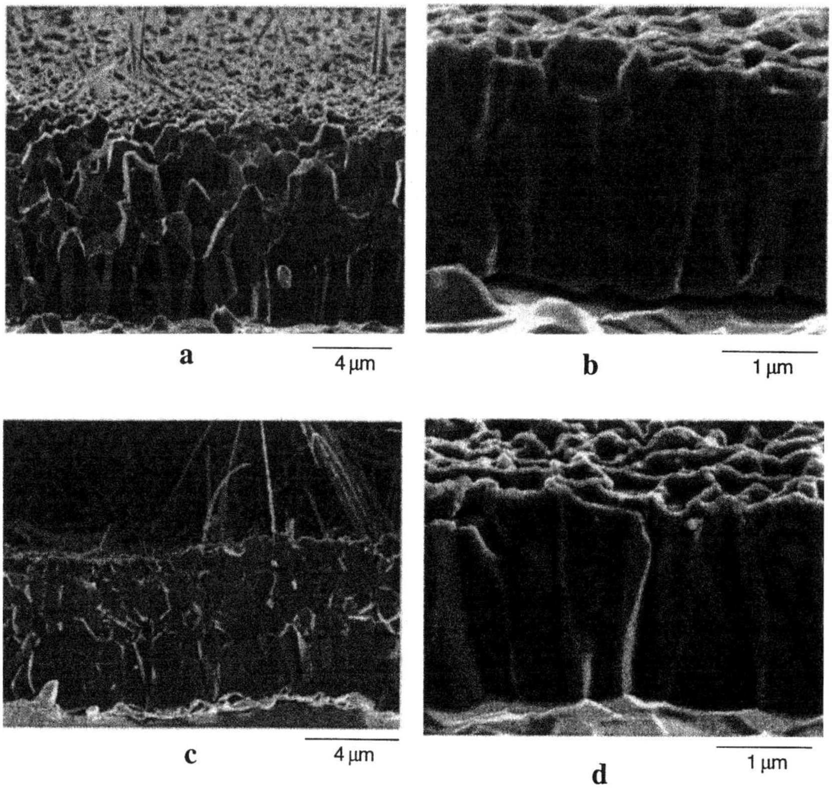

Fig. 8 SEM secondary electron cross-section images of the α-Al2O3 scale formed at 1200°C on (a) undoped NiPtAl after 100 h, (b) β-NiAl+Hf after 100 h, (c) PtAl after 100 h and (d) NiPtAl+Hf after 100 h.16 With the addition of a reactive element such as Hf, the oxide whiskers disappear and a more columnar grain structure develops.

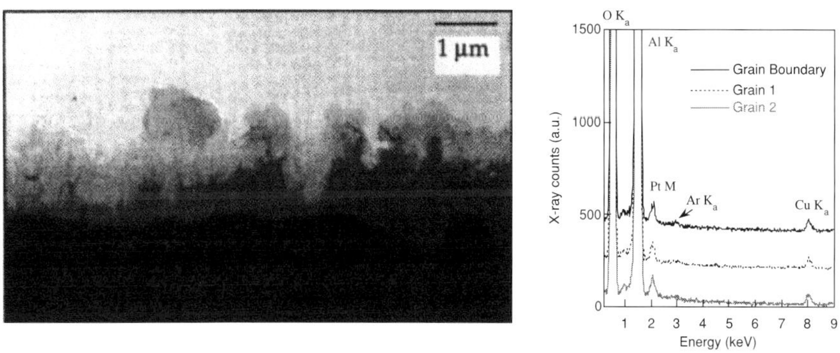

Fig. 9 (a) Bright-field TEM image of isothermally-grown Al_2O_3 scale on PtAl and (b) EDS spectra from Al_2O_3 grains and grain boundary showing no segregation to the grain boundaries. Note that the spectra intensities are offset for clarity. The Cu, Pt, and Ar peaks are artifacts from sample preparation.

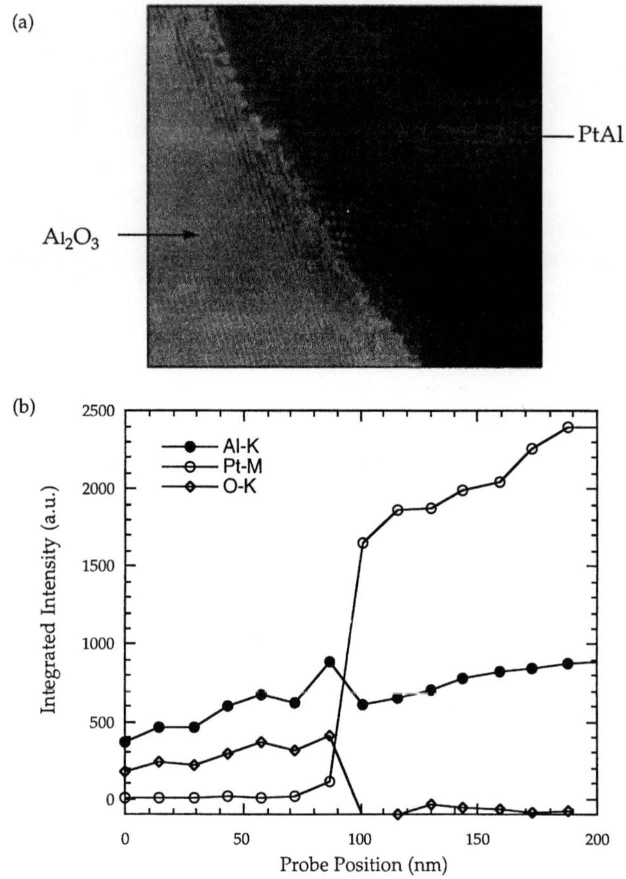

Fig. 10 (a) HREM image of PtAl–Al_2O_3 interface in isothermally oxidised (1200°C, 2 h) PtAl sample and (b) EDS profiles taken across the interface which show an abrupt transition between the metal and oxide scale.

STEM and EDS analyses were performed on PtAl and PtAl+Zr after oxidation for 2hr at 1200°C. At this time no scale spallation was observed. Figure 9(a) shows a bright-field TEM image of the $A1_20_3$ scale grown on undoped PtAl at 1200°C. EDS spectra, Fig. 5(b), were acquired from the interior of the $A1_20_3$ grains as well as the $A1_20_3$ grain boundaries. In these undoped PtAl specimens, no evidence was found for segregation of Pt or impurity elements, such as S, to the grain boundaries. EDS analysis was also performed across the PtAl–$A1_20_3$ interfaces. Figure 10(a) is a high resolution electron microscopy (HREM) image of this PtAl–$A1_20_3$ interface. The lattice image shows that the interface is abrupt with no other phases present at the PtAl–$A1_20_3$ boundary. EDS profiles, taken across the same boundary in 5nm steps, Fig. 10(b), show an abrupt chemical transition across the interface. EDS also showed no evidence for impurities, such as S, segregated to the interface.

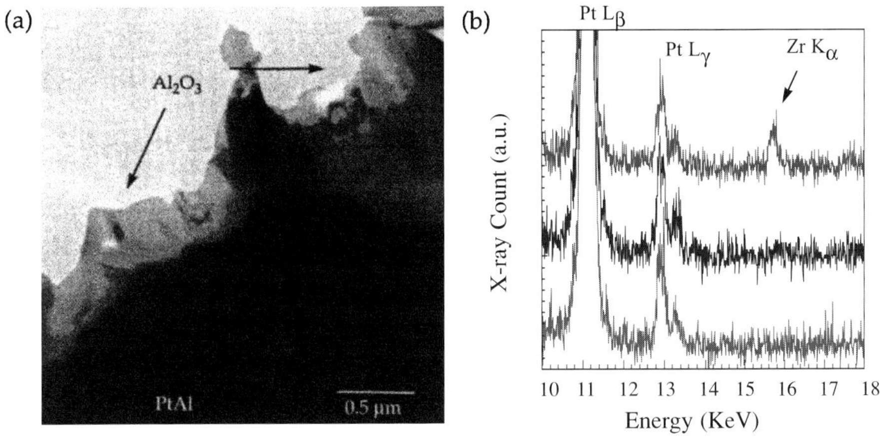

Fig. 11 (a) Bright-field TEM image of columnar alumina scale on isothermally oxidised PtAl+Zr and (b) EDS spectra from alumina grains and grain boundary showing segregation of Zr to grain boundaries. The intensities of the three spectra are offset for clarity. Traces of Pt are also detected in the alumina grains and grain boundaries.

Similar analytical analyses were performed on the scale formed on PtAl+Zr. Figure 11(a) is a bright-field TEM image of a portion of the $A1_20_3$ scale. EDS spectra, Fig. 11(b) shows that Zr is segregated to the $A1_20_3$ grain boundaries. This observation is consistent with numerous previous observation of RE ionic segregation (see literature reviews in Refs 12 and 24). The interface between the PtAl+Zr and $A1_20_3$ scale was also examined by HREM and EDS to look for interphases or segregation at the metal-oxide interface. Figure 12(a) is an HREM image of the interface between PtAl+Zr and $A1_20_3$. As in the undoped case, the metal-oxide interface is free of third phases and appears to be structurally abrupt. EDS profiles taken across the boundary, Fig. 12(b), show that the interface is chemically abrupt and that Zr segregation was not detected in this sample.

Additional TEM work on the NiPtAl alloys is not yet complete. These alloys are more brittle and thus more difficult to thin for TEM analysis.

While Pt is not as effective as RE-doping in improving alumina scale adhesion, the addition of Pt to a bond coat on a Ni-base single-crystal superalloy may have additional benefits on the oxide behaviour, not reproducible in these cast alloys. A comparison using TEM/EDS of the scale grown on uncoated Rena N5 with that grown on a Pt aluminide-coated Rena N5 substrate indicated that the scale on the uncoated alloy contained significant amounts of Ta and Ta ions segregated to every α-Al_2O_3 grain boundary.[25] Initial results with a Pt-aluminide bond coat show far less Ta in the scale and little segregation of Ta on the grain boundaries. Limiting the outward diffusion of elements such as Ta from the substrate into the external alumina scale may be an additional beneficial effect of Pt when present in an aluminide coating. In general, it has been observed that, while a small amount of RE-doping is beneficial to oxidation

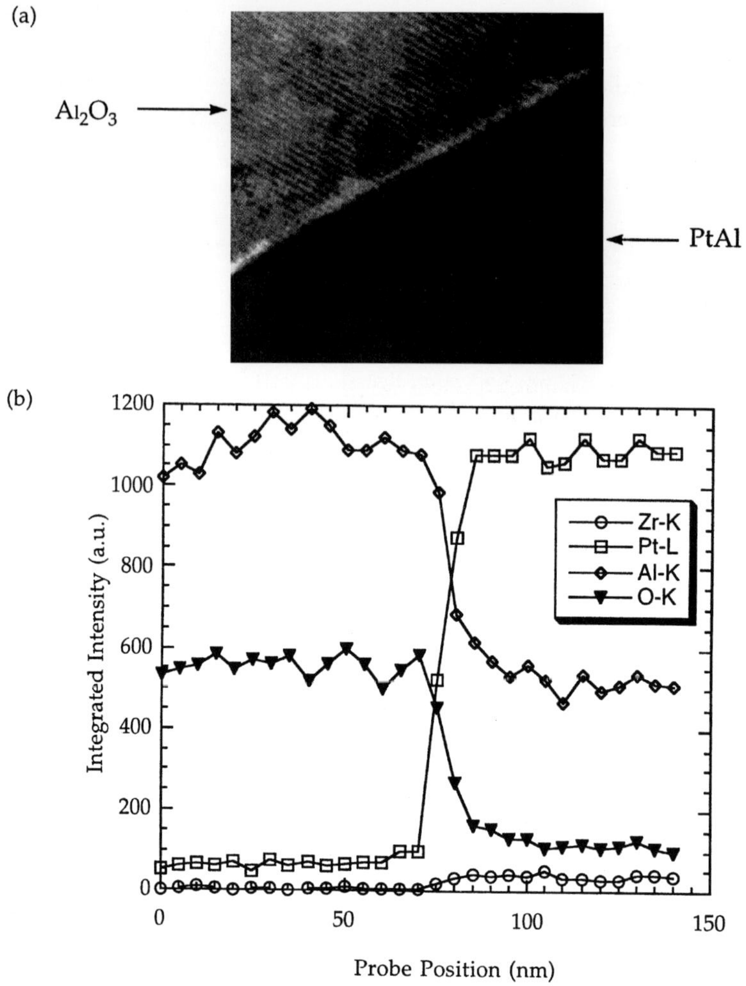

Fig. 12 (a) HREM image of PtAl+Zr–Al2O3 interface in isothermally oxidised (1200°C, 2 h) PtAl+Zr sample and (b) EDS profiles taken across the interface which show an abrupt transition between the metal and oxide scale.

performance, additional dopants in the scale are detrimental in the long term. Thus, Pt may have an additional beneficial effect on oxidation by limiting the diffusion of extraneous elements from the substrate into the alumina scale.

CONCLUSIONS

1. The addition of Pt to aluminides is beneficial to oxide adhesion but not to the same extent as reactive element (Zr, Hf, etc.) doping.

2. The addition of Pt does not change the scale growth rate or scale microstructure.
3. Using TEM/EDS, neither Pt or S is detected as segregated to the α-Al_2O_3 grain boundaries or the metal-scale interface on PtAl. For the scale on PtAl±Zr, Zr is observed as a segregant on the α-Al_2O_3 grain boundaries.

ACKNOWLEDGEMENTS

The authors would like to thank M. Brady and D. Wilson at ORNL for their comments. T. Geer and J. W. Jones at ORNL assisted with the metallography and TEM sample preparation. J. D. Whittenberger and J. Doychak at NASA Lewis Research Center provided some NiAl substrates. This research was sponsored by the U. S. Department of Energy, Assistant Secretary for Energy Efficiency and Renewable Energy, Office of Industrial Technologies, as part of the Advanced Turbine Systems Program (ORNL program manager, M. A. Karnitz) under contract DE-ACO5-960R22464 with Lockheed Martin Energy Research Corporation.

REFERENCES

1. R. A. MILLER: 'Oxidation-Based Model for Thermal Barrier Coating Life', *J. Amer. Ceram. Soc.,* 1984, **67**, 517–21.
2. A. BENNETT: 'Properties of Thermal Barrier Coatings', *Mat. Sci Tech.,* 1986, **2**, 257–61.
3. S. M. MEIER, D. M. NISSLEY, K. D. SHEFFLER and T. A. CRUSE: 'Thermal Barrier Coating Life Prediction Model Development', *J. Eng. Gas Turb. & Power,* 1992, **114**, 258–63.
4. *The Role of Active Elements in the Oxidation Behavior of High Temperature Metals and Alloys*, E. Lang, ed., Elsevier Applied Science, London, 1989.
5. *The Reactive Element Effect on High Temperature Oxidation – After Fifty Years,* W. E. King, ed., Materials Science Forum 43, Trans. Tech Publications, Switzerland, 1989.
6. D. P. MOON: 'Role of Reactive Elements in Alloy Protection', *Mat. Sci. Tech.,* 1989, **5**, 754–64.
7. A. STRAWBRIDGE and P. Y. HOU: 'The Rolc of Rcactive Elements in Oxide Scale Adhesion', *Mat. High Temp.,* 1994, **12**, 177–81.
8. B. A. PINT, A. J. GARRATT-REED and L. W. HOBBS: 'The Reactive Element Effect in Commercial ODS FeCrAl Alloys', *Mat. High Temp.,* 1995, **13**, 3–16.
9. B. A. PINT: 'Experimental Observations in Support of the Dynamic Segregation Theory to Explain the Reactive Element Effect', *Oxid. Met.,* 1996, **45**, 1–37.
10. E. J. FELTEN: 'Use of Platinum and Rhodium to Improve Oxide Adherence on Ni–8Cr–6A1 Alloys', *Oxid. Met.,* 1976, **10**, 23–8.
11. E. J. FELTEN and F. S. PETTIT: 'Development, Growth, and Adhesion of Al_2O_3 on Platinum-Aluminum Alloys', *Oxid. Met.,* 1976, **10**, 189–223.
12. J. G. FOUNTAIN, F. A. GOLIGHTLY, F. H. STOTT and G. C. WOOD: 'The

Influence of Platinum on the Maintenance of α-Al_2O_3 as a Protective Scale', *Oxid. Met.,* 1976 **10**, 341–5.

13. D. Lowrie and D. H. Boone: 'Composite Coatings of CoCrAlY Plus Platinum', Thin Solid Films, 1977, **45**, 491–8.
14. I. M. ALLAM, H. C. AKUEZLTE and D. P. WHITTLE: 'Influence of Small Pt Additions on $A1_20_3$ Scale Adherence', *Oxid. Met.,* 1980 **14**, 517–30.
15. J. P. ROUX, M. W. BRUMM and H. J. GRABKE: 'Effects of Orientation and Doping with Platinum on the Oxidation of β-NiAl', *Fresenius J. Anal. Chem.,* 1993, **346**, 265–8.
16. B. A. PINT and I. G. WRIGHT: 'The Effect of Pt on the Oxidation Behavior of Aluminides', manuscript in progress for submission to *Oxid. Met.*
17. B. A. PINT, J. R. MARTIN and L. W. HOBBS: '^{18}0/SIMS Characterization of the Growth Mechanism of Doped and Undoped α-$A1_20_3$', *Oxid. Met.,* 1993, **39**, 167–95.
18. B. A. PINT: 'The Oxidation Behavior of Oxide-Dispersed β-NiAl: I. Short-Term Performance at 1200°C', *Oxid. Met.,* 1998, in press.
19. *Binary Alloy Phase Diagrams,* 1st edition, T. B. Massalski *et al.* eds, ASM International, Metals Park, OH, 1986.
20. H. E. EVANS: 'Stress Effects in High Temperature Oxidation of Metals', *Int. Mat. Rev*. 1995, **40**, 1–40.
21. H. J. GRABKE, D. WEIMER and H. VIEFHAUS: 'Segregation of Sulfur During Growth of Oxide Scales', *Appl. Surf. Sci.*, 1991, **47**, 243–50.
22. B. A. PINT: 'On the Formation of Interfacial and Internal Voids in α-$A1_20_3$ Scales', *Oxid. Met.,* 1997, **48**, 303–33.
23. F. A. GOLIGHTLY, F. H. STOTT and G. C. WOOD: 'The Relationship Between Oxide Grain Morphology and Growth Mechanisms for Fe-Cr-Al and Fe-Cr-Al-Y Alloys', *J. Electrochem. Soc.*, 1979, **126**, 1035–42.
24. B. A. PINT, A. J. GARRATT-REED and L. W. HOBBS: 'A Possible Role of the Oxygen Potential Gradient in Enhancing Diffusion of Foreign Ions on α-Al_2O_3 Grain Boundaries', *J. Amer. Ceram. Soc.*, 1998, **81**, 305–14.
25. B. A. PINT, I. G. WRIGHT, W. Y. LEE, Y. ZHANG, K. PRÜßNER and K. B. ALEXANDER: 'Substrate and Bond Coat Compositions: Factors Affecting Alumina Scale Adhesion', *Mat. Sci. Eng*., 1998, **A245**, 201-211.

The Application of Glow Discharge Optical Emission Spectroscopy to the Study of Thermal Barrier and Environmental Coatings

A. HOQUE, J. HIGGINS*, D. RICKERBY*, J. CAWLEY
and M. IVES
Materials Research Institute, Sheffield Hallam University, Sheffield S1 1WB
**Surface and Coatings Technology Group, Rolls Royce Plc, Derby*

ABSTRACT

Glow discharge optical emission spectroscopy (GDOES) has been developed to investigate thermal barrier and environmental coatings. Qualitative analysis of thermal barrier and environmental coating has shown that this technique exhibits considerable potential. Work has been conducted to quantify the results. This involved the preparation of well documented calibration standards for alloy systems containing up to fifteen elements. The advantages of the technique lies in its ability to analyze up to 44 elements ranging from hydrogen to uranium with a lateral resolution in the order of tens of nanometers.

Results will be presented identifying the elemental distribution across the coating interface which will be compared with results from quantitative analysis conducted using scanning electron microscopy and X-ray fluorescence technique. The results will be discussed in terms of the GDOES technique sensitivity/limitations and elemental interference encountered and possible optimisation in developing this technique for understanding degradation mechanisms involved in thermal barrier and environmental coatings.

INTRODUCTION

Glow discharge optical emission spectroscopy (GDOES) is a powerful and versatile technique for the bulk analysis of ferrous and non-ferrous alloys with a sensitivity down to approximately 50 ppm; the technique is also capable of quantitative depth profile analysis.[1,2] This paper describes the work currently underway to apply the GDOES technique to understand the oxide bonding mechanisms and degradation modes for thermal barrier and environmental protection coating systems.

Overlay coatings of the MCrAlX type in which M represents a base metal of Ni, Co, Fe or some combination of elements, and X represents either a rare earth element or some other element such as Y or Zr with a high chemical reactivity for oxygen, have many applications in the fields of high temperature and environmental protection. In addition variants of the coatings are used to create the bond between turbine blades and thermal barrier coatings, which are usually stabilised zirconia. The mechanisms of bonding and the effects on the performance of the blades are not completely understood as yet.

Several analytical methods are available for the analysis of bonding mechanism and degradation process. Electron probe microanalysis is often used to characterise the composition of turbine blade coatings by analysing the cross section of the specimen. Scanning electron microscopy is used for microstructural assessment and X-ray diffraction for structural analysis. Other more sensitive techniques available include SIMS and AES but they are less convenient due to the high cost and the need to handle the specimen in ultrahigh vacuum. Thus, a more convenient method such as GDOES would be useful for understanding the role of low concentration elements.

THEORY

Glow Discharge Optical Emission Spectroscopy

Glow discharge spectroscopy is an atomic emission technique that relies on the excitation of atoms within the sample being analysed. It allows both qualitative and quantitative determination of metal and non-metal elemental composition as a function of depth.

If sufficient energy is transferred to an atom from the sample by an argon atom from a glow discharge, an increase in electron energy can occur such that a higher energy level orbital transition takes place to provide an excited state and a lower energy level vacancy. This state is unstable and the atom relaxes back to the ground state by the transition of an electron to fill the vacancy, normally accompanied by the liberation of the excess energy in the form of a photon of wavelength (Fig. 1) defined by the equation:

$$\lambda = hc/E1\text{-}E2$$

where; $E1$ and $E2$ are energy states, h is Planck's constant, c is the Speed of light and λ is wavelength.

Each element has different electron orbital energy levels, the wavelength of the photon emissions will therefore be characteristic of the elements present within the sample and the intensities of the emission lines will be proportional to the number of emission quanta and hence elemental concentration.

To produce clean line spectra without significant contribution from continuum energy in the form of thermal radiation or lattice vibrations, the atoms must be in the form of a low pressure gas, volatilised from a solid within a vacuum chamber.[3]

Glow discharge spectroscopy has a number of benefits over the more commonly used analytical techniques:[4]

1. The layer-by-layer removal of sample material allows both qualitative and quantitative depth profile analysis.
2. Very little or no sample preparation is required for bulk solids.

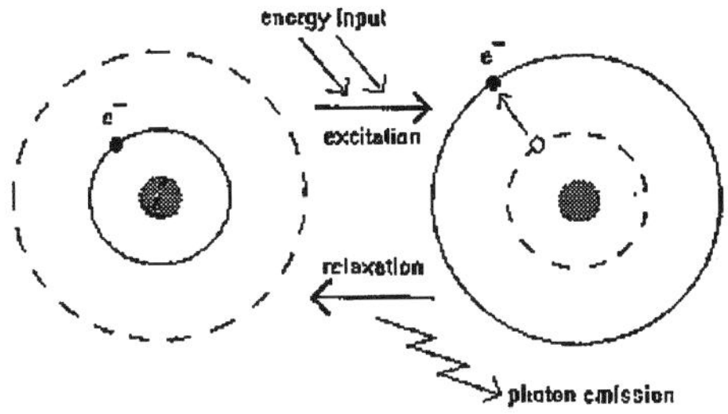

Fig. 1 Principle of excitation and emission.[4]

3. It naturally separates the removal of atoms from the solid surface (atomisation), from the subsequent excitation and ionisation of those atoms, thus reducing matrix effects and normalising sample-to-sample inconsistencies.
4. Due to the small cathode to anode distance, sputtering occurs over a narrow volume resulting in uniform sample erosion over the analysed area giving good resolution when used for depth profiling.
5. The excitation and emission layer is also thin, minimising self-absorption and giving a linear relationship between intensity and concentration, with sharp emission lines.

The glow discharge lamp utilised in this technique is based on a design by Grimm (Fig. 2). Basically, it consists of a small vacuum chamber with a hollow cylindrical anode 4 mm in diameter with the sample to be analysed acting as the cathode at one end. The outside of the anode cylinder is vacuum pumped and bled with a working gas (argon). The other end of the cylinder contains a lens through which the photon emission is detected via a polychromator. Typically operating conditions are -500 to -1500 V and 20 to 200mA , giving a sputtering rate in the range 10 to 100 nm per second.

Sputtering Process

The glow discharge sputtering process operates as follows (Fig. 3). The glow discharge lamp provides a low pressure argon environment over the sample. A high negative potential 500V to -1500 V is typically applied to the low pressure environment which causes breakdown of the discharge gas, normally argon, whose ions are then accelerated across the dark space and impact on the cathode (sample) surface, thereby ejecting primarily neutral atoms and electrons, although some ions and polyatomic species are also released. In addition to the ions striking the cathode, the charge-exchange processes in the dark space create fast atoms that also can cause sputter ablation. Through these sputtering steps, the solid sample yields an atom population that diffuses across a thin dark space into the adjacent negative glow, where collisions with electrons metastable atoms and other energetic species cause excitation and ionisation of a fraction of the sample atom, creating species that are analytically useful.[5]

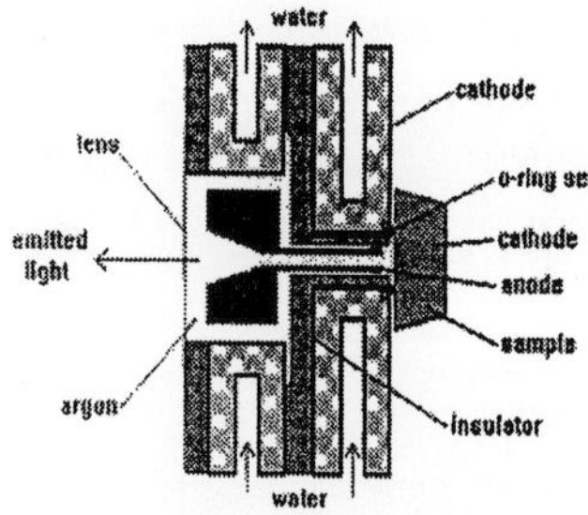

Fig. 2 Glow discharge lamp.[4]

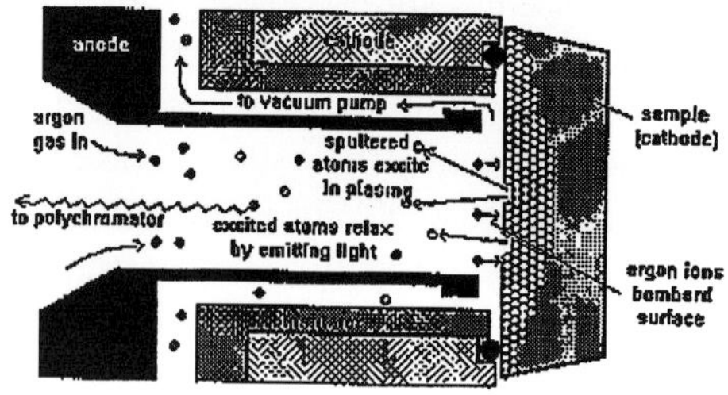

Fig. 3 Sputtering in the glow discharge lamp.[4]

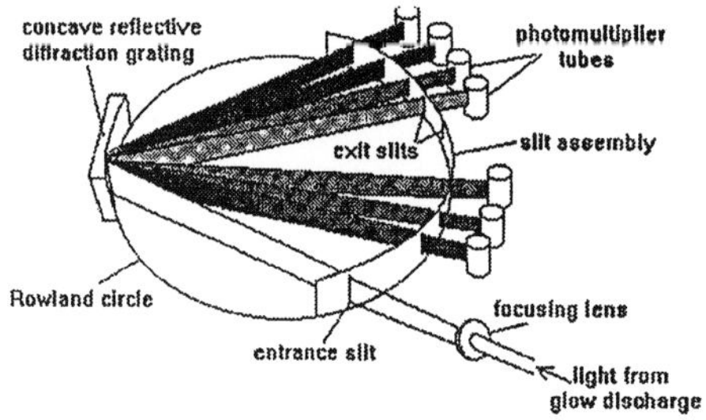

Fig. 4 Light emission and measurement.

Light Emission and Measurements

A polychromator (Fig. 4) based on the Rowland focusing property, separated by a window at the end of a glow discharge lamp is used to detect emitted photons.[4]

The basic arrangement of a polychromator is as follows:

- i a collimator or focusing lens to render the light rays parallel.
- ii an entrance slit that provides a narrow optical image.
- iii a concave reflective diffraction grating to disperse the light.
- iv exit slits at the appropriate locations to isolate particular spectral wavelengths.
- v photomultiplier tubes behind these slits to detect the light, amplify and convert it into an electrical signal for monitoring.

Glow discharge sputtering is strongly dependent on the working gas, sample composition and the applied parameters: current, voltage and pressure. Careful adjustment and monitoring of these parameters is vital for controlling surface analysis or rate of depth profiling and for quantification of collected results.

GDOES can be used to depth profile non-conductive coatings but the main requirements is to have a radio frequency (RF) source.

EXPERIMENTAL

To assess the capability of the GDOES technique to analyse complex materials, CMSX-4, a nickel-based superalloy used in turbine blade manufacture, and a CoNiCrAlY overlay coating (RT31), used for environmental protection and as a bondcoat for ceramic thermal barriers, were assessed by GDOES, electron probe microanalysis (EPMA) and X-ray fluorescence (XRF).

The glow discharge spectroscopy was carried out on a LECO GDS-750 QDP glow discharge spectrometer. The following optimised operating conditions shown in Table 1 were used. Both qualitatitative and quantitative studies were performed. Qualitative studies produced a plot of the intensity over time for given elements.

Table 1 Operating conditions for GDOES

Evacuation time	10 sec
Argon flushing time	20 sec
Preburn time	60 sec
Interval delay	0
Burn time	0
Numner of repititions	0
Compensation	5 sec
Integration time	10 sec
GDL operating time	11
Preburn excitation	700 V
Integration excitation	700 V; 30 mA
Flow rate in digits	400
Evacuate until	10E-3 mBar
Anode diameter	4 mm
Profile duration (depends on sample)	300–1000 sec

GDOES quantification was split into two parts: determination of the sputtered depth from the sputtering time and sputter rates of individual elements; and determination of the chemical content from the emission line intensities. The sputter rates can be calibrated from sputter depth measurements. The emission line intensities are dependent on the voltage, current and sputtering rates, as well as the chemical composition. Normally the intensity calibration is performed using certified reference materials of known chemical composition. During depth profile quantification the analysed composition is normalised to 100%. There are quite a number of steps involved in setting up a quantification program. Figure 5 gives an outline of the steps involved for superalloy turbine blade coatings.

A total of 49 calibration standards were used to calibrate the superalloy turbine blade coatings quantitative program covering must of the elements in the coatings. In order to ensure reliable results were obtained, the standards were selected so as to cover the elemental concentration range in the coatings. Table 2 shows an overview of the calibration elements and analytical range with the selected standards.

EPMA using WDS was used to perform chemical analyses on both the CMSX-4 and RT31 samples to enable comparisons to be made with GDOES and XRF results (the XRF results were supplied by Rolls-Royce).

As the substrate alloys and coatings were multiphased systems a straight forward single line scan would contain a lot of scatter as different phase compositions were crossed by the scanning electron beam. In order to reduce this problem a series of 'point' analysis at one micron intervals along a straight line were taken. At each point the beam of the probe was 'rastered' along a 50 μm line normal to the line produced by the series of analyses to form a 'ladder' scan. This enabled an average band (50 μm wide) to be assessed.

RESULTS AND DISCUSSION

The time-intensity curves and calculated quantification depth concentration profiles for CMSX-4 superalloy substrate are shown in Figs 6 and 7. Only elements Ni, Co, Al (major ones) and Ti, Hf, W (minor ones) have been shown for the sake of clarity. The tendency for the lines to rise as time and depth increases is due to matrix effects resulting from differences in elemental sputtering rates.

The crater produced during the analysis of CMSX-4 after 60 and 1000 seconds are shown in Figs 8(a) and (b), and their measurement by laser profiling is shown in Figs 9(a) and (b). It can be seen that the bottom of the crater in Fig. 8(a) is quite planar, while some re-sputtering of material has created a lip around the rim after 1000 seconds. The laser profile depth measurement shows a good correlation to the calibrated depth values obtained from Figs 7(a) and(b), with errors of the order of 10–15%. The roughness of the bottom of the crater normally increases linearly with time,[6] which is one of the factors limiting the depth resolution. This necessitates a compromise between good depth resolution and high intensity spectral lines when selecting operating parameters.

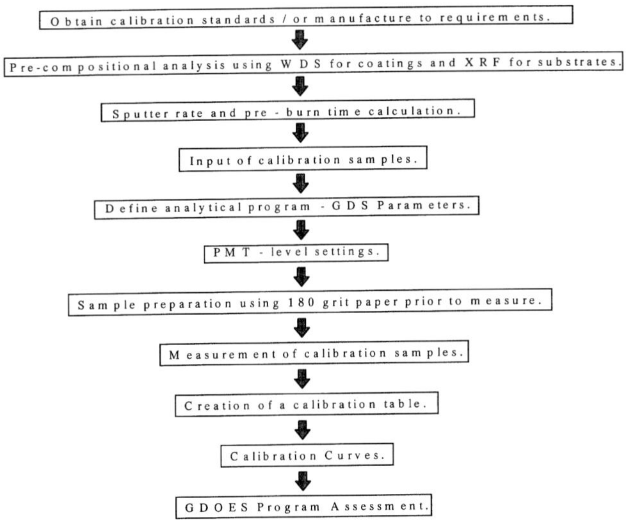

Fig. 5 GDOES quantification program set-up.

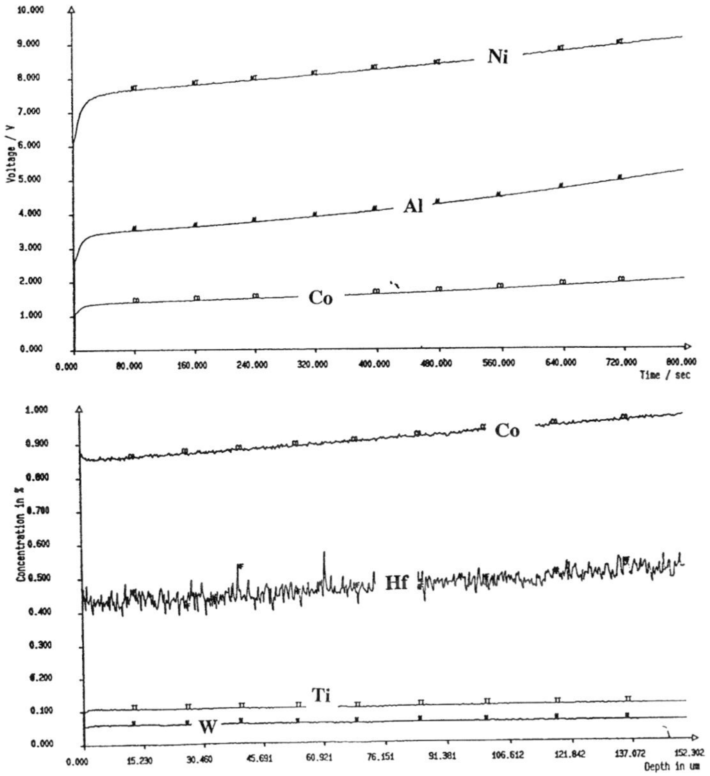

Fig. 6 (a) Time–intensity trace on CMSX-4 showing major elements Ni, Al and Co, (b) Time–intensity trace on CMSX-4 showing Co and minor elements Ti, Hf and W.

Table 2 Calibration standards overview

Analyte	Analytical range	Matrix Element in	Calibration points used
Ni	1.99–64.69%	Ni Fe grades; Aluminide coatings; Pure Ni; MCrAlY	29
Cr	0.23–8.78%	Superalloy: Marmoo2, CMSX-4 Low alloy steel; Pure Co	7
Al	0.00–17.91%	Aluminide coating; Superalloys Pure Al	16
Ti	0.11–1.44%	Superalloys; Aluminide coatings Pure Ti	10
Pt	0.00–99.99%	Pure Pt; Aluminide coatings	4
Y	0.00–99.99%	Pure Y MCrAlY coating	2
Co	0.38–9.42%	Pure Co; High alloy steel; Aluminide coating	19
Mo	0.30–9.41%	Low alloy steel grades High alloy steel grades	11
W	0.20–9.65%	Superalloys Aluminide coatings	9
Ta	1.13–6.48%	Superalloys Aluminide coatings	4
Hf	0.07–1.27%	Superalloy grades	2
Zr	0.00–97.67%	Pure Zirconia	1
Si	0.05–2.69%	Cast iron; High alloy steel Superalloy grades	17
P	45.2–225.8 ppm	Low alloy steel grades High alloy steel grades	13
B	4.27–27.12%	BTiAlN Ni–B	2
S	30.1–378.6 ppm	High alloy steel grades Low alloy steel grades	8
C	0.00–3.35%	Cast iron grades; High alloy grades; Low alloy steel grades	14
O	0.00–26.60%	FeO	1

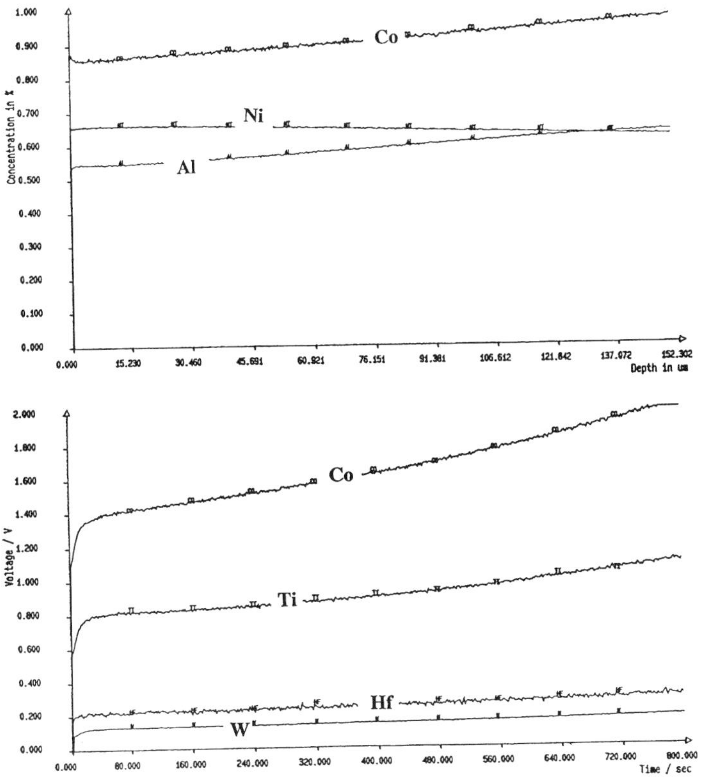

Fig. 7 (a) Depth–concentration profile on CMSX-4 showing major elements Ni, Al and Co, (b) Depth–concentration profile on CMSX-4 showing Co and minor elements Ti, Hf and W.

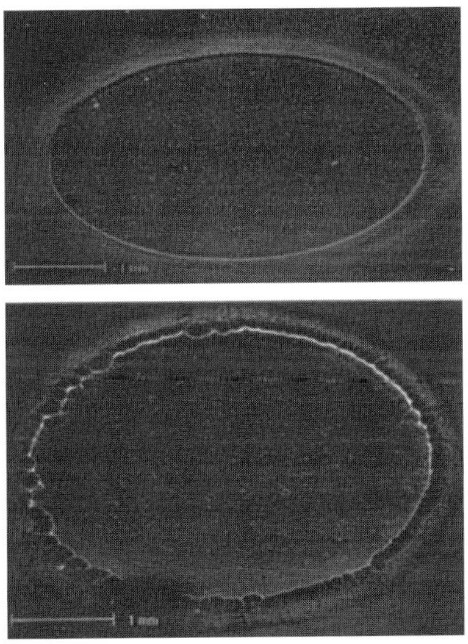

Fig. 8 (a) SEM micrograph of crater at 60 sec, (b) SEM micrograph of crater at 1000 sec.

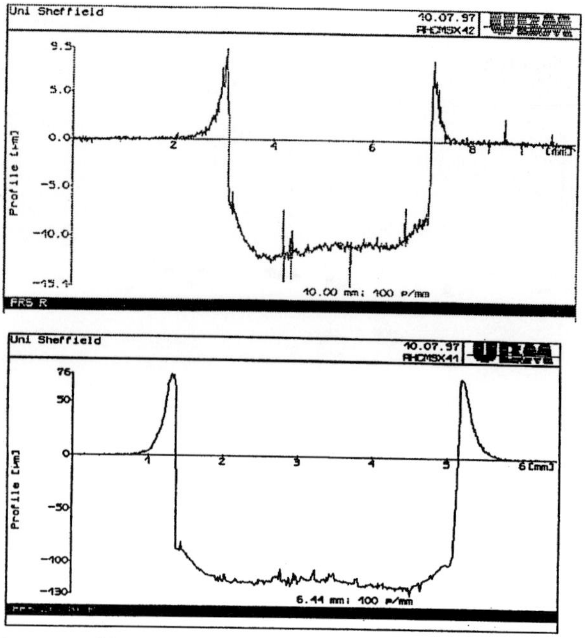

Fig. 9 (a) Laser profile of crater at 60 sec, (b) laser profile of crater at 1000 sec.

Table 3 Comparison of XRT, EPMA and GDOES compositions for CMSX-4.

	Ni	Cr	Al	Ti	Pt	Co	Mo	W	Ta	Hf	Zr	Si	Y	B	O	Re	C	P
XRF analysis	60.82	6.3	5.83	0.98	0.012	9.46	0.6	6.32	6.52	0.1	0.005	0.03				2.71		
EPMA analysis	59.9	6.7	6.3	1.1		9.6	0.6	6.1	6.4	0.1						3.7		
GDOES analysis	64.51	6.39	5.89	1.09	0.431	9.11	0.55	6.07	5.4	0.47	0.009	0.029	0.02	0	0	–	0.001	0.007

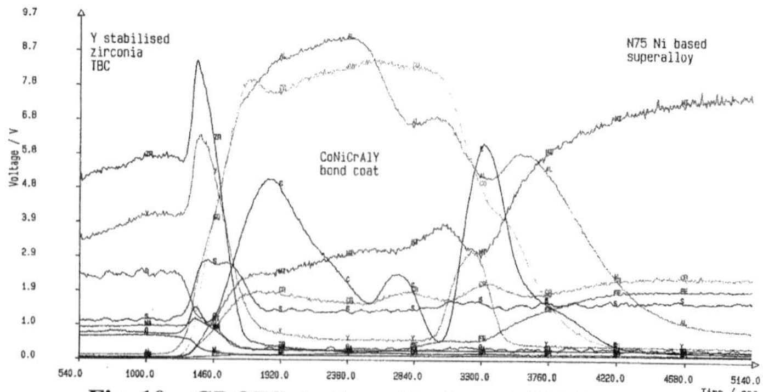

Fig. 10 GDOES depth profile through TBC system.

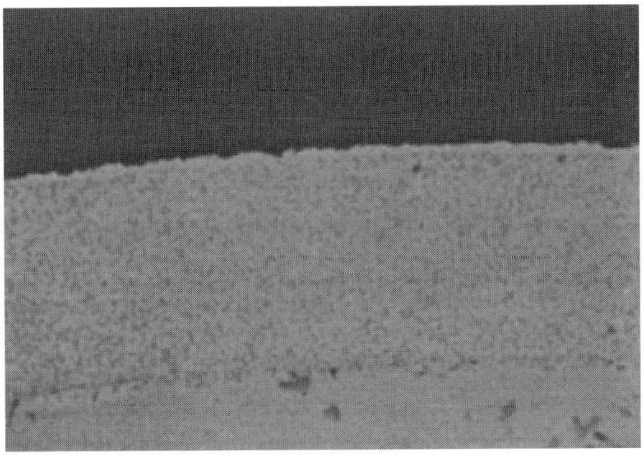

Fig. 11 Cross-sectional view of TBC.

The results of the GDOES and EPMA analyses are compared to the composition determined by XRF in Table 3. Both results showed a good correlation, with the GDOES results generally closer than those of EPMA, although there were some obvious discrepancies which require investigation; the values obtained for nickel, tantalum, hafnium and platinum are inaccurate and may be due to interference between the spectral lines. The GDOES equipment is also not currently able to detect rhenium.

The trace obtained from the thermal barrier system (Fig. 10) clearly distinguishes the different interfacial regions seen in the cross-sectional view (Fig. 11). The elements identified within these regions agree with the EPMA analysis (Fig. 9), with both techniques highlighting the yttrium concentrated at the ceramic/bondcoat and bondcoat/substrate interfaces. GDOES does have the advantage of better sensitivity to very low concentrations of elements such as sulphur, boron, carbon and phosphorus. The results can also be obtained more rapidly with minimum sample preparation.

The current semi-quantitative program set up for turbine blade coating analysis is limited by the number of standards and calibration points used (see Table 2). For example with a minimum of two calibration points it is thus possible to obtain a straight calibration line. It is however, recommended to use a minimum of 3 to 5 calibration samples. The more calibration samples are used, then the easier to detect an inaccurate calibration sample and if necessary to make appropriate correction or deletion. In addition extra elemental channel such as Re and additional elements needs to be installed.

A study was performed on a few selected superalloy turbine blade coatings to assess elemental interference in GDOES traces. Cross-over of emitted wavelengths can result in 'ghost' profiles being produced. Currently further investigations are underway to identify and overcome any interference problems.

The work conducted so far has shown GDOES to be a very useful analytical tool for the characterisation of turbine blade coatings. The large sample area analysis with GDOES, as compared to EPMA, SIMS and AES where microscopic areas are in-

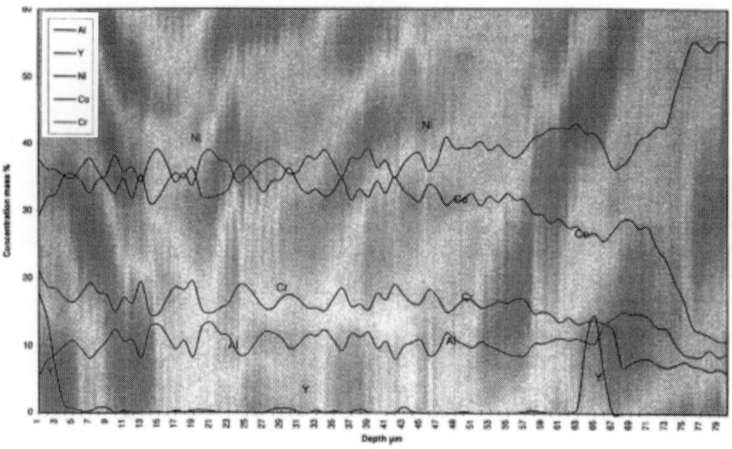

Fig. 12 EPMA analysis of TBC system.

volved is good for obtaining a better representative average result. A semi-quantitative GDOES program has been developed and with further work to resolve the problems seen with elemental interferences, an improved quantitative method will enable a much broader understanding of the role of trace elements in the bonding and degradation mechanisms of both thermal barrier and environmental coating systems.

ACKNOWLEDGMENT

The authors would like to thank Mr R. Jones (RR), Dr M.D. Bramhall (SHU), Mr P. Slingsby (SHU) and Mrs C. Shaw (SHU) for their help and advice towards this project.

REFERENCES

1. W. GRIMM: *Spectrochimica Acta*', 1968, **23B**, 443.
2. O. DESSENE, A. QUENTMEIER and H. BUBERT: *Fresenius – J. Analytical Chemistry*, 1993, **346**, 340–345.
3. L. A. DONOHUE: *Thesis: Deposition and Characterisation of (Ti,Zr) based hard compound and multilayer pvd films*, Sheffield Hallam University, 1995.
4. M. IVES: *Thesis: Fundamental studies of the PVD technique*, 1994.
5. W. W. HARRISON: *J. Analytical Atomic Spectrometry*, 1992, **7**(3), 75.
6. R. K. MARCUS: *Glow discharge spectroscopies*, Plenum Press, 1993.

Mechanical Properties of an Air-Plasma-Sprayed Thermal Barrier Coating and its Effects on Hastelloy X Substrates

XIN-HAI LI, L. JOHANSSON, M. BRODDEGÅRD,
D. ABRAHAMSSON, and B. GUDMUNDSSON
ABB STAL AB, Dept. RT, SE–612 82 Finspong, Sweden

ABSTRACT

Tension, compression, high cycle fatigue (HCF), and low cycle fatigue (LCF) behaviours of an air plasma sprayed thermal barrier coating (APS TBC) have been studied. Test specimens were TBC coated Hastelloy X. The tension and HCF tests were also performed on the uncoated Hastelloy X specimens. An acoustic emission system was employed in the room temperature tension tests to detect coating cracking and delamination. The tension and compression tests were performed in a temperature range from 20 to 840°C, while the HCF and LCF tests were performed at 800 and 850°C respectively. Results have shown that the tension strain to TBC cracking ranges from 0.7 to 1 % and the compression strain to TBC cracking varies from 1.3 to 2.5%, depending on TBC thickness, temperature, and heat treatment. During both LCF and HCF tests there was no TBC spallation found until the base material fractured. As compared to existing LCF results for the uncoated Hastelloy X, TBC has shown no effect on LCF life of the base material. But, TBC does reduce the HCF life of the base material. Failed specimens have been microscopically examined and failure mechanisms have been discussed.

1. INTRODUCTION

Thermal barrier coatings, TBCs, have been applied on the hot components, such as combustors, blades, and vanes, in gas turbines to increase the turbine efficiency and power and to reduce emissions. TBC development has been world-wide recognised as a key element for gas turbine development and particularly the future advanced land-based gas turbines will depend on TBCs.[1]

To effectively utilise TBC's benefits, TBC properties need to be provided to the turbine design and its limitations should be understood and considered. Although great efforts have been made on investigations and material testing of TBCs by NASA projects,[2,3] it is still necessary to investigate and test individual coating systems since the coating properties and behaviours strongly depend on applying method, process parameters, material chemical composition, structure, and even substrates. Coating properties can be investigated in two ways, one is to test the free-standing coating[4] and the other is to test coating on substrate. Apart from coating own properties, coating effects on base materials are interesting to explore. Coating could affect the strength and life of base materials negatively or positively,[5,6] depending on coating/substrate combination, coating process, and so on. The coating effects, especially the negative effects, have to be considered.

In this work an air plasma sprayed TBC applied on a Hastelloy X substrate which is a combustor material in gas turbines has been studied. Tension and compression tests have been carried out. Due to vibration of the combustor and thermally cyclic operation of the gas turbines, HCF and LCF tests have also been performed. TBC effects on the base material have been investigated.

2. EXPERIMENTS

Cylindrical bar substrates were fabricated for all tests and made of a Ni-based alloy, Hastelloy X. TBC was applied on the waisted part of the substrates by air plasma spraying, APS. The TBC system consists of two layers, a ZrO_2+7%Y_2O_3 ceramic top coat (TC) and a Ni23Co17Cr12Al0.5Y bond coat (BC). The bond-coat thickness was 0.15 mm, and the topcoat thickness was 0.4 mm except that the top coat on the RT tensile specimens had a thickness of 0.4, 0.6, and 1 mm.

Four types of HCF specimens were fabricated and tested in order to investigate the effects of coating, annealing, and grit blasting. They were as-coated, coated and annealed at 1080°C for 4 hr in vacuum, uncoated and polished, and uncoated and grit blasted. The waisted part and the inner hole of all the specimens were polished to a surface roughness $R_a \leq 0.2$ µm prior to the grit blasting and spraying.

A post annealing was carried out on some of RT tensile test specimens in Ar at 1000°C for 4 hr for studying annealing effects.

A Schenk-Trebel RSA 100 tensile testing machine was used for all the RT tensile tests. An acoustic emission (AE) system (supplied by Alvetec AB) was utilised to detect sounds generated by cracking and delamination of the coatings during RT tensile testing (the details of the testing method and the data evaluation can be seen in Ref 7). This method was not used during the HT testing because of the difficulty of cooling AE transducers. The recorded AE data were: time of pulse arrival, load (from the testing machine), strain, rise time, counts, energy, duration, and amplitude. Two acoustic emission sensors were attached to two ends of the specimen respectively. If a signal arrived at the two sensors in different time a corresponding event position was calculated from the time difference.

An Instron 1275 servo hydraulic testing system equipped with an infra-red radiation furnace was used for HT tensile, compressive, and LCF testing. It was possible to visually inspect TBC cracking through a hole on the infra-red furnace by a x30 microscope. The tests were controlled manually and a computer assisted data acquisition system was used to log the data of the tests. The HT tension and compression tests were performed at 600, 700, and 840°C and the LCF tests at 850°C. The compressive tests were performed by load control with a rate of 0.1 kN/s, while all the tensile tests were carried out by applying EN 100002 standards. The LCF tests were performed with three strain amplitudes 0.2, 0.3, and 0.6 % and the straining rate was 0.6 %min^{-1}. The compressive tests were terminated as the coating failed. The LCF failure cycle was defined as the cycle at which the maximum stress decreases to 50 % of the maximum stress in the first cycle.

HCF tests were performed with an Amsler UHF 422/589 resonance testing machine. The mean stress was set to zero for all the tests, the testing temperature was 800°C, and the frequency of the dynamic loading was around 100 Hz. The tests were terminated as soon as the load amplitude changed more than 10 % from the setting value.

Some of tested specimens were sectioned diametrically or axially for studying structures and failures appearance. Optical microscopes and a Joel JSM-5800 scanning electron microscope, SEM, were used to examine the samples.

3. RESULTS AND DISCUSSION

3.1 Tension Testing

TBC on all the tensile testing specimens was found to fail in two steps. Circumference cracks appeared first in the coating, and then coating delamination occurred. The coating delaminated at the interface between the top and bond coats and sometimes at both top/bond and bond/substrate interfaces, depending on the thermal history. After the final failure of the specimens the TBC became a segmented tube.

For the room temperature tensile tests, the strain values to both first cracking and delamination were determined by AE method. Figure 1 shows a typical relation between the AE events position and the strain value. It is seen that most of the AE events occuring within the first 0.2% strain come from the specimen holders and at two strain levels of 0.12% and 0.8% the events are located in the coating region as the first events and coating cracking respectively. The first events may be associated with microcracking

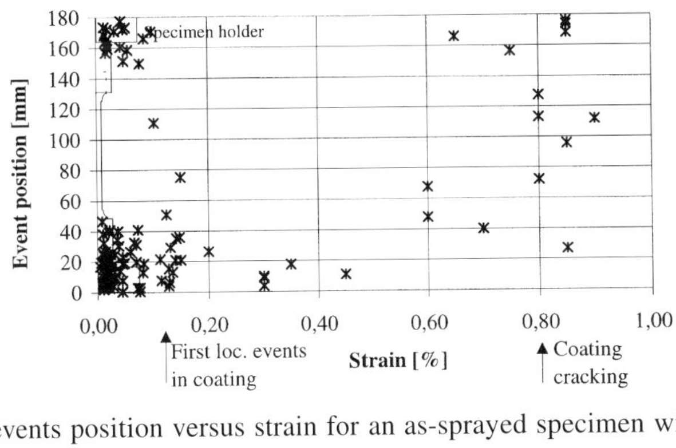

Fig. 1 AE events position versus strain for an as-sprayed specimen with 0.4 mm TC (and drawing of specimen and holders).

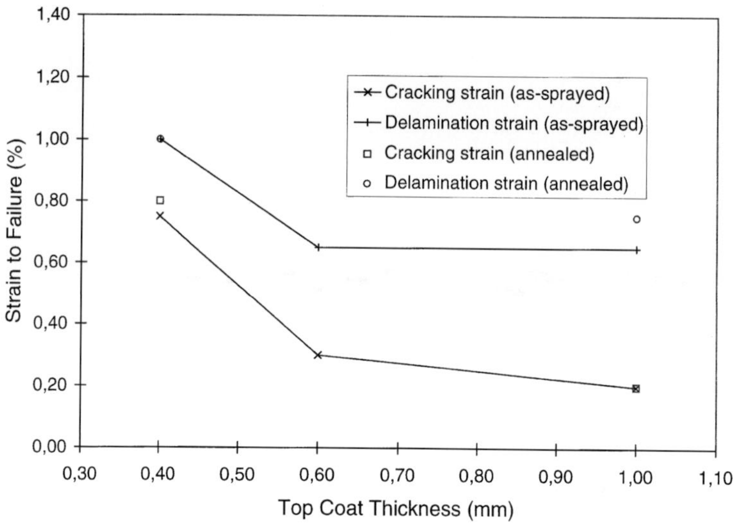

Fig. 2 Tensile strains to TBC delamination and initial cracking versus top coat thickness.

in the ceramic coating. The coating cracking was defined as the strain level at which more than one event was localised in the coated area. Using this detection method it was not possible to localise the delamination events since the signal overlaps resulted in a duration overflow. The delamination strains were determined from plots of energy and duration vs. strain and defined as onsets of energy peaks or onsets of the duration overflow.[7]

From the RT tension testing the strains to TBC cracking and delamination were found to be a function of the top coat thickness. Figure 2 shows that both cracking and delamination strains decrease with increasing top coat thickness with TBC delamination occurring at a strain value 0.2 % higher than the cracking. Both strain values decrease rapidly when the TBC thickness increases from 0.4 to 0.6 mm and then become stable when the thickness is above 0.6 mm. This may indicate that the value around 0.6 mm is a critical thickness above which the coating becomes much less elastic. According to the results, to achieve a reliable thick TBC, e.g. 2 mm thick for improved thermal insulation, the approach of changing TBC structure in order to increase the coating strain tolerance could be considered. For instance, increasing porosity or segmentation of top coat[8] has been employed for spraying thick TBC.

Figure 2 also shows the data for the specimens annealed at 1000°C for 4 hr in Ar. The annealing effect is not apparent, the strains increase only slightly after annealing. However, the BC delamination was found to be influenced by annealing. As shown in Fig. 3(a), in the ascoated specimen the delamination occurred at both BC/substrate and TC/BC interfaces. In the annealed specimens no delamination at BC/substrate interface was observed as seen in Fig. 3(b) but only the delamination at TC/BC interface, indicating that the bond coat adhesion to substrate was improved by the anneal-

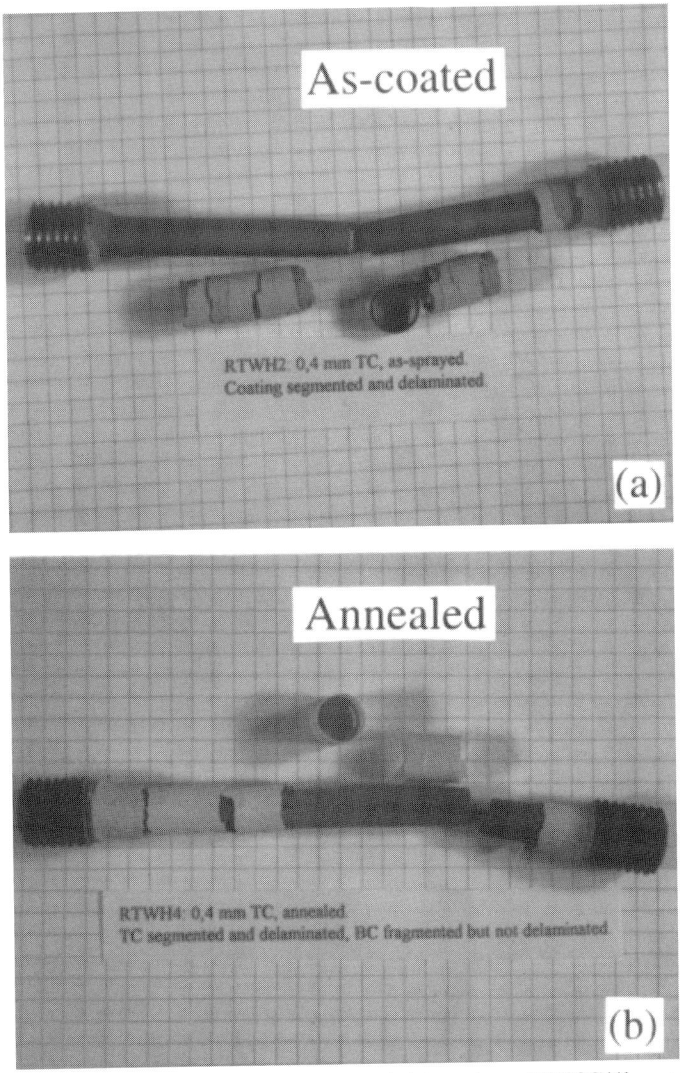

Fig. 3 Photographs taken from an as-coated (a) and an 1000°C/4hr annealed (b) specimens showing coating delamination.

ing. This effect is due to the fact that a diffusion zone formed at the BC/substrate interface during the annealing, resulting in a strong metallurgical bonding between the BC and the substrate.

For the high temperature tensile tests it was only possible to determine the strain of which resulted in TBC cracking. Figure 4 presents the mechanical strain to TBC initial cracking as a function of testing temperature (the solid line) measured from the specimens coated with 0.4 mm TBC and tested at RT, 600, 700, and 840°C. It is seen

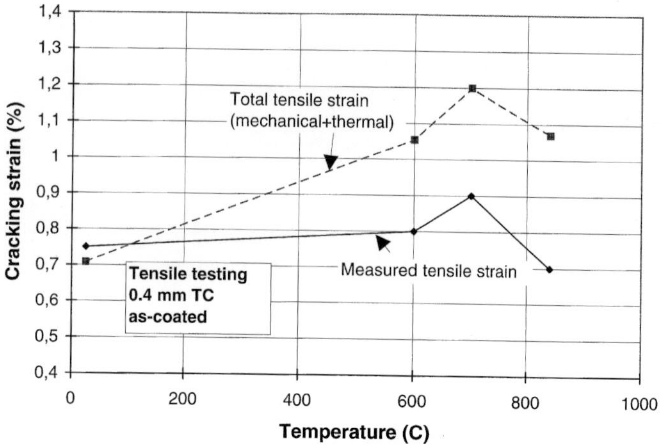

Fig. 4 Tensile strain to TBC cracking as a function of testing temperature.

that the strain value varies from 0.7 to 0.9 % and the plot does not show clear temperature dependence, in agreement with the findings of Ref. 9. However, during the HT testing the top coat actually suffered an extra thermal strain originated from a large difference in coefficient of thermal expansion (CTE) between the ceramic coat (~10x10^{-6} °C^{-1}) and metallic parts (bond coat + substrate, ~15x10^{-6} °C^{-1}). Assuming that the substrate temperature during spraying is 100°C and that the residual stress in the coating at this temperature is zero, the strain due to the difference in CTE can be estimated. The total tensile strain to cracking, i.e., the measured and the estimated thermal strains, of the top coat is plotted as a dashed line in Fig. 4. In contrast to the measured strain, the total strain increases with temperature rapidly, showing a strong temperature dependence. According to the plot the ceramic coat can endure a larger strain (1.05–1.2 %) at higher temperatures (600–840°C) before initial cracking.

The bond coat delamination during the HT tensile testing was found to be affected by testing temperature, though the specimens were all as-coated. The delamination occurred at the both interfaces when tested at 600°C but only between TBC and bondcoat at 700 and 840°C. It is unlikely that it is the annealing effect mentioned above since the diffusion at such low temperatures and for such a short testing time (<1 hr) can be neglected. Rather, ductile to brittle transition of the bond coat may contribute to the difference. At the higher temperatures the bond coat changes from brittle to ductile and can follow the plastic deformation of substrate to final failure. Unfortunately, there is no data of ductile to brittle transition temperature (DBTT) for the NiCoCrAlY coating but the DBTT of NiCrAlY is just between 600 and 700°C.[6]

Yield strength $R_{p0\cdot 2}$ and the tensile strength R_m data were measured at different temperatures for both coated and uncoated Hastelloy X. There is no distinct difference in both $R_{p0\cdot 2}$ and R_m between coated and uncoated specimens, indicating that the tensile properties of Hastelloy X were not affected by TBC or the coating process.

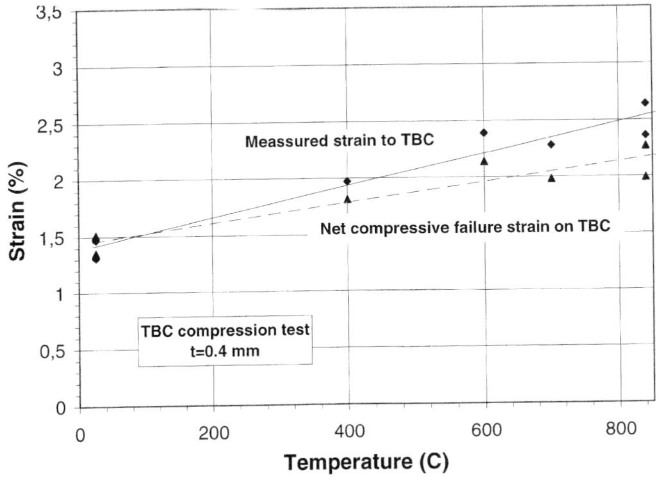

Fig. 5 Compression strain to TBC failure versus testing temperature.

3.2 Compression

TBC failure strain in compression is much larger than that in tension. Figure 5 shows the measured failure strain versus testing temperature (solid line). The compression failure strain has a strong temperature dependence and increases monotonically from 1.4 to 2.5 % with increasing temperature from RT to 840°C.

Unlike tension tests, TBC failed during the compression tests in a way that the ceramic coat buckled and spalled off instantly. No cracking was observed prior to the final TBC failure and the top coat cracking and spallation occurred simultaneously. In contrast to the circumference cracks observed on the tensile specimens, during both RT and HT compression tests the cracks developed axially in the top coat. During the compression tests delamination was always observed in the top coat near the TC/BC interface. This can be seen from two diametrical cross-section micrographs in Fig. 6 taken from the specimens tested at RT and 840°C, respectively.

During thc tcsts the strain to TBC failure was only able to be determined when the failure appeared on the surface. It was suspected that the TBC delamination and microcracking might be occurring at a strain level much lower than the failure strain determined by this approach. In order to clarify this a specimen was tested up to 1.9 % strain at 600°C. A cross-section of this specimen was examined microscopically and no delamination or cracking was observed.

The thermal strain due to the difference in CTEs was also taken into account to estimate the actual compression failure strain in the coating. The dashed line in Fig. 5 shows the net strain to the TBC failure and it is smaller than the measured strain.

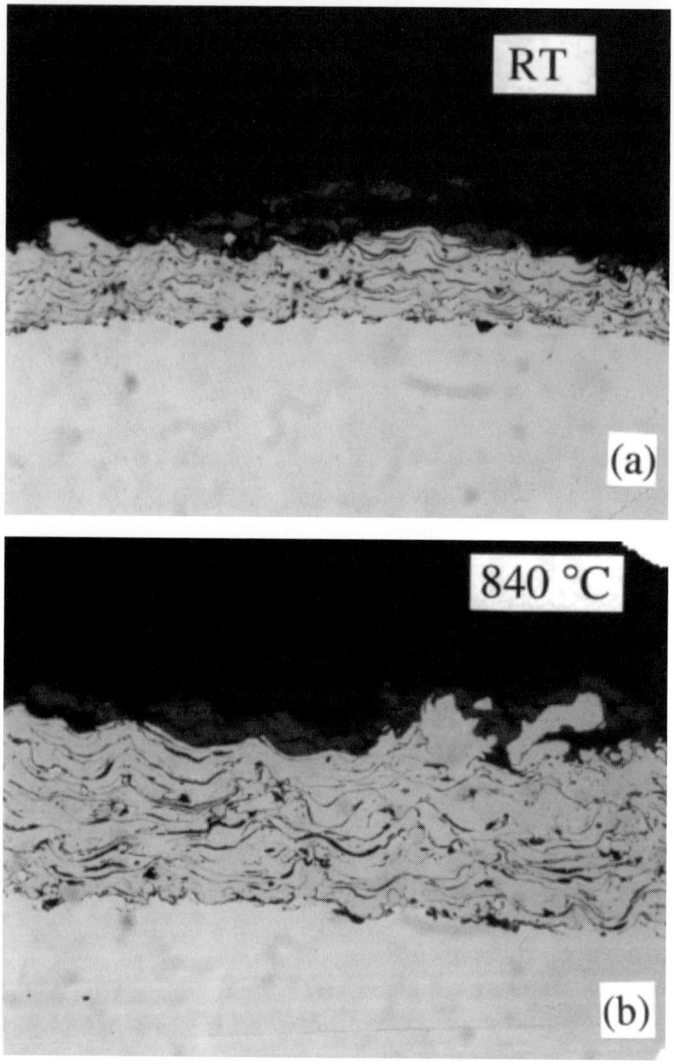

Fig. 6 Micrographs taken from specimens tested at (a) RT and (b) 840°C, showing only the TC delamination.

3.3 High cycle fatigue

Figure 7 is a histogram showing the fatigue strengths determined from the coated, uncoated, coated and 1080°C/4hr annealed, and uncoated but grit blasted test series. Fatigue strengths at 10^7 cycles were evaluated by employing staircase method[10] and the strength was determined from the total load divided by the cross-sectional area of the substrates. From the Fig. 7 it is seen that the coated series have a fatigue strength about 12% lower than the uncoated series do. Between the grit blasted and polished

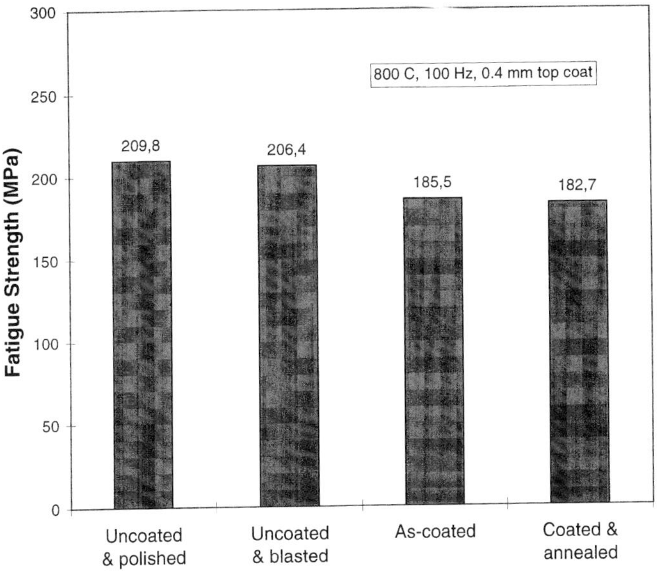

Fig. 7 A histogram, showing fatigue strengths determined from four test series performed on specimens which are as-coated, coated and annealed, uncoated and uncoated and grit blasted, respectively.

series the strength difference is only within the deviation of <±3%, indicating that the grit blasting has no distinct effect on the HCF strength of Hastelloy X material at the testing temperature of 800°C. The negligible strength difference between the as-coated and annealed series also implies that the annealing does not affect the HCF life of the coated Hastelloy X.

Calculation of the fatigue strength from the total load and the cross-sectional area of the substrate was based on the assumption that the load carried by the coating was negligible compared to the load on the substrate. However, if one calculates carefully the load carried by each layer by using elastic moduli for the bond coat and top coat (~130 and 20–30 GPa, respectively) found from the Ref. 2, it will be found that the TBC layers do carry about 8% of the total load in the elastic range and the load on the base material is smaller than the total load. This means that for the coated specimens the actual fatigue strength of the Hastelloy X is smaller than that shown in Fig 7 and this also confirms that the coating does reduce the fatigue strength of the base material.

TBC spallation and visible cracks were not observed on all of the runout specimens nor on the failed specimens which were not broken up (3 of 5 failed specimens). As shown in Fig. 8, the TBC spallation and crack were only found on the side of the area B in the specimens failed with breaking up. The fracture pattern in Fig. 8 shows clearly that fatigue crack was initiated from the area A and the fracture in the area B was caused by the forced separation at the last stage of the testing or during demounting.

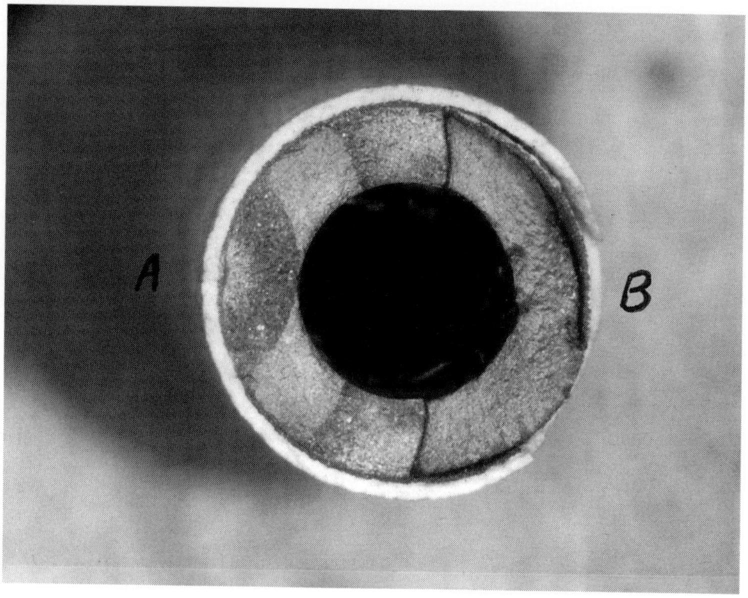

Fig. 8 HCF specimen tested to rupture, showing the fatigue fracture area (A) and the final separation area (B) in which TBC spalled.

To further investigate the cracks, both runout and failed specimens were sectioned axially (since the fracture is diametrical) and examined metallographically. These studies revealed no cracking in the run-out specimens, and for the failed specimens there was no cracking (>50 μm in length, and note that microcracks ≤20 μm in length always exist in the top coat[11]) found in the top coat on the side of the area A in Fig. 8, except for a crack going through the substrate and the coating and leading to the final fracture (see Fig. 9).

It is unlikely that the fatigue crack was initiated from the top coat, since no more cracks than the final one were found in the top coat in the area A. According to our previous experience the cracks in the top coat do not normally propagate into bond coat and substrate. Another reason is that the top coat which has a low elastic modulus of 20–30 GPa is very 'elastic' and can follow the elastic deformation of substrate. One possible contributing factor was grit blasting prior to spraying since this roughens the substrate surface and causes a notch effect, since a microcrack was discovered the substrate near the surface. But, according to the results shown in Fig. 7, the blasting can be excluded, a finding supported by the data published in Ref. 5 also show insignificant blasting effect on IN738 HCF life. The roughnesses on the blasted substrate surface and the interface between the bond and top coats were measured to be 3 and 16 μm, respectively. The rough bond coat interface which is desired for achieving a good mechanical bonding between the ceramic and bond coat is a possible cause of the

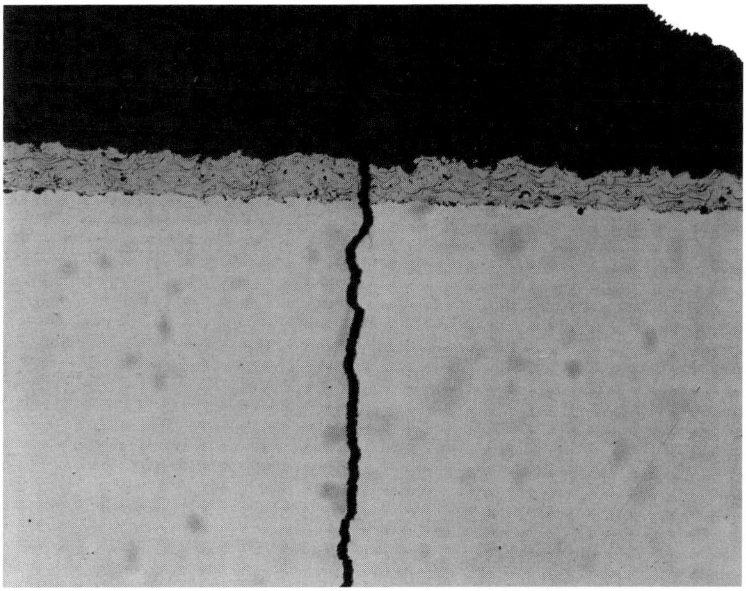

Fig. 9 A crack lead to final fracture in a HCF specimen which did not fall apart.

coating effect on the substrate HCF life. Smith[5] and Schneider *et al.*[6] have studied the effects of different MCrAlY overlays on base material HCF and their results show that the coatings have either no effect or positive effects on the base materials. However, the overlay coatings they investigated may have a smoother surface (6.4 μm peak height for as-coated and 0.5 μm for polished)[5] than the bond coat in our specimens. To confirm the roughness effect, further tests are needed.

3.4 Low Cycle Fatigue

The number of cycles to failure for the TBC coated specimens was determined to be 9200, 2220, and 596 for the strain amplitudes of 0.2, 0.3, and 0.6 %, respectively, as seen the strain-life diagram in Fig. 10. The LCF life data for the uncoated Hastelloy X material obtained from Ref. 12 (871°C) and internal results (850°C) are also plotted in the figure. It is seen from Fig. 10 that there is insignificant difference in LCF life between the coated and uncoated Hastelloy X. Figure 11 shows the monotonic and cyclic stress-strain curves obtained from the LCF results with evidence for a greater degree of the cyclic stress. This behaviour is also similar to that of the uncoated Hastelloy X tested at 871°C as seen the stress-strain curves in Fig. 12(a) from Ref. 12. As a comparison, Fig. 12(b) presents the curves for Hastelloy X tested at 695°C from the same reference, showing a different hardening behaviour between two different temperatures. The comparison between the coated and uncoated Hastelloy X reveals that the TBC has an indistinct effect on the LCF life of Hastelloy X.

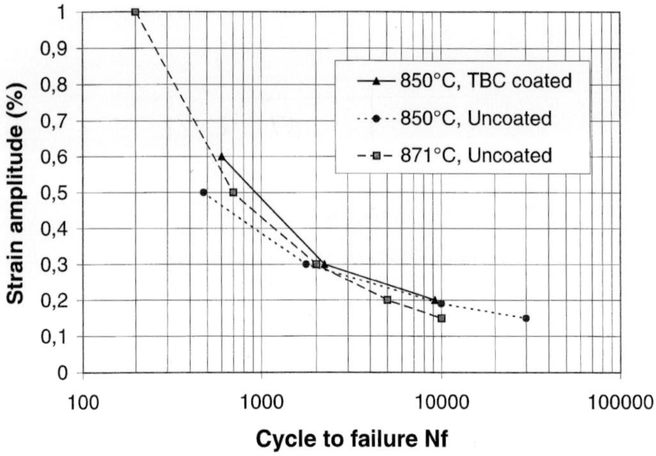

Fig. 10 LCF strain-life curves for TBC coated and uncoated Hastelloy X.

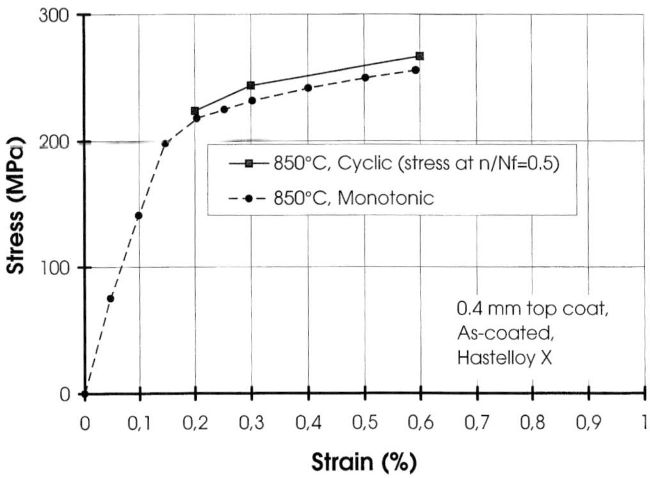

Fig. 11 Monotonic (dashed line) and cyclic (solid line) stress/strain curves for the TBC coated Hastelloy X tested at 850°C.

During the LCF tests no TBC spallation appeared and only one visible circumference crack on each specimen was observed only when the substrate started to crack, i.e., when the maximum stress started to decrease sharply. This indicates that the TBC coated specimens had a LCF life at least equivalent to that of the substrate material Hastelloy X. Some microcracks on the TBC surface which were almost not able to be seen by a x30 microscope appeared at the later stage of the testing.

Tested LCF bars were sectioned both axially and diametrically for microscopic ex-

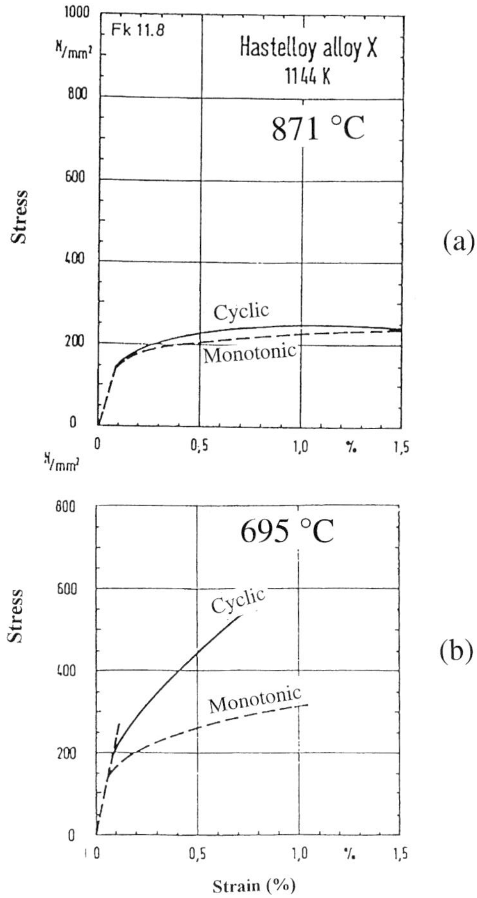

Fig. 12 Monotonic (dashed line) and cyclic (solid line) stress/strain curves for the uncoated Hastelloy X tested at (a) 871°C and (b) 695°C from Ref. 12.

amination. Fig 13(a) and 13(b) show respectively the axial and diametrical cross sections of the specimen tested with the strain amplitude of 0.6 %. As seen in Fig 13(a), a large crack which is associated with the final fracture is circumferential and goes through the TBC and the major part of the substrate. In the diametrical cross-section shown in Fig 13(b), cracks were observed in both top coat and substrate but they are not connected to each other. The substrate cracks was not observed in the other diametrical cross sections tested with the lower strain ranges.

In the axial cross section of the 0.2 % specimen, besides the large crack associated with the final failure, several small cracks or microcracks were discovered in the substrate, but only a few found in the other specimens with high strains. This means

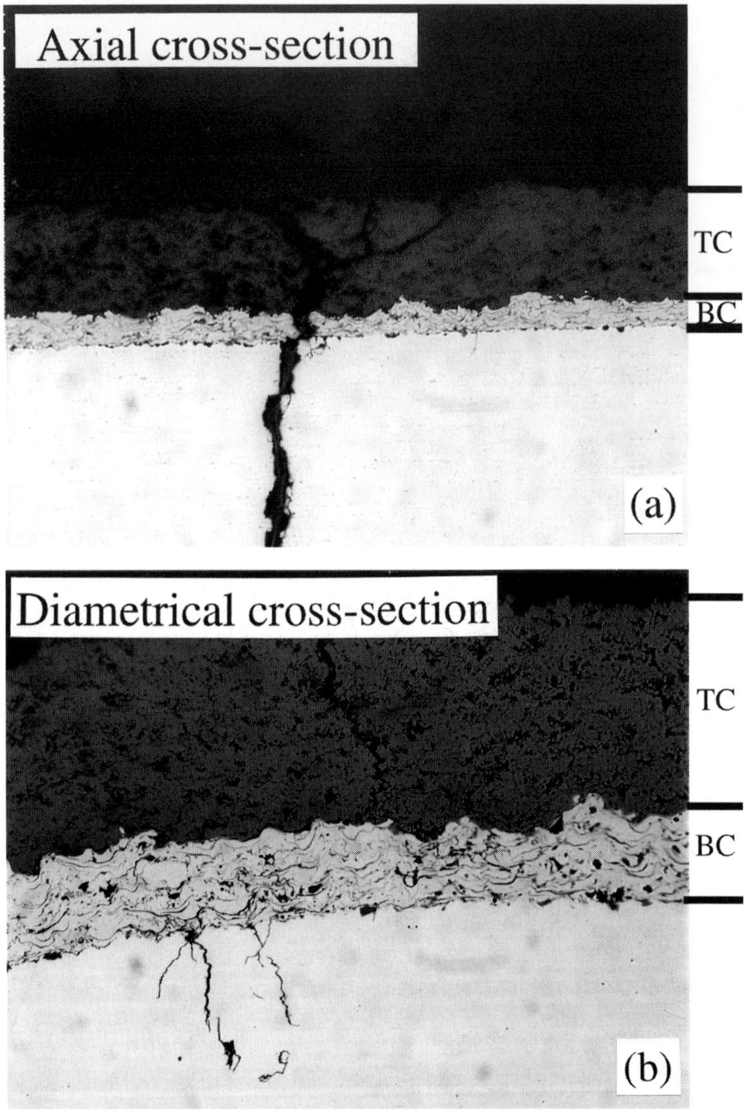

Fig. 13 Micrographs taken from the (a) axial and (b) diametrical cross-sections of the specimens LCF-tested with a 0.6% strain amplitude.

that the larger number of cycles (due to the small strain amplitudes) gives rise to more circumferential substrate cracks. The interesting observation is that most of the cracks start from the substrate surface and there are no cracks in the coating which link to substrate as seen in Fig. 14. The cracks start from the bottom of the valleys of the rough surface created by grit blasting. The valleys are filled with the residual of the

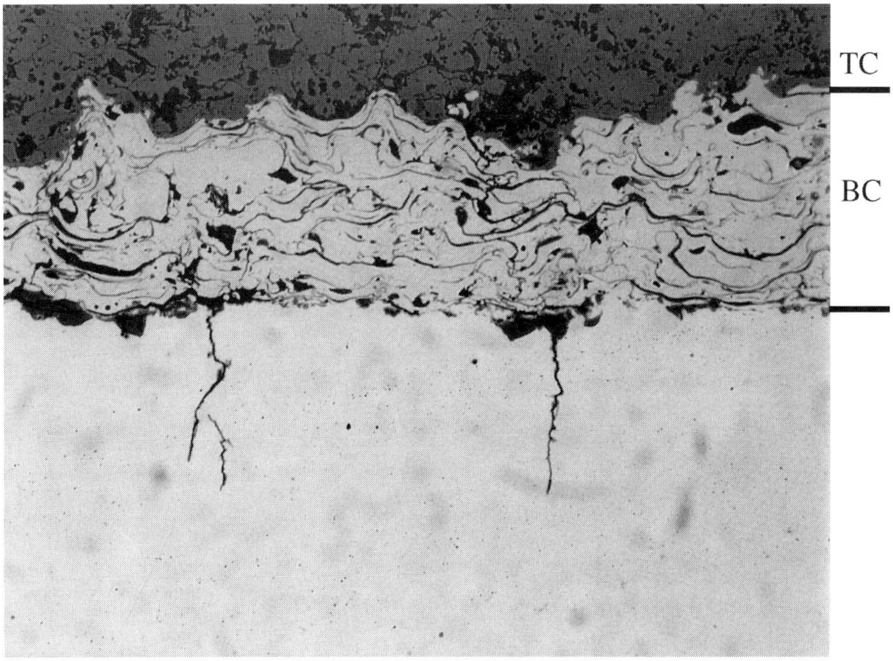

Fig. 14 Axial cross-section micrograph taken from the specimen LCF-tested with a 0.2% strain amplitude, showing microcracks initiated from the substrate surface.

ceramic grit. These imply that the LCF failure of the specimens was caused by the cracking initiated in the substrate, rather than in the coating. This could be the reason that substrate LCF life was not influenced by the coating.

4. CONCLUSIONS

- According to the RT tension tests, the tensile strains to TBC delamination and to first cracking, and TBC strain tolerance are influenced by the TC thickness. The cracking strain decreases from 0.75 to 0.2 % and the delamination one decreases from 1 to 0.65 % with increasing the TC thickness from 0.4 to 1 mm.
- Bondcoat adhesion to the substrate during the RT tensile tests has been improved by the post annealing at 1000^{0}C for 4 hr. It also varies with the testing temperature probably due to the ductile to brittle transition.
- Compared to the tensile strains, the TBC has a much larger value of compression strain to failure, ranging from 1.4 to 2.5 % depending on testing temperature (RT–840°C). Both compression and tension strains to TBC failure have a tendency to increase with the testing temperature. Comparably, the compression strain has a stronger temperature dependence.

- The TBC coating does not affect significantly the tension properties, yield strength and tensile strength of the Hastelloy X.
- No TBC spallation has occurred before the substrate failure during both HCF and LCF tests. The LCF life of the TBC is at least as long as the Hastelloy X material.
- The LCF life of Hastelloy X has not been affected by the TBC while the HCF life has been reduced by about 12 % by the TBC. The TBC effect on the HCF life of Hastelloy X is not caused by the grit blasting prior to the spraying but probably by the rough bond coat surface. A further investigation is needed to confirm this.

REFERENCES

1. W. P. PARKS, E. E. HOFFMAN, W. Y. LEE, and I. G. WRIGHT: 'Thermal Barrier Coatings Issues in Advanced Land-Based Gas Turbines', *Proceedings of Thermal Barrier Coatings Workshop 1995,* Cleveland, 1995, 35.
2. R. V. Hillery and B. H. Pilsner: 'Predicting Life of Thermal Barrier Coating', NASA Report, *NASA CR 180807*, 1989.
3. T. A. CRUSE, A. NAGY, and C. F. POPELAR: *Mechanical Testing of Advanced Coating System,* Final Report, NASA Cooperative Agreement NCC3-89, Southwest Research Institute, San Antonio, Texas, 1990.
4. K. F. WESLING, D. F. SOCIE, and B. BEARDSLEY: 'Fatigue of Thick Thermal Barrier Coatings', *J. Am. Ceram.* Soc., 1994, **77**(7), 1863–68.
5. R. W. SMITH: 'Mechanical Properties of a Low-Pressure-Plasma-Applied Co Cr–Al–Y Coating', *Thin Solid Films,* 1981, **84**, 59–72.
6. K. SCHNEIDER and H. W. GRUNLING: 'Mechanical Aspects of High Temperature Coatings', *Thin Solid Films,* 1983, **107**, 395–416.
7. M. BRODDEGÅRD: *Tensil Properties of a Thermal Barrier Coating on Superalloy Hastelloy X Substrate,* MS Thesis, No. LiTH-IKP-Ex-1299, Linköping, 1996.
8. T. A. TAYLOR, D. L. APPLEBY, A. E. WWATHERILL and J. GRIFFITHS: 'PlasmaSprayed Yttria-Stabilized Zirconia Coatings: Structure-Property Relationships', *Surface and Coatings Technology,* 1990, **43/44**, 470–480.
9. A. R. Nicoll: 'A Survey of Methods Used for the Performance Evaluation of High Temperature Coatings', *Coatings for high temperature applications,* E. Lang ed., Elsevier, London, 1983, 307.
10. R. C. PICE, 'Fatigue Data Analysis', *AMS Metals Handbook, 9th edition,* JR. Newby *et al.*, eds, 1985, **8**, 695.
11. P. BENGTSSON: *Microstructural Studies and Residual Stress Evaluation in Zircon ia-Based Thermal Barrier Coatings,* Thesis No. 524, Linköping, 1995, 54.
12. S. BHONGBHIBHAT: *Dehnungswechselkurven,* Vorhaben Nr. 224, Germany, 1980.

Thermo-Mechanical Fatigue Tests of Coatings for Turbine Blades

YU. A. TAMARIN*, V. G. SUNDYRIN†¶ and N. G. BYCHKOV†
*Scientific Consulting Centre 'VIAM-REMAVIA', 11–6–15, Molostovih St., 111555, Moscow, Russia
†Central Institute of Aviation Motors, 2, Aviamotornaya St., 111250, Moscow, Russia
(¶ABB Uniturbo Ltd at present)

ABSTRACT

Cracking of protective coatings is one of the most common types of aircraft engine blade surface damage. Investigations and calculations have shown that crack initition in the coating is caused by thermal stresses in the system of 'coating-superalloy'.[1] These thermal stresses are caused by the difference in the temperature expansion coefficients and modules of elasticity of the coating and the superalloy.

Coated turbine blade design techniques[1–4] allow the calculation of thermal stresses in the blade's surface layers and the coating service life under the thermocyclic loading. However, to predict the coated blade's service life it is necessary to compare the results of the calculations and the coating thermal fatigue test data.

This paper presents the results of thermo-mechanical fatigue tests of the specimens made with a CMSX-4 superalloy substrate and LCO-22 coating under conditions of elastic deformation of the substrate. Special load-adjusted thermo-mechanical fatigue-test techniques were developed for this purpose.

The suggested test method has shown that the coating service life is considerably shorter than that of the uncoated alloy under conditions which result in high temperature oxidation.

BASIC PRINCIPLES AND METHODS

Load-adjusted thermo-mechanical fatigue-tests (LA TMF-test) and strain-control thermo-mechanical fatigue-tests (SC TMF-test) are the main types of TMF-tests for high-temperature Ni-base alloys used in turbine blades. With the LA TMF-test, a constant stress range $\Delta\sigma$ = const and with the SC TMF-test a constant mechanical strain range $\Delta\varepsilon$ = const have usually been used.

Each of the LA and SC TMF-tests, mentioned above, makes it possible to only approximately simulate the deformation process of blade materials under operational conditions, because in a real blade complicated cyclic deformation takes place. This deformation is a combination of $\Delta\sigma$ = const. load (effect of centrifugal and gas loads) and of $\Delta\varepsilon$ = const. load (effect of thermal stresses). It is considered that the SC TMF-test is the most similar to real deformation conditions of blade airfoil materials. However, the last statement is true under conditions of elastic–plastic straining of blade materials only, when thermal stresses in blade airfoils are noticeable higher than mechanical stresses caused by centrifugal and gas loads. In addition irreversible creep strains take place in blade airfoils under operational conditions. This reason is in addition to the complicated real deformation picture.

Both methods mentioned above for TMF-tests could be used for the experimental investigations of thermo-mechanical fatigue of corrosion-resistant coatings for turbine blades. In such a case coated bars are used with different designs, depending on the test method and laboratory units. In the case of elastic straining of the coated bar both LA and SC TMF-test types are basicaly similar and produce similar results.

The following basic principles are important, when experimental TMF analysis of the corrosion-resistant coatings are being considered.

1. Basicaly, thermal fatigue cracking starts in the coating and then penetrates into the substrate. This is very typical for the single crystal blades cast with the primary crystallographic orientation <001> because of the low elasticity modulus of the material and therefore low thermal strain ranges for thermal cyclic loading.
2. It is necessary to take into account bi-axial stress states in the coating during analysis of the TMF test results of coated bars (see Fig. 1).
3. Cracking of the coating on a cylindrical bar during TMF-testing can be both diametrical and longitudinal.
4. The stress–strain state of the coating can not be directly measured by existing control methods during TMF-tests of coated bars. For analysis of non-elastic straining of the coating under these conditions it is necessary to use suitable constitutive models.
5. Strains in coatings are basicly dependant on the difference between thermal expansion coefficients of the substrate and the coating. These strains in the coatings are usually non-linear (plastic) even when linear (elastic) straining of the substrate occurs. Coating strain range values are as a rule higher than the corresponding strain range values of surface layers of blade (or bar) during thermomechanical cyclic loading especially for the single crystal and directionally solidified materials (see Fig. 2).
6. The number of cycles to the first crack initiation and their distribution on the outer surface of the single crystal blade airfoil cast with primary crystallographic orientation <001> depends on the secondary orientation of the crystallographic lattice in the airfoil cross-section. This is greatly dependant on the circumferential thermal stresses σ_s in the blade, resulting in longitudional coating cracking, and the value of circumferential elasticity modulus E_s for the single crystal superalloy. So, when TMF cyclic testing of single crystal coated bar, it is necessary to know the positioning of the axis of the secondary orientations <010> and <100>, for each bar (see Fig. 1(b)), because distribution of the first crack on the coated surface can depend on the axis of orientation of the bar.

So, the following experimental method of TMF investigations of the protective coatings for turbine blades is suggested.

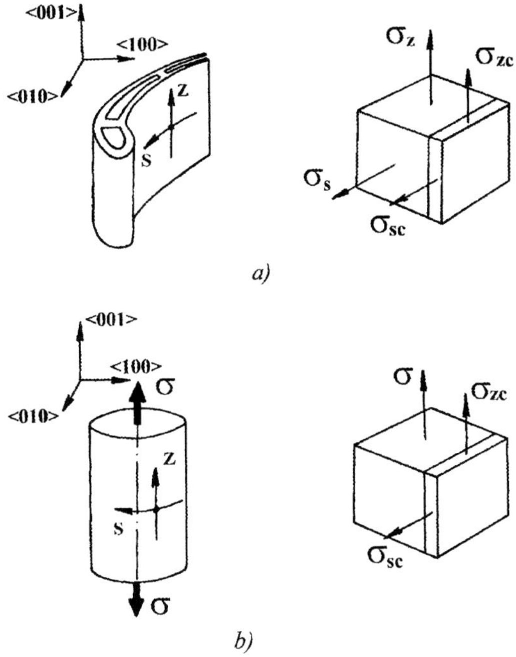

Fig. 1 Stress state of (a) coated single-crystal cooled blade and (b) coated whole cylindrical bar.

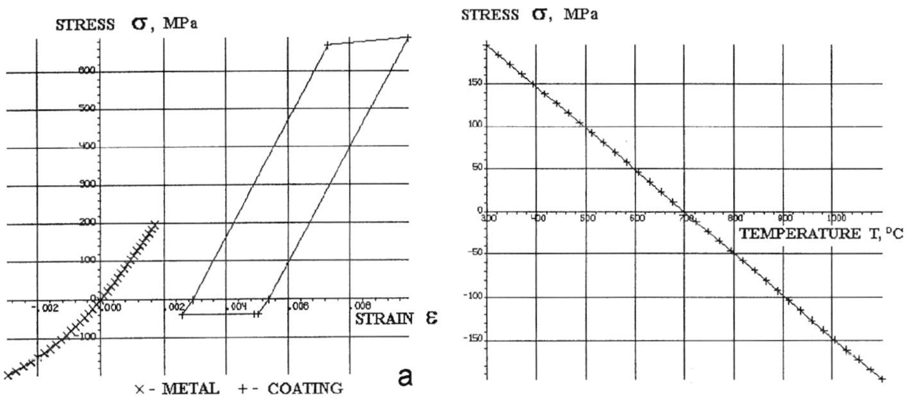

Fig. 2 Hysteresis loops for (a) Ni20Co20Cr12AlY coating and bar material MAR–M–002 and (b) of whole cylindrical coated bar when out-of-phase TMF-tests.

1. As elastic–plastic straining of the coating is possible even under elastic straining of the substrate, it is suggested to carry out TMF testing of the coated bars under conditions of elastic straining of the base material, only. This method is expedient for single-crystal turbine blades with primary crystallographic orientation <001>,

because there are usually no plastic strains in the outer layers of such blades at operating conditions. This is the result of the low elastic modulus and high yield strength of the blade material for the <001> orientation. The exceptions to the rule are small stress concentration zones near perforation holes at the cooled blade walls where plastic strains of blade metal are possible.

2. In accordance with the above mentioned, the LA TMF-test is recommended for TMF testing of coated bars. The results of this method are similar to the SC TMF-tests under conditions of elastic straining of the bar material, but is simpler in practice. The main problem with the SC TMF-test is the necessity for temperature uniformity of the metal in the extensometer-controlled zone at stationary and non-stationary thermal conditions. This is not necessary for the LA TMF-test, because during the test the loading of the bar is controlled. Moreover, non-uniformity of temperature distribution along the central work zone with a temperature maximum is preferrable as it helps to fix the location of maximum failure in the bar's cross-section.
3. It is recommended to use a standard whole cylindrical bar with constant square cross-sectional work zone for the LA TMF-test (see Fig. 3). The stress state of the coating applied to the outer surface of the bar is bi-axial, although the stress state of the bar itself during this test is uni-axial (see Fig.1(b)).
4. It is recommended to use a control system with two independent channels: first – to control load, second – to control temperature of the bar during LA TMF-testing. This control system gives the possibility to model changes in stress and maximum temperature of the bar's work zone vs time with any phase shift among these independently controlled parameters (for example, out-of-phase or in-phase LA TMF-tests, see Fig. 2). The possibility of stress asymmetry $R_\sigma = \sigma_{min}/\sigma_{max}$ varying for all types of LA TMF-test must be also limited (σ_{min} and σ_{max} are correspondingly minimum and maximum values of stress during the cycle).
5. The stress–strain state of the coatings on the bars is calculated for the TMF-test conditions with the use of the developed constitutive model[1,3] based on the deformational theory of plasticity. The calculational hysteresis loop for the overlay coating Ni20Co20Cr12AlY, obtained on the basis of this constitutive model for the out-of-phase LA TMF-test of the coated bar casted from MAR–M–002 superalloy is shown in Fig. 2.

Calculations of the strain ranges in the coating for the test conditions are necessary to plot the TMF curves

$$\Delta\varepsilon_{ic} = \Delta\varepsilon_{ic}(N_{Rc})$$

directly for the coating, using the method described in Ref. 3 and experimental TMF curves

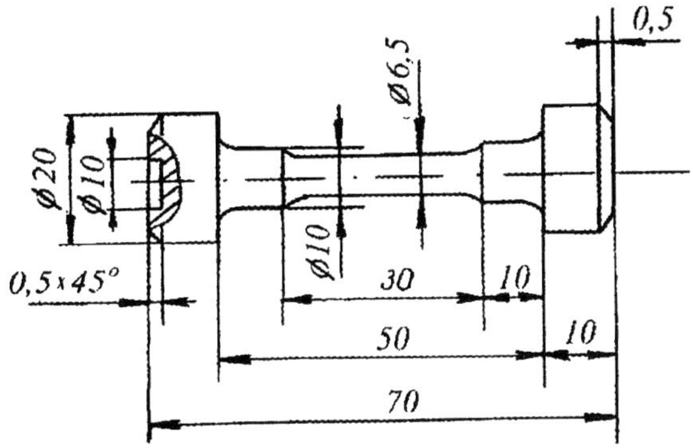

Fig. 3 Specimen for TMF-tests.

$$\Delta\varepsilon_z = \Delta\varepsilon_z(N_{Rc})$$

for the coated bars.

In this expressions:

$\Delta\varepsilon_z$ – elastic axial strain range of the bar controlled during LA TMF-test,

N_{Rc} – number of cycles of load up to the first crack initiation in the coating with depth 0.1 mm and length ~0.5 mm,

$\Delta\varepsilon_z = |\, \mathrm{sign}(\varepsilon_{1c}{}^{max})\cdot\varepsilon_{ic}{}^{max} - \mathrm{sign}(\varepsilon_{1c}{}^{min})\cdot\varepsilon_{ic}{}^{min}|$
– elastic–plastic v.Mises strain range in the coating,

$\varepsilon_{ic}{}^{max}$, $\varepsilon_{ic}{}^{min}$ – correspondingly maximum and minimum values of v.Mises strain $\varepsilon_{ic} = \sqrt{2/3}\sqrt{(\varepsilon_{zc} - \varepsilon_{sc})^2 + (\varepsilon_{sc} - \varepsilon_{nc})^2 + (\varepsilon_{zc} - \varepsilon_{nc})^2}$ of the coating during cycle of loading,

$\mathrm{sign}(\varepsilon_{1c}{}^{max})$,$\mathrm{sign}(\varepsilon_{1c}{}^{min})$ – signs of the maximum principal strain ε_{1c} for the maxi mum ($\varepsilon_{ic}{}^{max}$) and maximum ($\varepsilon_{ic}{}^{min}$) values of the v.Mises strain in coating, correspondingly,

ε_{zc}, ε_{sc}, ε_{nc} – strain components in coating (n-is the direction normal to the surface of the tested coated bar).

TMF TESTS OF COATED SPECIMENS

As an example the LA TMF-test results for the LCO–22 overlay coating applied on the specimens cast of CMSX–4 superalloy are presented below. The design of specimens is shown in Fig. 3.

During the test the following conditions were observed:

- out-of-phase temperature and load change condition;
- asymmetric load (stress) cycle without keeping the specimen at the maximum temperature of cycle.

The selection of these test conditions is dictated by the following reasons:

- out-of-phase thermal and load cycles correspond to the actual stress and temperature difference in the leading edge of the blade, wherein the temperature rise causes the compressive stresses to rise as well;
- during the out-of-phase test the load cyclic change results in the raising of additional substantial strains caused by the elongation of the specimen under load. During the heating semi-cycle substantial strains caused by the specimen compression are added to the compressive thermal stresses.

Thus during the tests a substantial expansion of the coating strain range with temperature and load cyclic change is observed. This makes the advent of a first thermofatigue crack in the coating more probable, but not in the substrate.

The maximum temperature, T_{max} of the specimen working part with the temperature cyclicly changing, varying from 100 to 1100°C. The specimens are heated by applying an electric current with a density of 20 Amm^{-2}. In synchronism with the temperature cycle the specimen load is applied with the use of a hydraulic loading device. The specimen and load change pattern is shown in Fig. 4.

During the test, the loading conditions for all the specimen cross-sections are similar. During heating and cooling the temperature varies along the specimen length. Heat is evenly dissipated while current runs through the specimen, but heat dissipation conditions for both butt ends are different. The butt end fastened in the holder cooled by water has a lower temperature than the butt end fastened near the hydraulic cylinder rod cooled by oil. Due to this, it is evident that the specimen is non-uniformly heated along its length.

To detect the actual temperature distribution along the length of the specimen it was thermometered. Seven thermocouples 0.2 mm in diameter were welded to the central area of the specimen. The location of the thermocouples is shown in Fig. 5. Also presented are the results of the specimen thermometering as the temperature on the working part reached its maximum, T = 1100°C.

The T_{max} zone moves from the specimen centre by approximately 2 mm towards the hydraulic cylinder cooled with oil. On the same side a sharper temperature drop is observed as one moves from the specimen centre. The temperature difference across the tested area, as the cycle temperature reaches its maximum, is 120°C.

To determine the specimen loading conditions during the test, cycle strains were measured in the area with T_{max}. Marks made up of wire 0.05 mm in diameter were spot-welded to the specimen surface. Length 'l' – the distance between the marks was measured with the use of an optical measuring device. The experimental error when measuring the distance between the marks does not exceed 0.005 mm. The distance (l_{min} and l_{max}) between the marks was measured at the maximum and minimum cycle temperatures without applying loads to the specimen. Then the thermal strain range in the maximum temperature area was calculated:

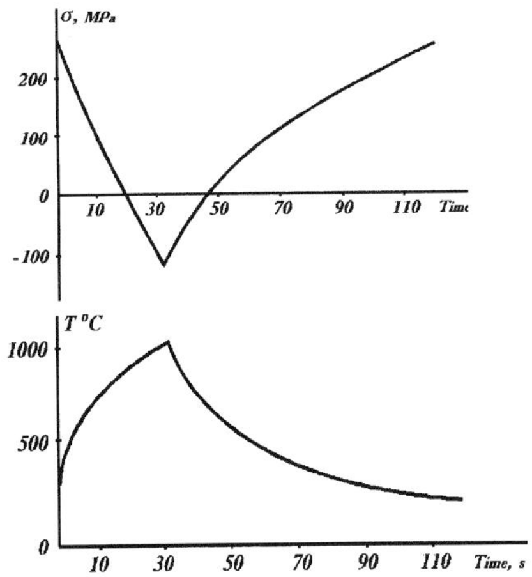

Fig. 4 Stress σ and maximum temperature T vs time distribution during out-of-phase LA TMF-test of coated bar.

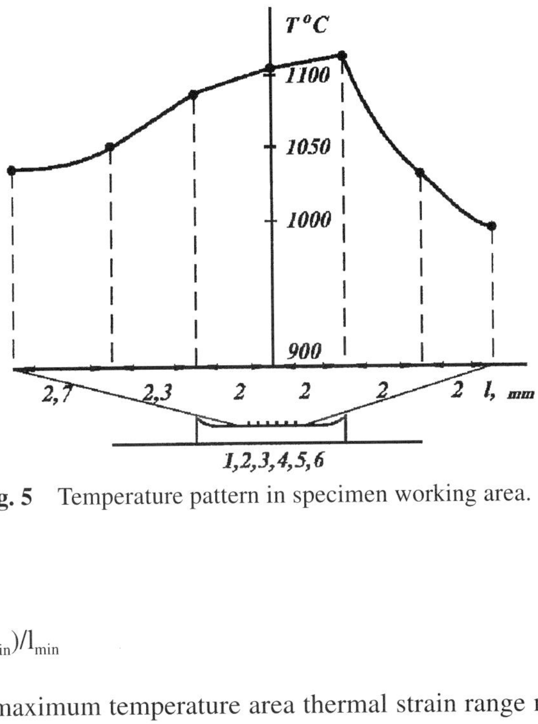

Fig. 5 Temperature pattern in specimen working area.

$$\Delta\varepsilon_T = (l_{max} - l_{min})/l_{min}$$

The results of the maximum temperature area thermal strain range measurement are shown in Fig. 6.

After measuring the thermal strain range, loads were applied to the specimens. The loading conditions being such to ensure that the following loads at the extreme points of the temperature cycle T_{min} = 100 and T_{max} = 1100°C were selected:

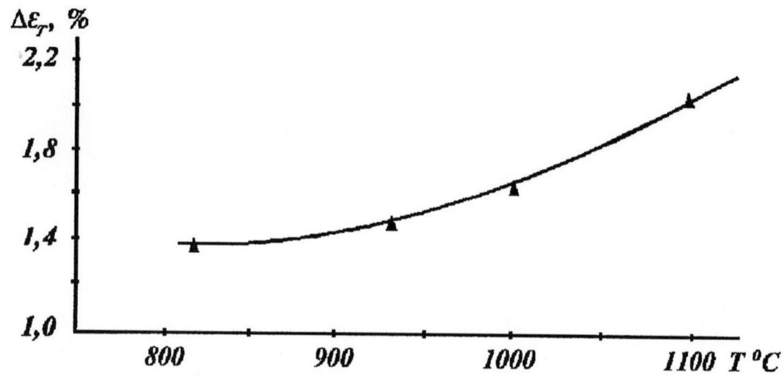

Fig. 6 Thermal strain range distribution in the specimen maximum temperature area.

1. $\sigma_{tensile}$ = +500 MPa, σ_{compr} = -150 MPa,
2. $\sigma_{tensile}$ = +200 MPa, σ_{compr} = -150 MPa;
3. $\sigma_{tensile}$ = +200 MPa, σ_{compr} = -100 MPa;
4. $\sigma_{tensile}$ = +200 MPa, σ_{compr} = -50 MPa.

In the process of loading the length of the section at temperatures T_{min} and T_{max} and strain range $\Delta\varepsilon$ in it caused by heating and loading of specimen was calculated. The strain range $\Delta\varepsilon$ in the tested area of specimen versus stress range $\Delta\sigma$ under thermo-mechanical loading conditions are shown in Fig. 7.

The obtained data characterise the cyclic stress–strain condition of the base material but not the coating. Due to the difference between the thermal expansion coefficients of the coating and specimen material additional stresses arise in the coating.

During the test the coating cyclic life was determined by the moment the first crack appears in the surface layer. The crack formation was checked by optical and non-destructive dye-penetrant test methods. These methods were able to check the crack length ≥ 0.5 mm and penetration depth to about ~0.1 mm. After initial cracking some specimens were tested until their failure.

The loading conditions which could have resulted in cracking of the LCO–22 coating were unknown. That is why the group of specimens used for selecting the test conditions under which it ensures cracking of the coating within 1000–3000 cycles of loading. The TMF-test results for the coated specimens are presented in Table 1.

The initially selected loading conditions at the extreme points of the temperature cycle are as follows:

1. Tensile stress at 100°C - $\sigma_{tensile}$ = +500 MPa,
   compressive stress at 1100°C σ_{compr} = -150 MPa, stress range $\Delta\sigma$ = 650 MPa.
2. Tensile stress at 100°C - $\sigma_{tensile}$ = +500 MPa,
   compressive stress at 1100°C σ_{compr} = -100 MPa, stress range $\Delta\sigma$ =600MPa.

The first tests showed cracking of the coating under these cyclic stress conditions

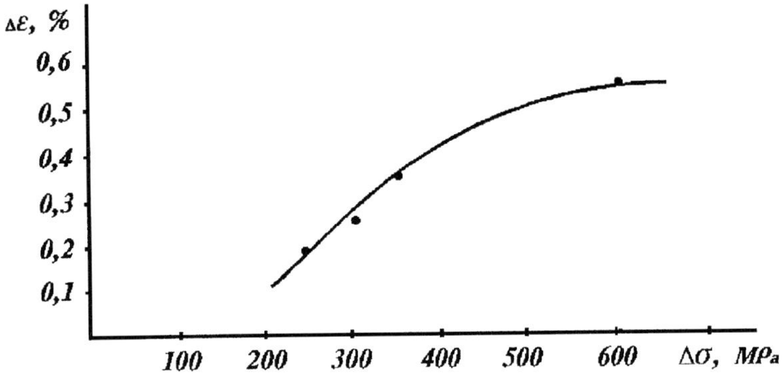

Fig. 7 Strain range Δε of specimen vs stress range Δσ in the maximum temperature area.

Table 1 TMF test results for specimens with coating LCO–22.

Specimen	Coating thickness, μm	Number of cycles until coating cracked	Number of cycles until specimen failure	Test condition, σ_{compr}, $\sigma_{tensile}$
FEP	120	383	2282	-150; +500 MPa
FEP	105	352	no destruction	-150; +500 MPa
FFR	105	381	4148	-100; +500 MPa
FFH	110	875	no destruction	-100; +200 MPa
FEV	110	958	no destruction	-100; +200 MPa
FEO	110	900	7368*	-100; +200 MPa
FER	105	2762	25850	-50; +200 MPa
FFL	105	1897	no destruction	-50; +200 MPa

*test condition until destruction 150–350 MPa

occuring in considerably less cycles than the preset arbitrary threshold of 1000 cycles. The reduction of compressive stress in the maximum temperature zone to σ_{compr} = -100 MPa had no effect on the coating cyclic life.

The appearance of the cracks and their closeness to the surface are similar to the coating TMF cracking (fragmentation) on the blade outer surfaces. The cracks in the coating are located all along the length of the working part of the specimen. Their advent does not depend upon the location of the maximum temperature zone. However the cracks causing the specimen failure are always located in the maximum temperature zone. A typical 'shrinkage' indicating that the coating is in strain at the maximum temperature is observed in the central part of the specimen.

To check the test selection method for observing the moment when cracking occurred was correct, the FET specimen was axis-cut after 550 cycles and the mechanism of crack initiation and propagation investigated. The cracks are all located along the specimen length and are observed in the areas where the maximum temperature does not exceed 1000°C . The initiation of cracks is preceded by an accumulation of plastic strain in the coating, the formation of pores and small surface tears which, then, gradually merge forming thermal fatigue cracks. The same manner of crack formation was also observed in the FFR specimen studied after its destruction.

The assay of the FFR specimen fracture failure showed that its structure is typical for the TMF-tests of the coated specimens. Failure of the coating starts at the specimen surface whereon there are numerous sites of failure. Development of cracks in the coating points to substantial strains in it. Numerous sites of failure appear all over the specimen surface. Thereafter a crack causing the specimen failure develops in the maximum temperature central zone.

Reduction of compressive stresses to σ_{compr} = -100 MPa did not change the coating cyclic life. That is why for further testing the loading conditions were changed. To increase the coating life to 1000 cycles tensile stresses applied during the cooling semicycle were reduced to $\sigma_{tensile}$ = +200 MPa. The tests proved that such a change of conditions resulted in a longer cyclic life of the coating, reaching the arbitrary threshold of 1000 cycles, typical for failure of coatings on turbine blades. The coating failure mechanism did not change with the reduction of the compressive stresses.

The first cracks in the FEV specimen were registered after 958 test cycles, testing was continued. After the first cracks the surface 'shrank' and cracks appeared at a certain angle to the specimen axis. The cracks then gradually expanded and were considerably extended in length. After 4120 test cycles the FEV specimen was cut axiswise to study the crack development. Over 3162 cycles past after the moment of initial cracking of the coating was registered, the crack had penetrated into the substrate by 100–120 μm.

To check strains in the specimen its elongation and diameter were measured after the tests were over. There was no plastic strain in the specimen during testing.

TMF TESTS OF UNCOATED SPECIMENS

Considering the idea of creating a coating with high TMF resistance – a coating with physical and mechanical properties similar to the protected superalloy may be assumed to be an ideal coating. It is hardly probable to realise such a coating since the thin layer of the polycrystalline coating can not have the same mechanical properties of a modern single crystal material.

However the value of cyclic life of the uncoated CMSX–4 superalloy which was tested according to the methods worked-out for the TMF conditions are interesting to be compared with the cyclic life of the coatings. As a result the uncoated CMSX–4

Table 2 Uncoated CMSX–4 superalloy specimen test results.

Specimen	Number of cycles until surface cracked	Number of cycles until specimen failure	Test condition, σ_{compr} - $\sigma_{tensile}$
FFO	2850	3628	-150; +500 MPa
FEX	6193	6508	-150; +500 MPa
EYQ	–	4840	-150; +500 MPa
EYI	15785	16953	-150; +300 MPa
EYX	20423	21821	-150; +200 MPa
EYL	48672	–	-100; +200 MPa
EYW	24122	25480	-100; +200 MPa
FFG	41271	44860	-50; +200 MPa
EYJ	–	15352*	-50; +200 MPa

*there is no cracking, the specimen remained intact.

specimens were subjected to the tests. The tests lasted until the specimens failed. The test results are presented in Table 2 and Figs 8 and 9.

The study of the FF0 specimen fracture showed that the nucleation and propagation of cracks on the uncoated specimen are similar to that of the tested coated specimens. The site of failure is located on the specimen surface. The cracks propagate inside the specimen and cause its failure. The specimen fracture is typical to the failure caused by the TMF test conditions characterised by narrow stress ranges.

The tests revealed the difference between the cyclic life of the uncoated and coated specimens. The cyclic life of the LCO–22 coating accounts for only 4–13% of the cyclic life of the CMSX–4 alloy. Thus if the blade has areas with a stress range close to the test ones, numerous TMF cracks failed to be predicted by the blade cyclic life calculations will appear on the blades, in the LCO–22 coating.

The selected test procedure for the specimens 7 mm in diameter was not supposed to be employed to study the effect of the coatings on the life of the CMSX–4 alloy specimens. Comparison of the number of cycles until the destruction of the uncoated specimens with that of the coated specimens shows the number of the destructed specimens and the obtained results are not sufficient for the objective assessment of this effect.

Compatibility of the results obtained for the cyclic life of the coatings on the speci-

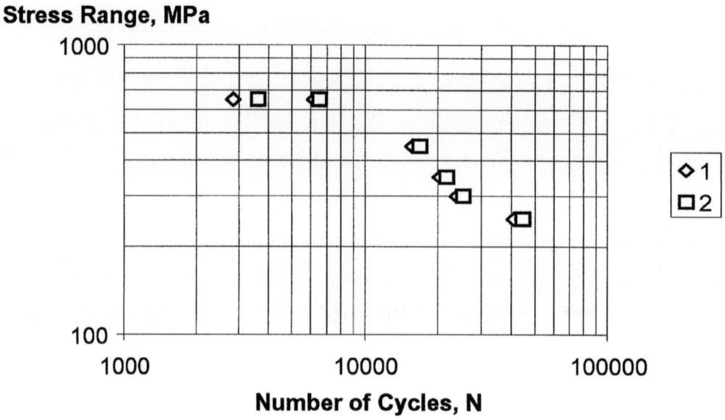

Fig. 8 Cyclic life of CSMX–4 alloy: 1 – cracks on surface and 2 – failure of specimen.

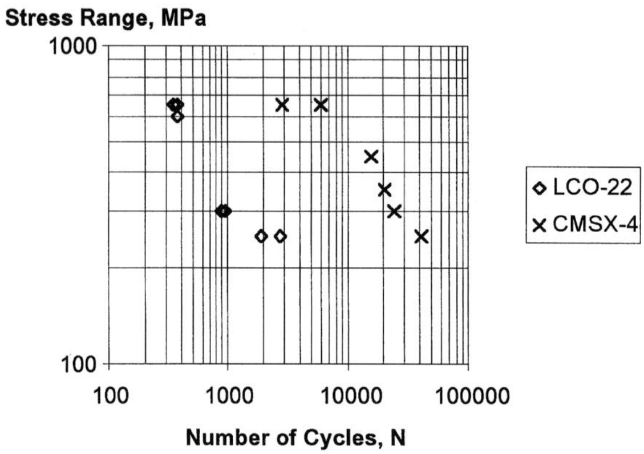

Fig. 9 Cyclic life of uncoated CMSX–4 alloy and CMSX–4 alloy with applied coating LCO–22.

mens with their lives on blades depends on the engine blade working conditions. During the engine cycle tests there is a great possibility for the coating cyclic lives to comply with the values obtained during the laboratory tests. In case of long tests and long operation a considerable effect on their results will be exerted by such characteristics as heat resistance and resistance to hot corrosion.

ACKNOWLEDGEMENT

The authors express their gratitude to Rolls-Royce plc for providing financial support of this research.

REFERENCES

1. Yu. A. TAMARIN, E. B. KACHANOV and V. G. SOUNDYRIN: 'Gas Corrosion and Thermal Fatigue of Protective Coatings for Turbine Blades', Proc. of Int. Symp. *High Temperature Corrosion and Protection*, June, 26–30, 1990, Shenyang, China.
2. V. G. SUNDYRIN, Yu. A. TAMARIN and V. Yu. KANAEV: 'Stress Calculation the Ceramic Thermal Barrier Coatings for Turbine Blades', Proc. of Int. Colloq. on *Mechanics of Creep Brittle Materials – 2*, Leicester, England, 1991.
3. Yu. A. NOZHNITSKY, R. A. DOULNEV and V. G. SUNDYRIN: 'Damage Mechanisms for Thermomechanical Fatigue of Aircraft Engines Materials', 81st Meeting of the AGARD SMIP, held in Banff, Canada from 2–4 October 1995, and published in CP–559.
4. G. R. HALFORD, T. G. MEYER, R. S. NELSON and D. M. NISSLEY: 'Fatigue Life Prediction Modeling for Turbine Hot Section Materials', NASA Technical Memorandum 100291, Prepared for the *33rd Intl Gas Turbine and Aeroengine Congress and Exposition* Sponsored by ASMIE, Amsterdam, The Netherlands, June 5-9, 1988.

TMF and LCF of Plasma Sprayed IN738 LC

L. LINDÉ and P. J. HENDERSON*
Swedish Institute for Metals Research, Drottning Kristinas vag 48, S–11428 Stockholm, Sweden
**Vattenfall Energisystem AB, P.O. BOX 528, S–16216 Stockholm, Sweden*

ABSTRACT

In the present work, coated IN738 LC was tested in thermo-mechanical fatigue (TMF) in-phase (IP) between 400 and 850°C as well as low cycle fatigue (LCF) at 850°C. The coating was of a plasma sprayed NiCoCrAlY type, PWA 1365–2. In addition LCF testing was performed on polished and heat treated specimens. When comparing the results for the two testing types it was found that the LCF tests had a longer fatigue life with a factor of two or more compared to the TMF tests. Polishing of the coating resulted in a minor increase in fatigue strength compared to as sprayed tests and the heat treated material had a lifetime similar to the polished specimens. Metallography after testing revealed distinct differences in crack initiation and propagation behaviour between LCF and TMF specimens. In the LCF specimens all cracks initiated in the coating surface and, probably because of the slow propagation rate, many cracks could initiate. TMF specimens exhibited fewer cracks in the coating, instead there was evidence of cracks initiating between the coating and the base metal in one case. These cracks initiated at defects in the interdiffusion zone and could propagate into the substrate alloy with the coating intact. In the same specimen there were internal intergranular cracks in the substrate indicating some creep damage as well. When comparing the results with data on uncoated IN738 LC from earlier work, it was found that the influence from the different cycle types was much larger for the coated material.

INTRODUCTION

Coatings are often used on components for gas turbines and aero engines as a protection against aggressive environments. The coating technique has been used for many years, first in aero and marine turbines and later in industrial turbines. Coatings are applied in order to protect the metal from oxidation and corrosion resulting in a prolonged life, increased reliability and reduced maintenance. They form a protective scale by interaction with the oxidising environment and thereby isolate the base metal from the environment. Relevant properties to consider are Youngs modulus E, the coefficient of thermal expansion P_{α}, the thermal conductivity P_{γ}, the yield strength R_e, the fracture strain A and also the toughness or fracture toughness of the different zones of a coated system.[1] The oxides that form effective scales are Cr_2O_3, SiO_2 and $A1_2O_3$ where only $A1_2O_3$ is effective above 900°C.[2] A Cr_2O_3 layer offers resistance to hot corrosion attack while Al_2O_3 offers resistance to oxidation.[3]

However, applying a coating to a substrate also introduces new factors to be considered. For instance, the coating must be compatible with the substrate alloy and be able to withstand the strains it is exposed to during service. Further, the coating will protect the base metal only if it is intact. Mechanical loads, *e.g.* during fatigue, may cause the coating to form cracks and thereby expose the base metal to environmental attacks. Cracks in the coating can also propagate into the substrate alloy causing premature fracture. The effect of coatings on the fatigue life is therefore an important field of study for the applications mentioned above since fatigue often is a major life limiting factor. However, the results from such studies are often non-conclusive.[4]

One important aspect of coatings which affect their mechanical properties is the ductile to brittle transition temperature (DBTT). Below this temperature the brittle behaviour reduces the ability of the coating to endure plastic deformation. It has been defined as the temperature corresponding to a fracture strain of 0.6%.[5] The DBTT depends *e.g.* on coating type, thickness and composition.[6–8] If the coating is strained below the DBTT cracks are more likely to initiate.

There are several types of coatings available which use different application techniques and compositions. In this study the effect from two different fatigue conditions on a plasma sprayed coating, which is one of the most widely used coating types, applied to a nickel base superalloy has been investigated. Previously, the fatigue behaviour under similar conditions has been investigated for aluminised material using the same substrate alloy.[9]

EXPERIMENTAL

Investigated material in this study was the conventionally cast Ni-base superalloy IN738 LC with a plasma sprayed NiCoCrAlY coating, PWA 1365-2. The chemical composition of the alloy and coating is given in Table 1.

Table 1 Chemical composition of the investigated IN738 LC alloy and NiCoCrAlY coating wt%.

	Al	B	C	Co	Cr	Fe	Mo	Nb	Ni	Ta	Ti	W	Y	Zr
IN738 LC	3.6	0.01	0.10	8.10	16.0	0.06	1.7	0.80	bal	1.7	3.4	2.6	–	0.03
Coating	12.5	–	0.01	22.5	17.1	–	–	–	bal	–	–	–	0.52	–

The coated material was tested for thermo-mechanical fatigue (TMF) between 400 and 850°C with temperature and strain cycled in-phase (IP). Isothermal low cycle fatigue (LCF) testing of the material at 850°C was also performed. Mechanical testing was carried out in a servohydraulic test machine where the specimen was heated by a high frequency generator equipped with an induction coil. The strain was measured using a side mounted axial extensometer and testing was performed under total axial strain control. For the TMF tests the total strain was equal to the sum of total mechanical and thermal strain, while for the LCF tests it was equal to the total mechanical strain. LCF was performed on as-sprayed material and on specimens where the coating had been polished. Before polishing, the R_a, value was 5.8 µm compared to 0.36 µm after polishing. One LCF test was also made on an as-sprayed specimen which was heat treated for 3 weeks (500 h) at 850°C prior to testing. This was made in order to investigate the effect of coating degradation on the fatigue life. TMF testing was only performed on as-sprayed material. Testing temperatures were chosen to be above the DBTT during LCF and cycled through during TMF. The transition temperature was then assumed to be less than 800°C.[8]

As a failure criterion, the number of cycles to 90% of linear maximum tensile stress, N_{90}, was used which was determined as a 10% drop from the extrapolated linear behaviour of the maximum tensile stress versus number of cycles.[9] For characterisation of the testing conditions half the numbers of cycles to crack initiation, N_i, was used. Load and force ranges were evaluated from the loop closest to $0.5{\cdot}N_i$.

In the analysis of the hysteresis loops from the TMF tests the following expression for Youngs modulus was used:

$$E = 191 + 4.32{\cdot}10^{-3}{\cdot}T - 6.50{\cdot}10^{-5}{\cdot}T^2 \tag{1}$$

where E and T are expressed in GPa and °C respectively.

After testing some specimens subjected to metallographic examination were cut longitudinally, polished to 1 µm and etched in glyceregia which consists of 10 ml HNO_3, 20 ml HCI and 20 ml glycerol. Before etching the solution was heated to 50–60°C. The specimens were then examined by light microscopy and/or scanning electron microscopy (SEM).

RESULTS

Fatigue Testing

The results from the LCF and TMF testing are shown in Table 2. As can be seen in Fig. 1, where the total mechanical strain range is plotted against N_{90}, there was a large difference in fatigue life between TMF and LCF tests performed at similar total mechanical strain ranges. One TMF test which is marked with parenthesis failed prematurely. It can also be seen that the polished LCF specimens had a slightly longer life compared to as sprayed tests. For the heat treated specimen, with an as sprayed surface, the fatigue life was similar to the polished ones. The lines in Fig. 1 correspond to a summation of the Coffin–Manson and Basquin equations, eqns 2 and 3 respectively. These equations are:

Coffin–Manson $$\Delta\varepsilon_p = A{\cdot}N_{90}^{\alpha} \tag{2}$$

Basquin $$\Delta\varepsilon_e = B{\cdot}N_{90}^{\beta} \tag{3}$$

Table 2 Results from LCF and TMF testing.

Test type	Temp [° C]	Test no.	Total strain range [%]	Plastic strain range [%]	Elastic strain range [%]	Stress range [%]	N_{90}	Notes
LCF	850	P6	0.92	0.27	0.65	965	980	
		P7	1.96	1.08	0.88	1300	80	
		P8	0.61	0.12	0.50	734	4425	
		P9	0.44	0.05	0.39	575	16350	
		P3	0.94	0.27	0.67	985	1468	polished
		P4	2.00	1.08	0.92	1364	107	polished
		P5	0.64	0.10	0.5	799	3530	polished
		P15	0.96	0.35	0.61	904	1163	heat treated
TMF-IP	400–850	P10	1.40	0.42	0.98	1661	32*	
		P11	0.92	0.14	0.78	1324	252	
		P13	0.66	0.06	0.60	1021	1050	
		P16	1.92	0.94	0.98	1669	30	

*Premature failure

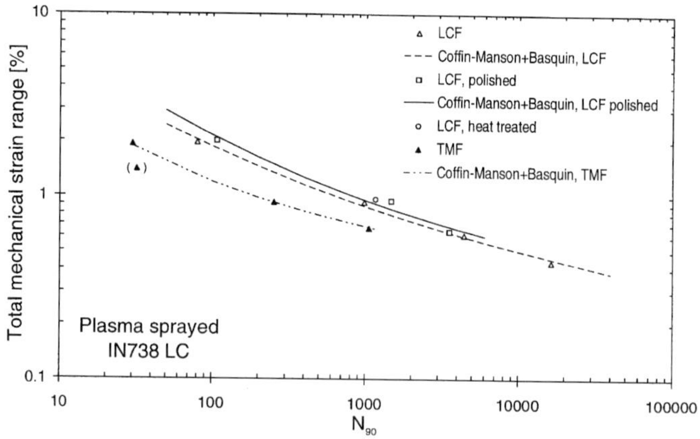

Fig. 1 Total mechanical strain range versus number of cycles to 10% stress reduction for plasma sprayed IN738 LC. The lines correspond to a summation of the Coffin–Manson and Basquin equations respectively.

where $\Delta\varepsilon_p$ and $\Delta\varepsilon_e$ are the plastic and elastic strain ranges respectively and A, α, B and β are material constants. The values of the constants are shown in Table 3. The TMF tests had a compressive mean stress due to the lower temperature in the compressive part of the cycle which could be compared with the symmetrical LCF loops as shown in Fig. 2. In the figure, loops from two tests performed at approximately 1.9% total mechanical strain, P7 and P16, are shown.

Table 3 Values of constants in equations 2 and 3.

Test series	A	α	B	β
LCF	14.2	-0.58	1.77	-0.15
LCF polished	23.6	-0.64	1.85	-0.15
TMF	11.6	-0.76	1.59	-0.14

Metallography after Testing

The thickness of the coating and interdiffusion zone was 100–130 and 15–35 μm respectively for the as-sprayed material, see Fig. 3. In the interdiffusion zone defects such as pores and inclusions were observed. Extensive crack initiation was generally found in the LCF tests except at the largest strain range and all observed cracks initiated at the outer coating surface. Only a minor fraction of the cracks propagated through the coating into the substrate alloy but the fraction of cracks that reached the substrate increased for specimens tested at lower strain ranges. There was a noticeable difference in crack density between the two test types. The number of cracks in the coating was generally much lower in the TMF specimens except at the largest strain where it was similar to the corresponding LCF test. Both test types showed an increase in the

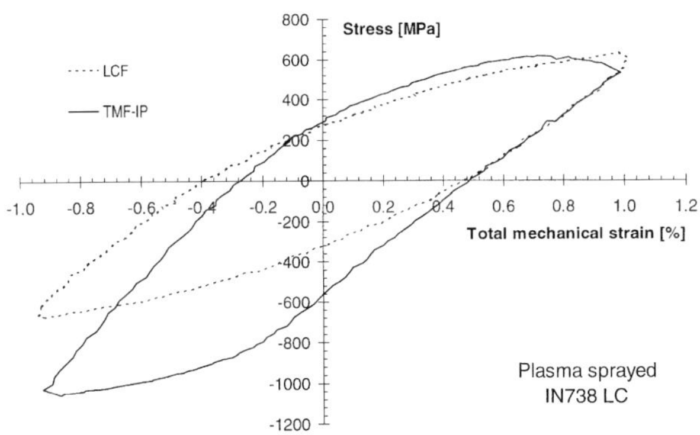

Fig. 2 Comparison between hysteresis loops recorded during LCF testing, specimen P7, and TMF-IP testing, specimen P16, at approximately 1.9% total mechanical strain.

Fig. 3 SEM picture showing the structure of the plasma sprayed coating. From top to bottom the coating, the interdiffusion zone and the substrate are shown. In the interdiffusion zone defects can be seen. Magnification 500X.

number of cracks with decreasing strain range, although the number of cracks showed much less variation with strain for the TMF tests compared to the LCF tests. In the TMF specimens some cracks probably initiated at the coating surface but there were no cracks found in the coating that had not propagated into the substrate. Further, in one of the TMF specimens tested at a total mechanical strain range of 0.92% cracks were also observed that clearly initiated in the interdiffusion zone, see Fig. 4. In the same specimen internal cracks were observed in the substrate, see Fig. 5.

DISCUSSION

Influence of Fatigue Conditions

In earlier investigations uncoated IN738 LC was tested under similar conditions as in the present work, *i.e.* LCF at 850°C and TMF-IP between 400 and 850°C.[10,11] The results are shown in Fig. 6, where the mechanical strain range is plotted against the number of cycles to crack initiation. It should be noted that the results for the uncoated material was obtained using diametrical strain measurements and may therefore not be directly comparable with the results for the plasma sprayed material in the present study. Nevertheless a few observations can be made, giving some indications regarding the influence of the plasma sprayed coating on fatigue life. As can be seen in Fig. 6, the LCF tests of the uncoated material exhibited a slightly longer fatigue life than for the TMF-IP tests. But the different cycle types did not have as large an effect on the fatigue life of the uncoated alloy as was observed for the plasma sprayed material in Fig. 1.

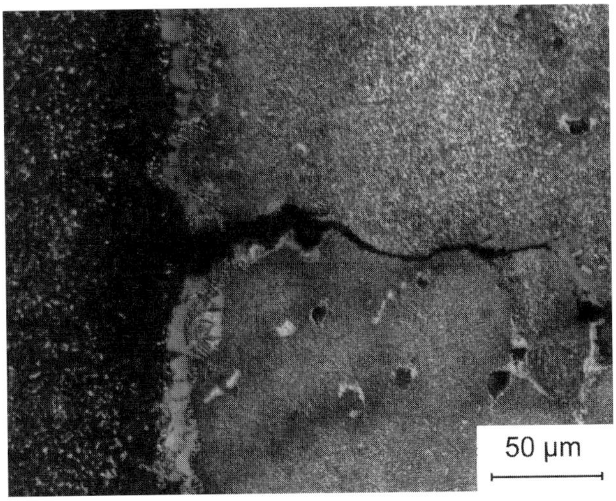

Fig. 4 Crack initiated in the interdiffusion zone of TMF-IP specimen P11 tested between 400°C and 850°C with a total mechanical strain range of 0.92%. Stress direction vertical. Magnification 400X.

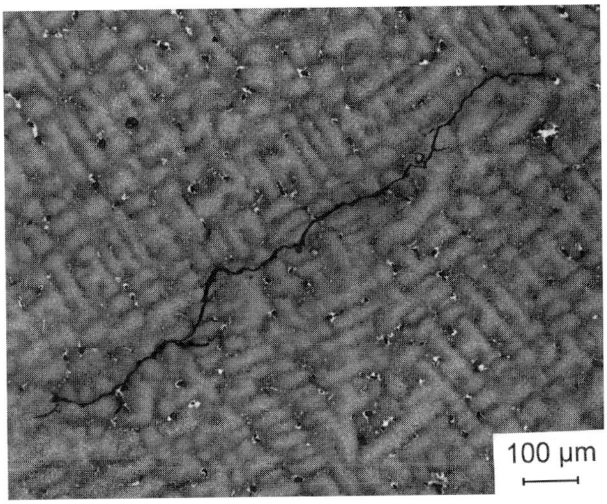

Fig. 5 Internal crack in TMF-IP specimen P11 tested between 400°C and 850°C with a total mechanical strain range of 0.92%. Stress direction vertical. Magnification 100X.

LCF testing of the coated material resulted in a fatigue life a factor of two or more greater than in TMF-IP. This difference was larger than for the tests of the uncoated material mentioned above and was at least partly associated to the fact that the temperature was cycled through the DBTT of the coating in the TMF-tests. At lower temperatures, where the coating was brittle, an applied strain could easily cause cracking of the coating. In addition, during the TMF tests thermal stresses and strains may have

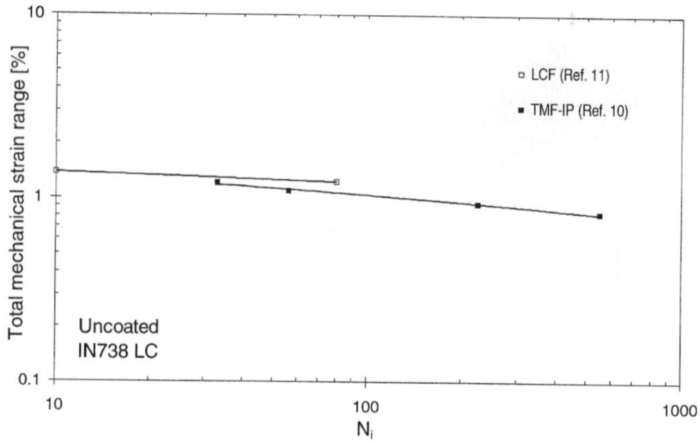

Fig. 6 Total mechanical strain range versus number of cycles to initiation for uncoated IN738 LC, from Refs 10 and 11.

been introduced in the coating due to the cyclic temperature. On the other hand, the isothermal LCF tests were performed above the DBTT where the coating was ductile and therefore less inclined to form cracks. The microstructural investigation showed a clear influence of the testing type on the cracking behaviour of the coating. During LCF, crack propagation was arrested by the coating, which was shown by the large fraction of cracks that did not propagate into the substrate alloy. The picture was the opposite for the TMF tests where the coating was not able to prevent the cracks from reaching and propagating into the substrate. When a crack was initiated in the coating during a TMF test it probably propagated quickly through the coating when subjected to strains under the DBTT, regardless of the strain range. During a LCF test the cracks probably propagated more slowly through the coating which prolonged the fatigue life. An indication of this was the larger amount of cracks which could initiate during the LCF tests. One exception was the specimen tested at the highest total strain range, 1.96%, which exhibited quite a few cracks. Compared to the other LCF specimens the number of cracks was much lower for this test and similar as for the TMF tests. This may indicate that the ductility of the coating above DBTT is somewhere between 0.5 and 1%, if one assumes that only strain in tension would cause crack initiation. Another difference between the LCF and TMF tests was that in one specimen there was evidence of fatigue cracks initiating at defects in the interdiffusion zone between the coating and substrate alloy. When cracks initiate in such a way the strength of the coating is irrelevant, the cracks will initiate and propagate through the substrate alloy regardless of whether the coating cracks or not. However, it should be noted that such initiation only occurred in the specimen tested at the intermediate strain range, neither specimen tested at higher or lower strain range exhibited interdiffusion zone cracking. This indicates that the observed interdiffusion zone cracking could be the result of scatter in the properties of the coating, possibly in combination with the testing parameters.

The internal cracks observed in one of the TMF specimen is not usually found in material after fatigue testing. Considering this and since the internal cracks did not show any signs of oxidation and were formed in an intergranular manner, it was assumed that they were related to creep damage rather than fatigue. Similar cracks were also found in a TMF test of uncoated IN738 LC.[10] The presence of intergranular cracks indicates that despite the compressive mean stress during a TMF-IP test, creep damage may occur.

In the present study the LCF tests had longer fatigue life compared to the TMF-IP tests. When testing the same substrate alloy as in the present work but with an aluminised coating, the difference between the two test types was much smaller and similar to what was found for the uncoated material.[9–11] The fatigue lives for LCF and TMF tests of the aluminised material were similar to the TMF results for the plasma sprayed material in this study. However, for the aluminised material there was a cross-over in fatigue life observed between the TMF and LCF tests. The cross-over corresponded to the strain range which was believed to cause cracking in the coating due to deformation below the DBTT. One reason for the better performance of the plasma sprayed material during LCF could be the fact that the plasma sprayed coating was more effective in arresting the growth of cracks.[12] It has been considered before that the most important mechanical property of a coating is the resistance to thermal-induced cracking.[13,14] These results emphasise the fact that isothermal fatigue testing cannot always include all factors that controls the component life and may result in lifetime predictions that are non-conservative. However, in a study of NiCoCrAlY coated PWA 1480 single crystal the opposite relationship was found.[15] During bithermal TMF-IP testing between 650 and 1050°C the fatigue life at high strains/short lives was similar to the LCF life at 650°C but at low strains/long lives the lifetime increased dramatically. This abrupt change in behaviour was connected to the fact that crack initiation in the coating was more difficult in the TMF test than in the LCF test thereby extending the fatigue life.

Polishing and Heat Treatment

Polishing of the rough coating surface increased the fatigue life during LCF, which would be expected. However, the increase was rather small which implies that the rough original surface did not have a very large effect on the fatigue properties. When applying a coating by plasma spraying, a high temperature plasma stream is created in a plasma gun. The coating alloy, which is in the form of a powder with a particle size less than 60 μm, is then injected into the stream where it melts and is accelerated towards the substrate.[16] As the particles hit the surface they spread forming a structure perpendicular to it. Although the result is a rough surface, it is in the form of pits and not grooves which would be much more detrimental to the fatigue performance. Also the bottom of the pits were not sharp but rounded, as can be seen in Fig. 3, which made the stress concentration smaller. In another work, a similar life increasing effect as the one observed in this study was found for pack aluminised Udimet 700 tested in LCF at 927°C, where polishing of the coating resulted in a slightly increased fatigue life.

Heat treatment of one plasma sprayed specimen at 850°C for 500 hours was performed prior to testing in order to investigate any possible effect from coating degradation on the fatigue life. Such a heat treatment results in a reduction of the interdiffusion zone and formation of a thin oxide layer on the surface.[12] It was found that such a heat treatment increased the fatigue life slightly, similar to the polished specimens. It is difficult to reach any certain conclusions based on a single test, but the result indicates that the heat treatment did not reduce the protective properties of the coating during LCF due to coating degradation, which could be expected.

CONCLUSIONS

In this investigation plasma sprayed IN738 LC, a nickel based superalloy, was tested in TMF in phase between 400 and 850°C and LCF at 850°C. It was found that the LCF tests had a fatigue life a factor of two or more longer than the TMF tests. The difference in fatigue strength could be associated to the ease by which fatigue cracks initiated and propagated during the two test types. In the LCF specimens only cracks initiated in the coating surface were observed. The cracks probably propagated slowly through the coating which had a beneficial effect on the fatigue strength. In the TMF specimens few cracks were observed in the coating and those had propagated into the substrate. Cracks were also observed that initiated in the interdiffusion zone between the coating and base metal. It could therefore be concluded that the coating had no beneficial effect during TMF. Further it was found that polishing of the coating, which reduced the surface roughness considerably, increased the fatigue life only slightly during LCF. This indicated that the as sprayed surface was not as detrimental as the high R_a, value might suggest. LCF testing was also made on one specimen that was heat treated for 500 hours at 850°C prior to testing. Since the lifetime of that test was similar to polished tests, the degradation of the coating obviously did not have a negative effect on the fatigue resistance.

ACKNOWLEDGEMENTS

The present work was performed at the Swedish Institute for Metals Research. The authors wish to thank Dr Jacek Komenda, SIMR, for metallographic assistance. Financing was provided by the Advanced Materials Group of Sweden's High Temperature Materials Committee (Varmhållfasthetskommittén). Participating companies are ABB STAL AB, Avesta Sheffield AB, CSM Materialteknik AB, AB Sandvik Steel, Volvo Aero AB and Värmeforsk. The support is gratefully acknowledged.

REFERENCES

1. H. W. GRÜNLING, K. SCHNEIDER and L. SINGHEISER: 'Mechanical properties of coated systems, Proc. *First International Symposium on High Temperature Corrosion of Metals and Coatings for Energy Systems and Turboengines,*

Université de Provence, France, July 7–11, 1986, Herman H *et al.* eds, *Mat. Sci. and Eng.*, 1987, **88**, 177–189.

2. *Research and Development of High Temperature Materials for Industry*, B. Bullock *et al*, eds, Elsevier Science Publishers Ltd, 1989.
3. G. H. MARIJNISSEN and T. v SCHAIK: *Study of gas phase coating mechanisms,* COST 50/II, Final Report, 1980.
4. L. LINDÉ: 'The fatigue properties of coated Ni-base alloys - a literature study', Swedish Institute for Metals Research, *Report IM-2984*, 1993.
5. R. LOWRIE and D. H. BOONE: *International conference on metallurgical coatings,* San Francisco, USA, 1977.
6. G. WAHL, F. SCHMADERER, M. METZGER and A. R. NICOLL: 'The CVD deposition of Ti-Si containing coatings on Ni-base superalloys', Proc. on *8th CVD International Conference,* Paris, 1981, 685–698.
7. W. BETZ, H. HUFF and W. Z. TRACK: 'On the assessment of protective coatings to combat hot gas corrosion on gas turbine blades', *Z. f Werkstofftech.,* 1976, **7**, 161–166.
8. D. H. BOONE: *Airco Temescal Data Sheets*, Airco Inc., Berkeley, California, 1977.
9. L. LINDÉ and P. J. HENDERSON: 'Thermo-mechanical and low cycle fatigue of an aluminised nickel-based superalloy', *8th CIMTEC,* Techna Srl., 1995, 313–320.
10. A. SAMUELSSON and L. E. LARSSON: 'Thermomechanical fatigue of the nickel base alloys IN738 LC and 1N597', Swedish Institute for Metals Research, *Report IM-1609*, 1981.
11. G. ENGBERG and L. E. LARSSON: 'Thermomechanical and low cycle fatigue properties and their relation to creep for the nickel base alloys IN738 LC and 1N597', Swedish Institute for Metals Research, *Report IM-1850*, 1984.
12. L. NICKLASSON: 'The effects of high temperature exposure on the microstructures of aluminised and plasma sprayed coatings on a nickel-base alloy substrate', Swedish Institute for Metals Research, *Report IM-3381*, 1996.
13. P. C. PATNAIK and J. P. IMMARIGEON: 'Protective coatings for aero engine hot section components', *Mat. & Man. Proc.,* 1989, **4,** 347–384.
14. A. STRANG and E. LANG: 'High temperature stability of pack aluminide coatings on IN738 LC', Proc. Symp. *High temperature alloys for gas turbines,* Reidel Publishing Co., 1982, 469–506.
15. J. GAYDA, T. P. GABB, R. V. MINER and G. R. HALFORD: 'Bithermal low-cycle fatigue behaviour of a NiCoCrAlY-coated single ciystal superalloy', Proc. Symp. *1987 TMS-AIME Annual Meeting* in Denver, Colorado, February 22–26, 1987, P. K. Liaw and T. Nicholas eds, The Metallurgical Society Inc., 1987, 179–198.
16. A. WERONSKI and T. HEJWOWSKI: *Thermal fatigue of metals*, Marcel Dekker, Inc., 1991.

High Temperature Fatigue Behaviour of a Thermal Barrier Coating on a Single Crystal Superalloy

A. BICKARD and L. REMY

Centre des Matériaux, Ecole Nationale Supérieure des Mines de Paris, U.M.R. C.N.R.S. No. 7633, BP 87–91003 EVRY Cedex, France

ABSTRACT

LCF tests were performed at 1100°C under strain control on a thermal barrier coating composed by a platinum–nickel modified aluminide bondcoat and an yttria partially stabilised zirconia outer layer. Continuous triangular wave cycle and tensile or compressive dwell cycles were used. Complex damage mechanisms were identified and a strain-oxidation interaction has been analysed for the alumina scale growth.

INTRODUCTION

Nickel aluminide coatings are currently used for cooled turbine blades in jet engines to improve the corrosion-oxidation resistance of superalloys during steady operation. Thermal barrier coatings (TBCs) are now considered to further reduce the temperature of the substrate while the bondcoat provides the corrosion-oxidation resistance like conventional coatings[1] as well as the accommodation of the strains developed between the ceramic coating and metallic substrate due to their mismatch in material properties. Rotating components may experience severe thermal transients that may damage the superalloy substrate, the coating or the interface.

The single crystal superalloy AM1, processed by SNECMA in advanced cooled engine parts, was coated with a thermal barrier coating, manufactured by electron beam physical vapour deposition[2] (EBPVD) of yttria partially stabilised zirconia (YPSZ) on a platinum modified nickel aluminide bondcoat. A comparison is made with the former coating used on AM1: C1A, a chromising and aluminising process by chemical vapour deposition.[3] In our TBC system, the chemical composition of the bondcoat is fairly similar to that of C1A. The external part of the layer (YPSZ side in case of the TBC) is β-NiAl and the internal part is constituted by a mixture of β-NiAl and precipitates which are enriched in heavy elements coming from the substrate. An alumina (Al_2O_3) scale grows between the bondcoat and the YPSZ of the TBC.

The thermal fatigue (TF) test is a useful tool to investigate the damage evolution in structures exposed to thermal transients but results depend on the specimen and rig geometries.[4] Moreover, predicted lifetime depend on the life prediction model and on constitutive equations as well.[5] Thermal-mechanical fatigue (TMF) tests are used to complement TF tests and to simulate the volume element behaviour of critical areas. Most turbine components are designed using simple low cycle fatigue (LCF) tests as an approximation of TMF.

This paper presents current results of an on-going investigation of damage mechanisms for a TBC deposited on AM1 under high temperature cyclic loadings. Results of LCF at the maximum temperature of the TMF cycle are depicted here as a first step of this investigation. The alumina scale thickness for different cycle shapes is found to be dependent of a strain-oxidation interaction and is tentatively rationalised

STUDUED MATERIALS

Superalloy Substrate AM1

The substrate is a high γ' volume fraction nickel-base single crystal superalloy with principal alloying additions of Co, Cr, Al. Cast single crystals of the substrate have a dendritic microstructure that is solution treated at 1060°C. Thus, a high creep resistance is achieved for using cuboidal microstructure given by a FCC γ-phase matrix strengthened by $Ni_3(Al,Ti,Ta)$ intermetallic γ' precipitates, Fig. 1(a).

Nickel Aluminide Coating C1A

To improve the corrosion-oxidation resistance of AM1, the C1A coating by SNECMA, results from a chromising followed by aluminising by a low-activity pack cementation process. The coating microstructure presents an outer layer of the β-NiAl intermetallic and an inner layer of elongated columnar chromium enriched precipitates in a β-NiAl matrix, Fig. 1(b). The superficial aluminide provides aluminium to form an alumina layer on the surface without affecting the substrate. Corrosion resistance is improved by chromium.

Thermal Barrier Coating (TBC)

The TBC used here allows a platinum–nickel modified aluminide bondcoat. Its microstructure is fairly similar to C 1As. Platinum-modified aluminide coatings have proved their efficiency against molten salt-induced high-temperature corrosion[6] but the presence of brittle $PtAl_2$ lowers their ductility.[7,8]

The topcoat is composed by Yttria partially stabilised zirconia processed by electron beam physical vapour deposition that confers a columnar structure, Fig. 1(c).

EXPERIMENTAL PROCEDURE

Thermal-mechanical tests were performed on AM1 coated by ClA with a thermo-mechanical cycle which simulates fairly severe thermal shocks on the leading edge of turbine blades during take-off and landing between 600°C and 1100°C, Fig. 2(a).

Fully reversed strain controlled ($\varepsilon_{min}/\varepsilon_{max} = -1$) isothermal fatigue tests were performed on the coated test pieces at 950°C for C1A only and at 1100°C for C1A and TBC. 950°C corresponds to the temperature for the minimum of mechanical strain in

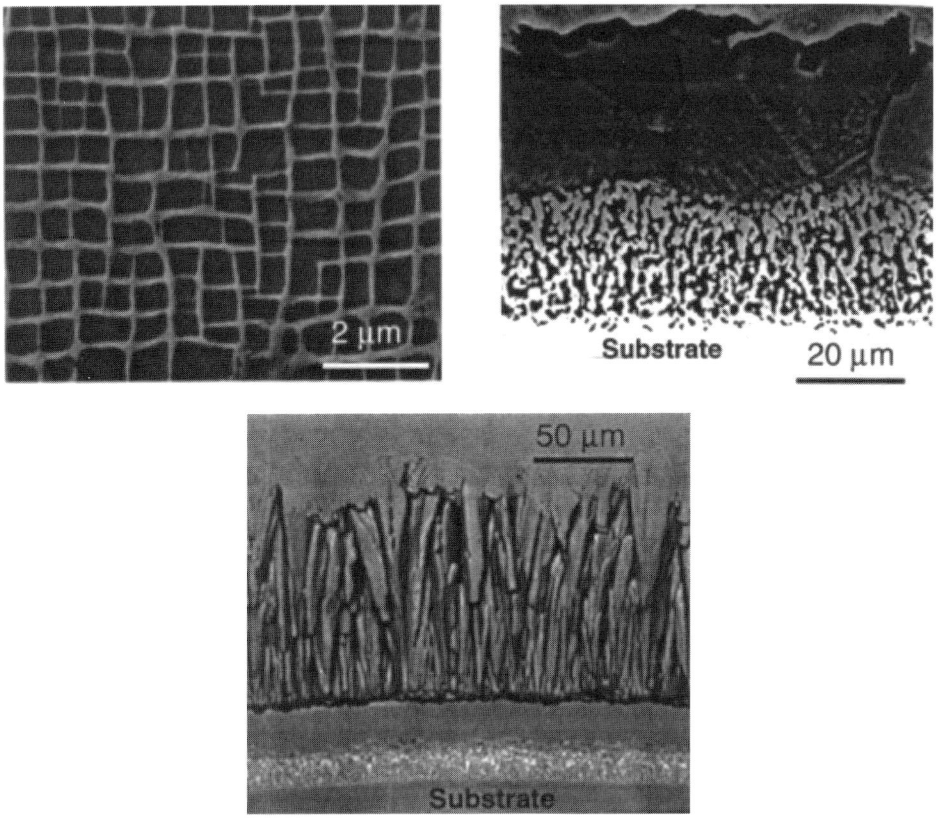

Fig. 1 (a) γ–γ′ microstructure of AM1, (b) microstructure of C1A coating and (c) microstructure of TBC.

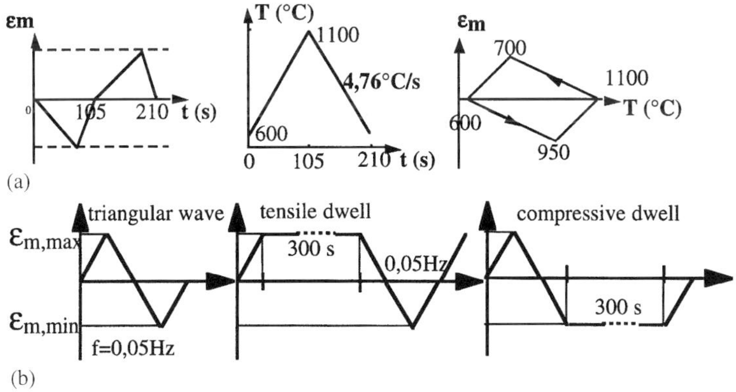

Fig. 2 (a) Thermal–mechanical cycle and (b) 1100°C low cycle fatigue cycles.

a thermal-mechanical cycle. The reason for choosing 1100°C is twofold: (i) it is the maximum temperature for turbine blades in service and therefore the maximum temperature of the thermal-mechanical cycle and (ii) for the TBC this temperature is the worst with respect to oxidation. Tests used a triangular strain waveform at the frequency 0.05 Hz for C1A and some tests on the TBC included a tensile dwell or a compressive dwell of 300 s, Fig. 2(b). Mechanical strain ranges investigated here were 0.75% and 1.6%.

Specimens were manufactured with their longitudinal axis parallel to the [001] cristallographic direction of AM1. Specimen temperature was maintained using a four-elliptical-lobe lamp furnace and was monitored using a K thermocouple located at the edge of the gauge length.

Load, mechanical strain and temperature were recorded through out the tests with a microcomputer. The damage was monitored using the DC potential drop technique. Tests were stopped either when a crack of 1 mm depth had grown in the specimens (in case of tests on C1A, the DC potential drop had been calibrated) or when the variation of potential drop exceeded 8% (in the case of tests on the TBC). We yet assume that the second criterion is obeyed after the first one.

The surface of the gauge length area as well as longitudinal sections with or without etching with a γ'-phase etch-stain (composition 66 vol.% hydrochloric acid, 33 vol.% nitric acid), were observed by scanning electron microscopy (SEM) with secondary emission electrons or backscattered electrons.

RESULTS

Comparison of the Lifetime under TMF and LCF for a Coated Superalloy

A previous study on AM1 coated by C1A[9] under TMF and LCF had led to the following conclusions. The TMF lifetime of specimens coated with C1A, was compared with that under LCF. According to previous experience on bare and aluminised superalloys using LCF, TMF and TF, the TMF life was in pretty good agreement with the LCF life, at 950°C for a given stress range and at 1100°C for a given mechanical strain range. Plots of applied strain range *versus* number of cycles at the end of the test at each temperature are reported in Fig. 3.

The strain cycle and the temperature cycle can be chosen independently in TMF tests and this allows to highlight damage mechanisms that cannot be activated in LCF.[10] In case of C1A under fairly severe thermal-mechanical cycles, LCF tests provide a good simulation of the lifetime and damage mechanisms occuring under TMF.

Therefore we started to study the TBC under LCF at 1100°C, before completing the analysis in TMF in a near future.

Lifetime of TBC Specimens under LCF

The lifetime of TBC tested specimens was compared with that of bare specimens under LCF at 1100°C *versus* mechanical strain range in Fig. 4. The comparison be-

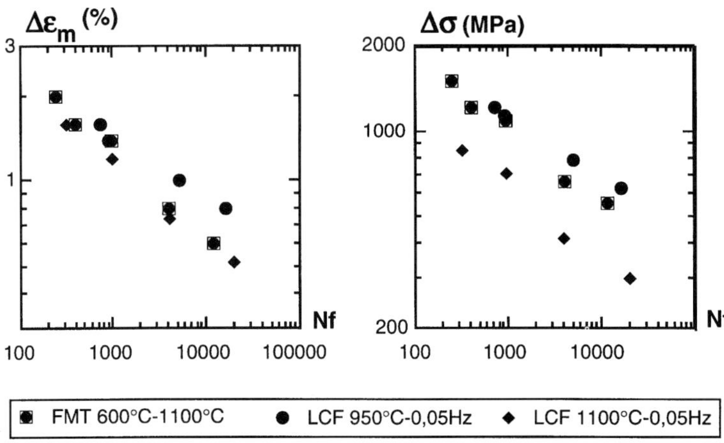

Fig. 3 TMF and LCF lifetime comparison for a 1 mm crack depth of C1A coated AM1.

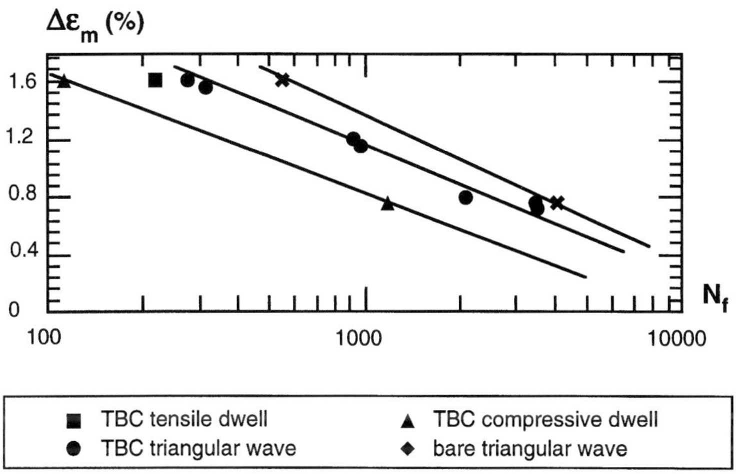

Fig. 4 Variation of the LCF lifetime at 1100°C with mechanical strain range for TBC.

tween bare and coated AM1 for a continuous triangular cycle can be made keeping in mind that the criteria to stop the tests were not the same. Thus, if we only consider the criterion used for TBC, the lifetime of bare AM1 should have probably been lengthened. The oxidation caused by the tensile or compressive dwell dramatically reduces the lifetime of coated specimens, with respects to the continuous triangular cycle. By comparing coated specimens for different kinds of cycle shape, the tensile dwell cycle was the most damaging but, as we shall see below, damage mechanisms were quite different and may not be considered as equally detrimental for the samples.

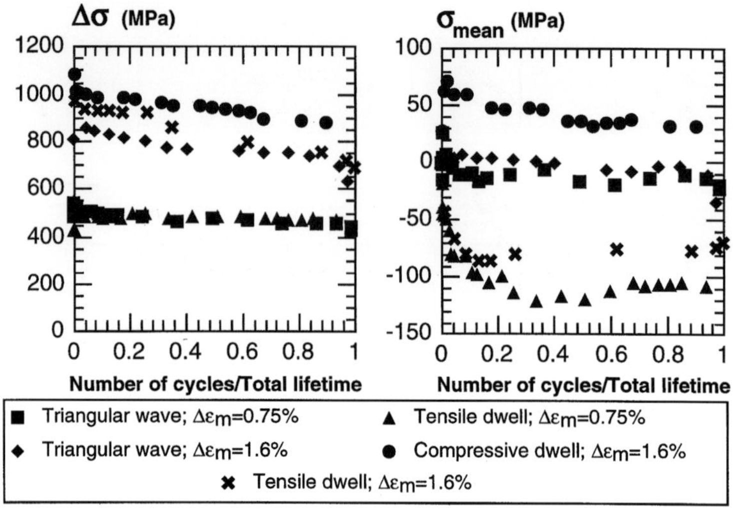

Fig. 5 (a) Variation of stress range in 1100°C LCF on the TBC and (b) variation of mean stress in 1100°C LCF on the TBC.

Stress Response

Whereas the DC potential drop slowly varied during the test, it increased rapidly within a few tens of cycles at the end. This period also corresponded to a rapid drop of the maximum and minimum stresses. Figures 5(a) and (b) respectively plot the stress range and the mean stress *versus* the fraction of lifetime. The slight decreasing of the stress range at the beginning of the test highlighted a slight softening due to rafting in AM1. The severe decreasing of stress range in the last tens of cycles corresponds to the last stage of a main crack propagation. The mean stress was significantly lowered when testing with dwell in tension or in compression because of a significant plasticity by relaxation.

The stress–mechanical strain hysteresis loops during the stable phase of the test showed an unusual behaviour. For the triangular cycle and the tensile dwell cycle, the loops always showed a variation of the compliance in compressive stresses and when the strain becomes negative, Figs 6(a) and (c). Obviously, specimens weren't buckled. Microstructural observations are going to explain that behaviour. For the compressive dwell cycle, this effect is less visible because of creep that reduces the minimum stress, Fig. 6(d).

Metallographic Observations

Continuous triangular wave cycle

The external observation for the continous triangular wave cycle showed a major crack in YPSZ perpendicular to the load direction and rather straight with few ramifications, Fig. 7(a). Secondary cracks, as well as crack branches are more numerous when the strain range increases, Fig. 7(b).

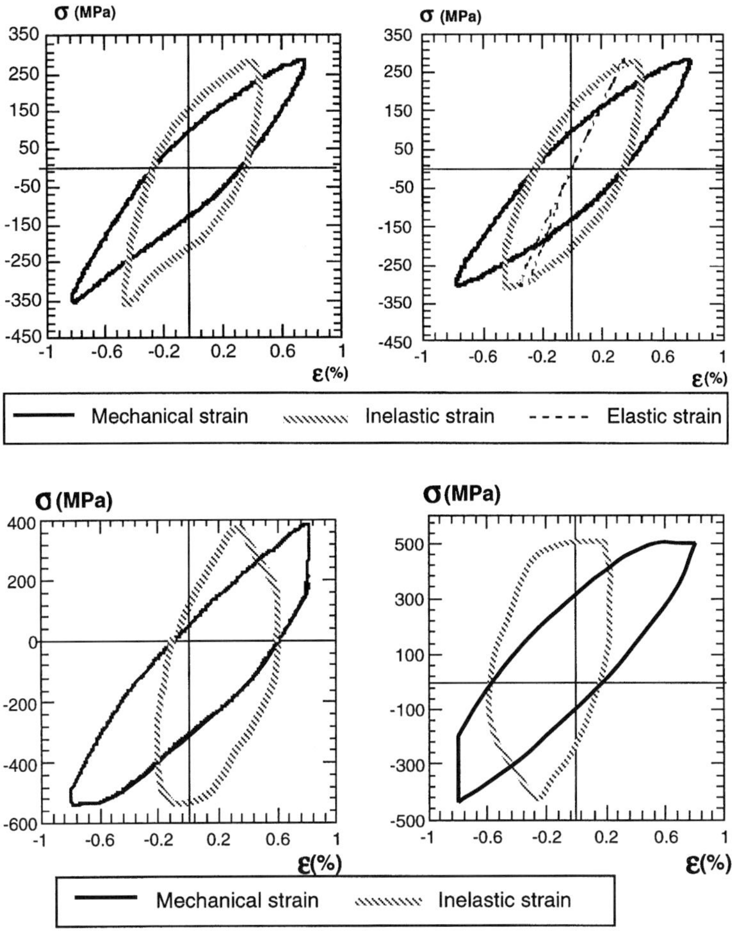

Fig. 6 Stress–strain hysteresis loop for LCF triangular wave cycling; N = 536, (b) stress–strain hysteresis loop with composite effect for LCF triangular wave cycling; N = 536, (c) stress–strain hysteresis loop for tensile dwell cycling; N = 30 and (d) stress–strain hysteresis loop for compressive dwell cycling; N = 100.

Sections showed the major crack that penetrated straight and deep in the substrate. The TBC bondcoat contains lots of transverse cracks located in the β-NiAl part of the bondcoat (external side). At low mechanical strain, cracks extended in a direction parallel to the load axis , mostly in the diffusion part of the bond coat, Fig. 8(a). Higher magnification, Fig. 8(b) highlighted the delamination between the alumina scale and the bondcoat. Delamination consisted in of holes at the alumina-boncoat interface and a complex cracking pattern of the alumina scale.

Some cracks, mainly at high mechanical strain range, interacted with cracks initiated in the substrate on casting micropores and then propagated in the YPSZ layer,

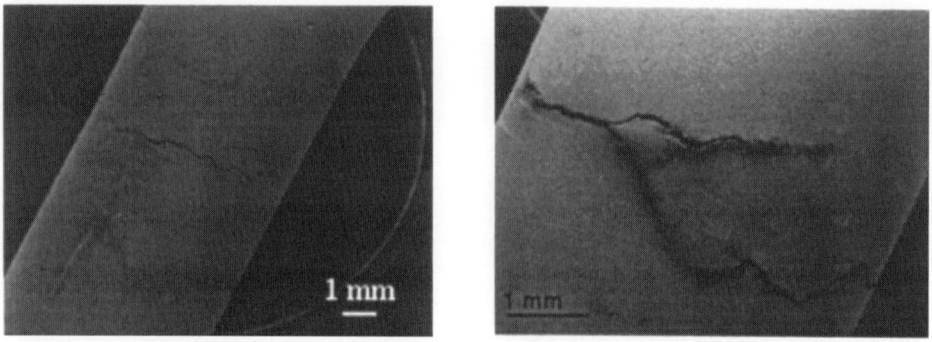

Fig. 7 (a) SEM image of gauge length after triangular wave cycling at $\Delta\varepsilon_m = 0.75\%$ and (b) SEM image of gauge length after triangular wave cycling at $\Delta\varepsilon_m = 1.6\%$.

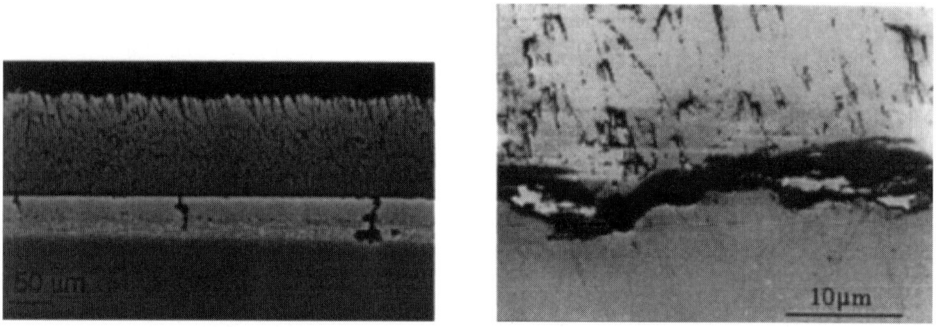

Fig. 8 (a) SEM image of bondcoat multicracking after triangular wave cycling at $\Delta\varepsilon_m = 0.75\%$ and (b) SEM image of alumina scale after triangular wave cycling at $\Delta\varepsilon_m = 0.75\%$.

becoming the secondary cracks mentioned above, Fig. 9(a). But in that case and although the alumina scale was thicker and full of holes, no spallation was observed, Fig. 9(b).

Tensile dwell cycle

The YPSZ surface for the tensile dwell cycle showed very numerous small cracks spread uniformly and perpendicular to the load axis, Fig. 10(a).

The growth of alumina scale is much increased. Therefore, the alumina scale penetrates in the γ-NiAl underlayer producing large cracks that stop when they reach the diffusion part of the underlayer. Moreover, alumina clusters, located between the β-NiAl part and the diffusion part of the underlayer extend in a direction parallel to the load axis, Fig. 10(b) and in a direction perpendicular through the diffusion part of the underlayer. This behaviour was strongly increased at high mechanical strain range and produced cracks in the substrate. Lower mechanical strain range fairly showed the same microstructural aspect but moreover, spallation occured, Fig. 10(c).

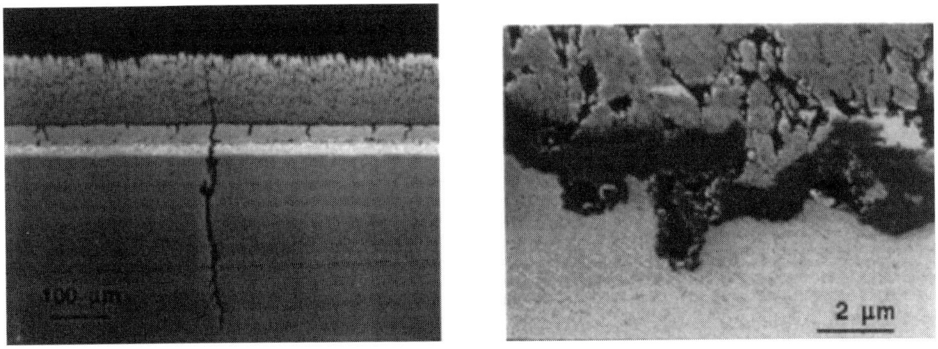

Fig. 9 SEM image of a secondary crack and bondcoat multicracking after triangular wave cycling at $\Delta\varepsilon_m = 1.6\%$ and (b) SEM image of alumina scale after triangular wave cycling at $\Delta\varepsilon_m = 1.6\%$.

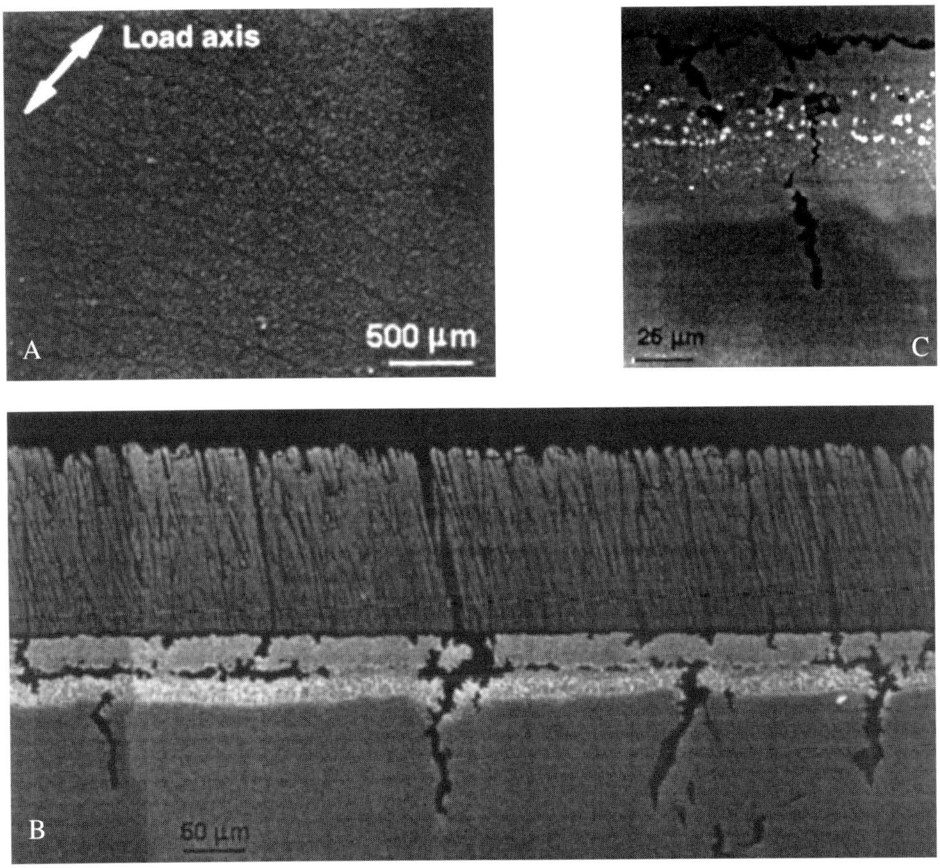

Fig. 10 (a) SEM image of gauge length after tensile well cycling at $\Delta\varepsilon_m = 1.6\%$, (b) SEM image of bondcoat after tensile dwell cycling at $\Delta\varepsilon_m = 1.6\%$ and (c) SEM image of bondcoat after tensile dwell cycling at $\Delta\varepsilon_m = 0.75\%$.

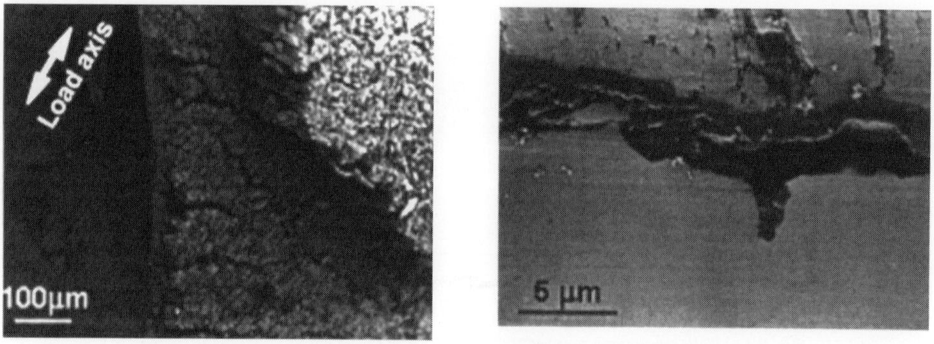

Fig. 11 (a) Fractographic SEM image of the transverse multicracking of the bondcoat after compressive dwell cycling at $\Delta\varepsilon_m = 1.6\%$ and (b) sectional SEM image of bondcoat spallation after compressive dwell cycling at $\Delta\varepsilon_m = 1.6\%$.

Compressive dwell cycle

With a compressive dwell cycle, the sample broke before reaching 8% variation in potential drop. On both sides of the rupture surface, YPSZ had totally spalled off. The bondcoat and the YPSZ that remain on the sample were dramatically bearing multiple cracks, Figs 11(a). Except for the major crack that produced final fracture, numerous cracks were located in the β-NiAI part of the bondcoat. Consequently to oxidation, cracks were largely opened. The discrete spallation under the remaining ceramic layer consisted in an alternate holes and complex cracking of the alumina scale, Fig. 11(b).

Rafting

For coated specimens, it is generally accepted that the rafting of γ′ precipitates is oriented in a direction parallel to the load axis near the coating and in a perpendicular direction in the bulk.[11] These observations made for C1A coating according to a triangular wave cycle, are also valid for the TBC, Fig. 12(a). On the contrary, the rafting was found to be uniformly oriented in the whole specimens with a tensile dwell or a compressive dwell cycle as for the bare AM1. Figures 12(b) and (c) respectively show the rafting near the interface for a tensile dwell cycle and for a compressive dwell cycle. The rafting orientation is the same in the whole section. This means that in triangular cycling, the local inelastic strain tensor principal direction was not the same in surface and bulk. But, in the dwell case, the direction is the same in the whole sample.

DISCUSSION

Analysis of Stress-Strain Hysteresis Loops

Assuming that the ceramic layer is damaged within a few cycles, the sample behaves as if it were not coated in tension yet all components bear the load in compression. The stress-strain loops are computed with an effective section of the sample so that at

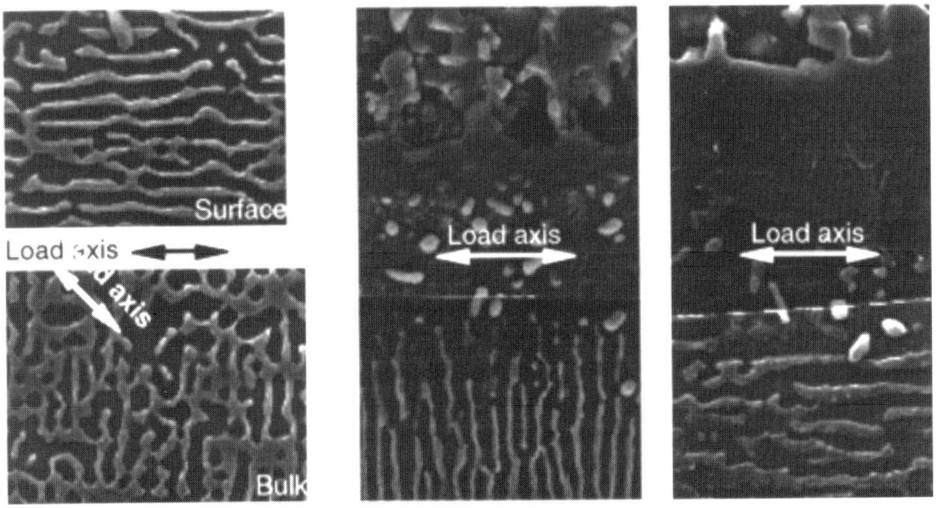

Fig. 12 (a) Rafting of γ' after triangular waver cycling at $\Delta\varepsilon_m = 1.6\%$, (b) rafting of γ' after tensile dwell cycling at $\Delta\varepsilon_m = 1.6\%$ and (c) rafting of γ' after compressive dwell cycling at $\Delta\varepsilon_m = 1.6\%$.

maximum compressive strain the load-bearing section includes the TBC while at maximum tensile strain, this is only on the AM1 substrate. Between these two stages, the load-bearing section gradually changes. This is depicted in Fig. 6(b) for the triangular wave cycle.

The elastic strain values are proportional to the section variations that we computed to obtain the regular loop shape.

Alumina Thickness

Oxidation kinetics

Figure 13(a) plots the mean thickness of the alumina scale *versus* the time spent at 1100°C for specimens exposed to temperature in a furnace as well as for LCF specimens. Specimens exposed to temperature without applied stress exhibit a usual parabolic oxidation kinetics, which reads in differential form as:

$$de_{OX}^2 = \alpha^2 dt \tag{1}$$

where e_{Ox} is the alumina scale thickness; t the exposure time; α an oxidation constant.

The oxide thickness of stressed/strained samples is higher than in steady oxidation in unstressed specimens whatever the cycle shape. Oxidation kinetics is thus triggered by strain cycling. This strain-oxidation interaction is strongly dependent upon the strain wave shape since a tensile dwell seems much more effective than a triangular cycle. On the opposite, a compressive dwell is almost equivalent to a triangular cycle.

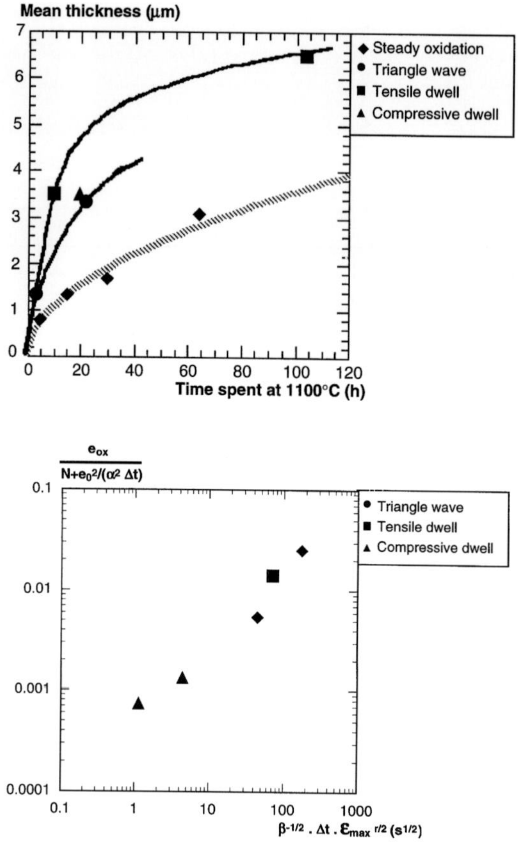

Fig. 13 (a) Variation of the alumina scale mean thickness with the time spent at 1100°C and (b) normalised alumina scale growth per cycle as a function of maximum strain range and cycle period in LCF at 1100°C.

This may be partly due to the peculiar mechanical behaviour of the YPSZ layer which has been evidenced by the microstructural observations and the shape of the stress-strain hysteresis loops. We assume that in compressive dwell, ceramic columns that are squeezed together and thus impede the inward diffusion of oxygen ingress to the bondcoat. This strain-oxidation interaction is analyzed further in the following.

Strain-oxidation interaction

The influence of straining on the growth of oxide scale on bare alloys has been studied in various steels and superalloys under creep or under fatigue (see for instance Ward *et al.*,[12] Hancock and Nicholls,[13] Rémy[14] for a review of these effects). Whatever the straining mode at sufficiently low stresses the scale thickness displayed a usual parabolic kinetics as for an unstressed alloy and at higher stresses, the scale thickness is much larger and shows an apparent linear variation with cycling time. According to eqn (1), if failure occurs for a given strain level and repeats for the same strain increment in creep, the oxidation rate is increased and can follow a pseudolinear kinetics.[14]

$$de_{OX} = e^* \frac{dt}{t^*} \tag{2}$$

where e_{Ox} is the alumina scale thickness; t^* is the time to fracture the oxide scale; e^* is the alumina scale thickness at t^*.

Thus, fracture of the oxide scale in our TBC-single crystal superalloy system may involve micro-cracks perpendicular to the load axis and/or local loss of adhesion between the alumina scale and the bondcoat, since the delamination plane is located here in our case. After fracture of the oxide scale, direct access of oxygen to the metal is achieved so that oxidation kinetics will resume from this new time origin. From this point the oxidation kinetics is given by the oxide growth law of eqn (1).

The quantitative application of eqn (2) relies on the failure criterion used to describe the time to fracture the oxide scale t^*. For monotonic loading a critical tensile ductility concept or a creep law can be used.[14]

For cyclic loading some authors[15] suggest to use a fatigue criterion to describe the time $t^*=N^*\cdot\Delta t$, where Δt is the cycle period and N^* the number of cycles to fracture the oxide scale. This number was thus related to mechanical strain range.

We assume in the present case that t^* depends upon the mechanical strain according to a creep rupture type law given by eqn (3).

$$t^* \varepsilon_{\max}^{r} = \lambda \tag{3}$$

where $\varepsilon_{\max}$ is the maximal strain of the cycle; r and λ are two constants.

We assume that only the tensile strains are damaging for the oxide scale, which is supported by the experimental observation of the influence of the strain wave cycle on oxidation kinetics. This equation obviously applies for a constant strain loading. For a given strain wave cycle, eqn (3) has to be used in a differential form and integrated over a strain-time cycle.

This results in the following equation for the alumina scale thickness after N fatigue cycles.

$$\frac{e_{OX}}{N + \dfrac{e_0^2}{\alpha^2 \Delta t}} = \frac{\alpha}{\lambda^{r/2}} \beta^{-1/2} \cdot \Delta t \cdot \varepsilon_{\max}^{r/2} \tag{4}$$

where e_0 is the initial scale thickness before strain cycling; β is a cycle shape factor which describes the effective creep strain component for the oxide layer; the other terms have been defined earlier.

Plotting the left member of eqn (4) *versus* $\beta^{-1/2} \cdot \Delta t \cdot \varepsilon_{max}^{r/2}$, Fig. 15(b) in a log–log chart displays a good linear fit.

The oxide scale growth and the interaction between oxidation kinetics and strain can thus be well described assuming scale cracking under strain cycling at regular time intervals and using a creep type law.

Damage Mechanisms

Let us summarise the three types of damage we observed.

For a triangular wave cycle, the damage is a main crack surrounded by secondary cracks with increasing mechanical strain range and a multicracking of the bondcoat.

For a tensile dwell cycle, the damage is quite diffuse in the bondcoat and towards the substrate. Eventually, in a compressive hold cycle, we had one main crack that propagate through the whole sample but diffuse cracking in the bondcoat.

The mean stress for a compression dwell, the mean stress is increased by the relaxation period. Thus the yield strain in tension is reached earlier and more tensile plasticity is created. The stress concentration at the crack tip increases the crack propagation. A tension dwell results in an inversed effect with much compressive plasticity. In addition to the compressive residual stresses in the coating, the stress-activated oxidation was the preponderant damage mechanism. And at last the triangular wave cycle allows a competition between cracks developing in the substrate and bondcoat damaging by transverse cracking and oxidation.

Spallation occured mainly at low mechanical strain ranges but this mechanism did not influence the criterion to stop the tests as the DC potential drop can only detect cracks in a direction perpendicular to the load axis. Nevertheless, spallation occured for a thick enough alumina scale, when oxidation times are long and which is favoured by the interaction between strain and oxidation, as discussed earlier. The criteria for TBC delamination and fracture are still an area for future work.

CONCLUSION

Complex damage mechanisms have been identified mostly in the bondcoat: alumina scale growth; delamination of the YPSZ-bondcoat interface; growth of alumina precipitates located in the bondcoat in a direction parallel to the load axis; transverse cracking of β-NiAl and of the diffusion part of bondcoat; localised cracking in the substrate or diffuse cracking of the bondcoat.

Among these mechanisms, those related to oxidation are deeply favoured by a tensile or compressive strain hold which reduces the lifetime. But the total lifetime of specimens depends on several damage mechanisms that interact with each other. A single parameter cannot describe the whole behaviour and a lifetime model should integrate mechanical and microstructural parametres thus showing the complexity of the problem.

We have shown an obvious effect of strain on the oxidation kinetic and described the interaction with an alumina scale creep law, assuming that the oxide scale breaks after a time of creep. Such a law can well describe the alumina scale growth in a phenomenological way.

Moreover, spallation should be further studied in a microstructural way to deduce physical mechanisms that occur and eventually, these statements will have to be validated under TMF in order to simulate the influence of thermal transients on the material.

ACKNOWLEDGEMENT

The authors are grateful to SNECMA for financial support and provision of coated material. Engineers of the research center of SNECMA are gratefully acknowledged for their interest in the present study.

REFERENCES

1. G. W. MEETHAM: 'The use of protective coatings in aero-gas turbine engines', *Materials Science and Technology,* 1986, **2**, 290–294.
2. H. LAMMERMANN and G. KIENEL: 'PVD coatings for aircraft turbine blades', *Advanced Materials and Processes,* 1991, **140,** 18–23.
3. R. MORBIOLI, Y. HONNORAT: 'Coatings for heat engines', *Proceedings of the 1st NATO advanced workshop,* J. Clark ed., Acquafradda, Italy, 1984, 179–191.
4. E. GLENNY and T. A. TAYLOR: 'Study of the thermal fatigue behaviour of metals, the effect of test conditions', *Journal of the Institute of Metals,* 1970, **88**, 449–461.
5. L. REMY; 'Recent developments in thermal fatigue', *High Temperature Fracture Mechanisms and Mechanics,* EGF 6, P. Bensussan, J. P. Mascarell, eds, 1990, Mechanical Engineering Publications, London, 353–377.
6. G. LEHNERT and H. W. MEINHARDT: *Electrodeposition Surface Treatment,* 1972, **1**, 189.
7. K. SCHNEIDER, H. VON ARNIM and H. W. GRUNLING: 'Influence of coatings and hot corrosion on the fatigue behaviour of Nickel-based superalloys', *Thin Solid Films,* 1981, **84**, 29–36.
8. S. ALPERINE, P. STEINMETZ, P. JOSSO and A. COSTANTINI: 'High tem perature-resistant Palladium-modified aluminide coatings for Nickel-based superalloys', *Materials Science and Engineering,* 1989, **A121**, 367–372.
9. E. CHATAIGNER and L. REMY: 'Thermomechanical fatigue of coated and bare nickel base superalloy single crystals', *Thermomechanical Fatigue Behaviour of Materials,* ASTM STP 1263, M. J.Verrilli and M. G.Castelli eds, American Society for Testing and Materials, Philadelphia, 1996, 3–23.
10. E. CHATAIGNER and L. REMY: 'Influence d'un revêtement protecteur C1A sur la tenue en fatigue d'aubes de turbines monocristallines', *Fatigue et traitements de surface*, comptes-rendus des Journées de Printemps, Paris, 1995, Commission de Fatigue des Métaux, Société Française de Métallurgie et de Matériaux, Editions de la Revue de Métallurgie, Paris, 11–22.

11. L. REMY, F. HANRIOT, E. CHATAIGNER, L. ESPIE and E. FLEURY: 'Fatigue anisotherme de monocristaux de superalliage a l'état nu ou revêtu', *Proceedings of the Colloque National des Superalliages Monocristallins',* Toulouse, France, 1995.
12. G. WARD, B. S. HOCKENHULL and P. HANCOCK, *Metallurgical Transactions,* 1974, **5**, 1451–1455.
13. P. HANCOCK and J. R. NICHOLLS, *Materials Science and Technology,* 1988, **4**, 398–406.
14. L. REMY: 'Oxidation effect in high temperature creep and fatigue of engineering alloys', *Proccedings of the Corrosion-Deformation Interaction International Conference,* Fontainebleau, France, 1992.
15. M. REGER and L. REMY: 'Fatigue oxidation interaction in IN100 superalloy', *Metallurgical Transactions A,* 1988, **19A**, 2259–2268.

Creep of Plasma Sprayed CoNiCrAlY and NiCrAlY Bond Coats and its Effects on Residual Stresses During Thermal Cycling of Thermal Barrier Coating Systems

J. A. THOMPSON, Y. C. TSUI, R. C. REED, D. S. RICKERBY* and T. W. CLYNE

Department of Materials Science and Metallurgy, University of Cambridge, Pembroke Street, Cambridge CB2 3QZ

**Rolls-Royce plc, P.O. Box 31, Derby DE24 8BJ*

1. INTRODUCTION

There has been intensive study recently[1–12] of the mechanical and microstructural stability of thermal barrier coating (TBC) systems. Particular attention[13–17] has been focussed on the behaviour of the bond coat under service conditions. It is well established that creep of MCrAlY-type bond coat material can be highly significant at temperatures encountered during service of thermal barrier coating (TBC) systems. Stress relaxation in the bond coat can substantially affect the stress state in a TBC system during prolonged exposure to high temperatures. Although this generally results in reduced stress levels at the creep temperature, it can give rise to enhanced residual stresses after cooling down to ambient temperature. This may tend to promote spallation during thermal cycling.

In this study, creep data and expansivities have been obtained for NiCrAlY and CoNiCrAlY bond coats produced by vacuum plasma spraying. A numerical process model has then been used to predict the changing residual stress distributions within TBC systems incorporating bond coats of these two materials.

2. EXPERIMENTAL PROCEDURES

2.1 Spray Deposition

The bond coat powders used in this work were supplied by Sulzer Metco (US) Inc. and are designated Amdry 962 and Amdry 995/C. The nominal compositions of these two powders arc Ni–22Cr–l0Al–1Y and Co–32Ni–21Cr–8Al–0.5Y respectively (all compositions in wt.%). Zirconia powder was supplied by MEL Chemicals (designation 619/04). Spraying was carried out with a Plasma Technik VPS unit, using the conditions shown in Table 1. For production of bond coat creep specimens, the bond

Table 1 Plasma spraying parameters

Spraying parameter	Deposited material		
	NiCrAlY	CoNiCrAlY	ZrO_2–8 wt.% Y_2O_3
Type of spraying	VPS	VPS	APS
Spraying distance (mm)	270	270	127
Arc current (A)	700	700	600
Voltage (V)	55	60	70
Argon plasma gas flow rate (l min^{-1})	50	50	60
Hydrogen plasma gas flow rate (l min^{-1})	6	10	9
Speed of the gun (mm s-1)	85	85	50
Chamber pressure (mbar)	200	200	–
Nozzle diameter (mm)	8	8	8

coat materials were sprayed onto a Nimonic 80A substrate, with a thickness 1.24 mm. The coatings were then detached by cutting a notch in the coating by spark erosion, before loading in three point bending such that the coating debonded from the substrate. For the case of the CoNiCrAlY material, the detached coating had a significant curvature (of about 5 m^{-1}). In order to produce an approximately planar specimen, suitable for creep testing, the detached coatine (of thickness about 0.8 mm) was used as the substrate for further deposition onto the convex surface. In this way, a flat specimen of thickness 1.4 mm was produced. For the NiCrAlY material, flat specimens of thickness about 1.4 mm were produced by a single spray deposition process, the detached coatings being approximately flat. Specimens were cut from the resulting sheets of bond coat material to the required shape by spark machining. To ensure even thickness, creep specimens were ground on both sides. This had the added benefit of improving the surface finish of the specimens.

2.2 Scanning Laser Extensometry

A scanning laser extensometer[18] was used to obtain steady state creep rates as a function of applied stress and temperature and also to measure the thermal expansivity over a range of temperature. For the expansivity measurements, the applied load was removed and the specimen was thermally cycled to ensure that there was no significant hysteresis (ie that the specimen did not creep under self-weight).

3. RESULTS

3.1 Creep Rates

For all of the creep tests, the strain histories showed primary, secondary and tertiary creep regimes, with a well-developed secondary stage from which a steady state creep rate could be derived. These creep rates are plotted in Figs 1–2 as a function of stress and temperature. The creep behaviour conforms for both bond coats to a power law expression of the form

$$\frac{d\varepsilon}{dt} = A\sigma^n \exp\left(\frac{-Q}{RT}\right) \tag{1}$$

The values of the pre-exponential constant, activation energy and stress exponent derived from these plots are shown in Table II. While the activation energies are similar, the NiCrAlY has a higher stress exponent. As a consequence of this, the NiCrAlY is more creep-resistant at relatively low stress levels (<~30 MPa), but less creep-resistant under high stresses (>~70 MPa).

Table II Creep parameters for the two bond coat materials.

	NiCrAlY	CoNiCrAlY
Activation energy, Q (kJ mole^{-1})	346	361
Stress exponent, n	4.5	2.9
Pre-exponential factor, A (s^{-1}Pa^{-n})	2.5 x 10^{-23}	4.6 x 10^{-10}

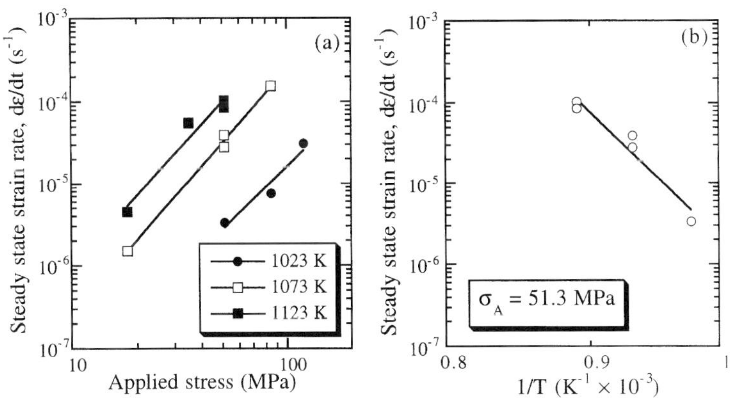

Fig. 1 Steady state creep rates for vacuum plasma sprayed CoNiCrAlY as a function of (a) applied stress and (b) temperature.

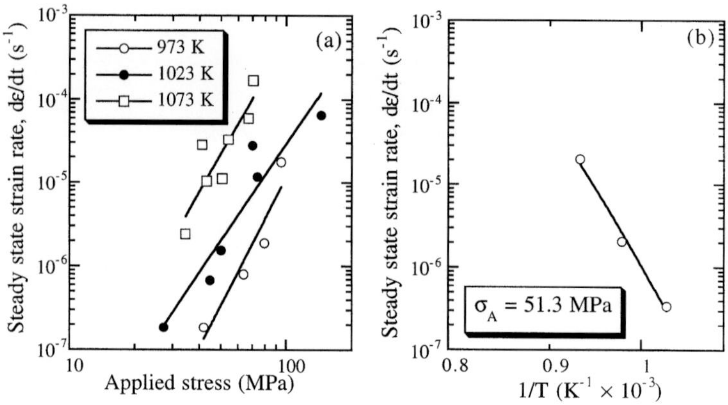

Fig. 2 Steady state creep rates for vacuum plasma sprayed NiCrAlY as a function of (a) applied stress and (b) temperature.

3.2 Expansivities

A typical strain history is shown in Fig. 3. Also shown in this figure is a second order polynomial which has been fitted to the experimental data. It can be seen that this provides a good fit, indicating that the expansivity is expected to be a linear function of temperature, although it should be noted that these measurements were only made up to about 800°C. Thermal expansivity values are shown in Fig. 4 as a function of temperature for the two bond coats and also for a superalloy substrate and a ZrO_2–8%Y_2O_3 top coat. Note that the expansivities of the bond coats increase faster with temperature than does that of the substrate, which thus has the lowest value of the three metal alloys at elevated temperature. However, while the net thermal misfit strain to be accommodated by a bond coat on a massive substrate ($=\int(\alpha d-\alpha s)\,dT$) after cooling from, say, 1000°C to room temperature would be close to zero for the NiCrAlY, it would have a substantial positive value for the CoNiCrAlY.

4. REPRESENTATION OF CREEP CHARACTERISTICS

These parameters are such that bond coats of both alloys are likely to exhibit creep relaxation of residual stresses during deposition of the TBC system and under service conditions. This can be quantified by considering the simple case of a bond coat on a non-creeping, massive substrate at a fixed temperature. The misfit strain, ε, must be accommodated entirely within the bond coat, which will therefore be subject to a stress, σ, given by the product of the misfit strain and the (biaxial) modulus, E′ (= E / (l-υ)). The decay of this stress with time as the bond coat creeps is described by the expression

$$\frac{d\sigma}{dt} = -E'\frac{d\varepsilon}{dt} = -E' A\sigma^n \exp\left(\frac{-Q}{RT}\right) \tag{2}$$

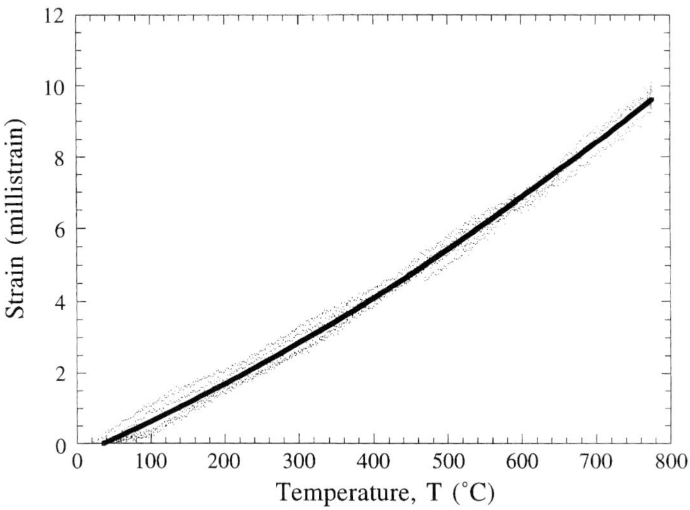

Fig. 3 Experimental strain data acquired during two thermal cycles of a NiCrAlY specimen with no applied load, together with a best fit second order polynomial expression.

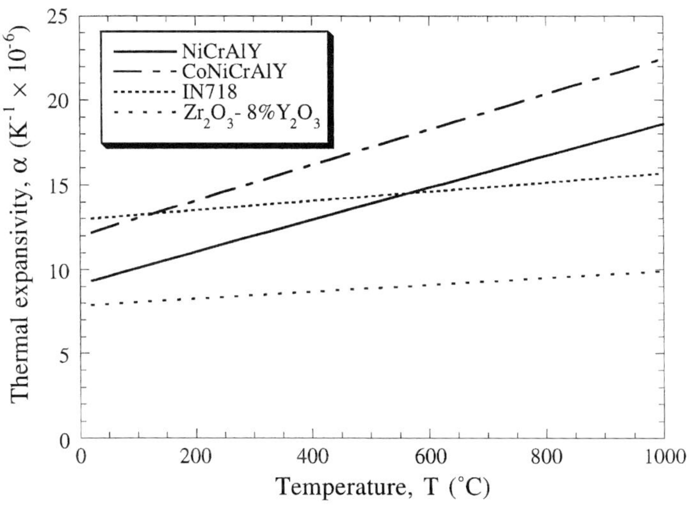

Fig. 4 Measured data for the thermal expansivity of the two bond coats. These plots were obtained by fitting a second order polynomial to the measured strain–temperature plots. Also shown are data from the literature[17] for the IN718 superalloy and for a sprayed ZrO_2–8%Y_2O_3 topcoat.

The time, t_{relax}, for this stress to decay away from an initial value, σ_0, to some negligibly small final value, σ_f, can therefore be obtained by integrating this equation between suitable limits.

This leads to the following expression for the relaxation time.

$$t_{relax} = -\frac{\sigma_f^{1-n} - \sigma_0^{1-n}}{E' A(n-1)\exp\left(\frac{-Q}{RT}\right)} \quad (4)$$

The behaviour predicted by eqn (4), using the data in Table II, is shown in Fig. 5. It can be seen that at high temperatures (>1000°C) relaxation occurs quite rapidly (t_{relax} < 100 s, depending on the assumed relaxation end stress) for the CoNiCrAlY. The NiCrAlY material is more creep-resistant at the low stress levels to which this calculation is sensitive. (Compare Fig. 1(a) and Fig. 2(a)). Nevertheless, even for NiCrAlY, a high initial stress will fall at 1000°C to a few MPa within a time of the order of 100 s.

While the above analysis may give a useful guide to the general effect of creep in the bond coat, a numerical model is needed in order to predict the changing stress distribution within a TBC during deposition and service. A finite difference model of this type,[19–22] based on unidirectional heat flow and neglecting through-thickness stresses, has been used in the present work to predict the changing stress distribution and associated specimen curvatures and strain energy release rates for interfacial debonding. This has been done both for massive substrates and for relatively thin ones, the latter being of interest because they give rise to curvature changes which are experimentally measurable.

5. MODELLING OF RESIDUAL STRESS DISTRIBUTIONS

Predictions have been obtained using the numerical model. In order to carry out this simulation, it is first necessary to measure the quenching (or deposition) stress.[23,24] This is the stress in a splat immediately after it has been deposited and quenched to the temperature of the underlying material. In the present work, the value has been obtained by comparisons between measured and predicted thermal and curvature histories during deposition onto a thin substrate. A typical comparison is shown in Fig. 6, which refers to an APS NiCrAlY coating. The magnitude of the positive increment of curvature which occurs during each spray cycle gives an indication of the value of the quenching stress. The value for the (VPS) NiCrAlY was found to be about 10 MPa. In contrast, the value of the (VPS) CoNiCrAlY coating was found to be significantly higher at about 80 MPa. This rather striking difference may be partly attributable to the higher expansivity and stiffness of the CoNiCrAlY material. Probably of more significance is the fact that the CoNiCrAlY is the more creep-resistant material in the high stress level – short timescale regime which is important in determining the value of the quenching stress. It is also likely that the CoNiCrAlY material has a higher yield stress at elevated temperature, although reliable data concerning this are in short supply.

Predictions are now presented for a TBC system composed of a massive (IN718)

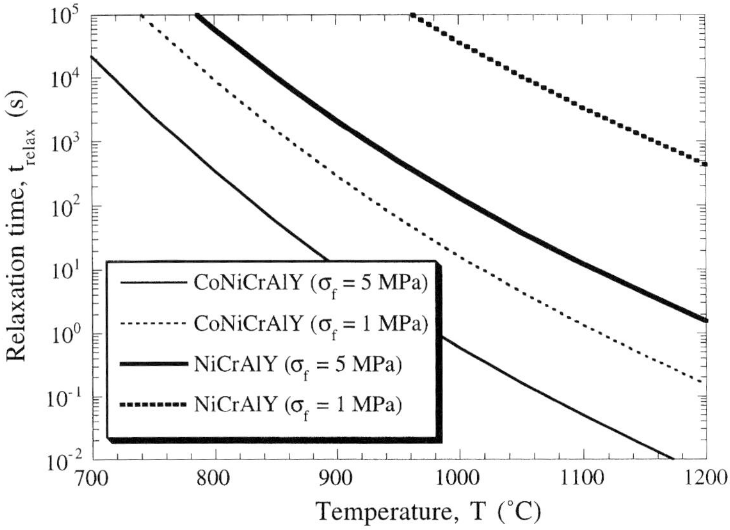

Fig. 5 Dependence of the stress relaxation period for the two bond coats on the temperature, according to eqn (3). The curves were obtained using the data in Table I, for an initial stress level, σ_0, of 50 MPa and a value for the biaxial modulus, E′, of 80 GPa.

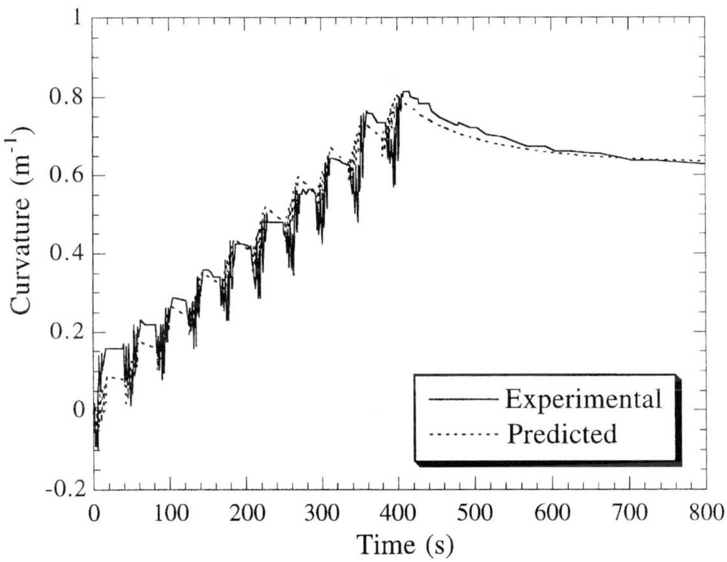

Fig. 6 Predicted and measured curvature histories during a 10 cycle spray raster for deposition of a NiCrAlY coating on a thin (1.6 mm) superalloy substrate.

substrate with a 100 μm bond coat (CoNiCrAlY or NiCrAlY) and a 400 μm ZrO_2 top coat. Temperature and stress histories during production of this were simulated using the numerical model. The system was then exposed to the thermal cycle shown in Fig. 7(a). Variations in the average stress level within bond coat and top coat, and the associated strain energy release rates for debonding at the two interfaces, have been calculated.

Consider first the case of the CoNiCrAlY bond coat. It can be seen from Fig. 7(b) that stresses in the top coat remain small throughout. (This is largely a consequence of the low stiffness of sprayed zirconia, which arises from the presence of a dense network of microcracks.) The stress in the bond coat becomes less tensile as differential thermal expansion occurs. (The thermal expansivity of the bond coat is greater than that of the substrate over almost the complete temperature range – see Fig. 4.) This is sufficient to put the bond coat into compression before the holding temperature of 900°C is reached. The stress in the bond coat then relaxes as creep occurs, becoming close to zero by the end of the holding period. Cooling then has the effect of regenerating a tensile stress, which becomes significantly larger than the original level. This is caused by the higher rate of temperature change in the range 600–900°C during cooling than during heating (limiting the degree of creep relaxation which can take place), which will commonly occur in practice.

Also shown in Fig. 7 is the changing strain energy release rate, G_r, for interfacial debonding. This is dependent on the residual stress distribution and the changes in it which would occur if debonding were to occur. It represents the driving force for debonding at the interface concerned. As expected, G_r for debonding of the bond coat falls as the stress in it drops and then rises again during cooling. In this case, it is predicted to reach relatively high values (~50 J m^{-2}). This might well be greater than the critical value for debonding at this interface (i.e. the interfacial fracture energy), although this interface is usually relatively tough.[17] However, also likely to be significant is the variation in G_r for detachment of the top coat from the bond coat. Although this never reaches high absolute values, the fracture energy is usually relatively low for this interface.[17] It is therefore of concern that the heating will cause an increase in the value of G_r for the top coat/bond coat interface. Note also the small peak at the start of the cooling period (which can be seen more clearly in Fig. 8(c)): this is associated with the thermal gradients through the specimen thickness which occur in this regime. It may be concluded that there is a particular danger of top coat spallation at the onset of rapid cooling after a period at high temperature.

In this context, note should also be taken of the importance of the thermally grown oxide at the bond coat/top coat interface. Since this is thin, the associated changes in strain energy release rate are expected to be relatively small.[25] However, it is probable that the toughness (resistance to fracture) of this region is degraded by the microstructural changes which occur as the oxide grows (possibly as a result of the formation of pores and microcracks) and it seems likely that it is this effect which is largely responsible for the common observation of top coat spallation after prolonged exposure to high temperature.

The corresponding plots are shown in Fig. 8 for the NiCrAlY bond coat. The stress levels in the bond coat are now much lower, both before and during the thermal cycle.

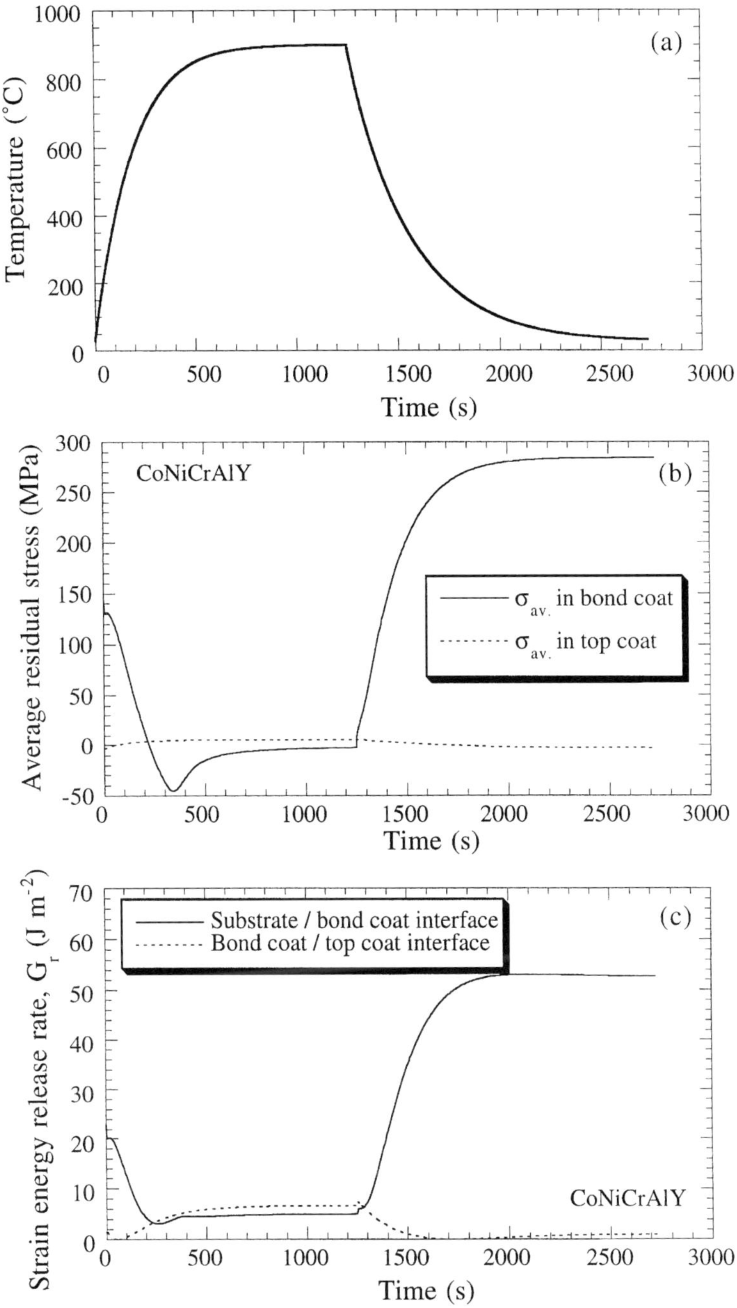

Fig. 7 Changes in parameters during a thermal cycle imposed on a Ni superalloy – CoNiCrAlY bond coat (100 μm) – ZrO_2 top coat (400 μm) system: (a) temperature, (b) average stresses in bond coat and top coat and (c) strain energy release rate for debonding at substrate/bond coat and bond coat/top coat interfaces.

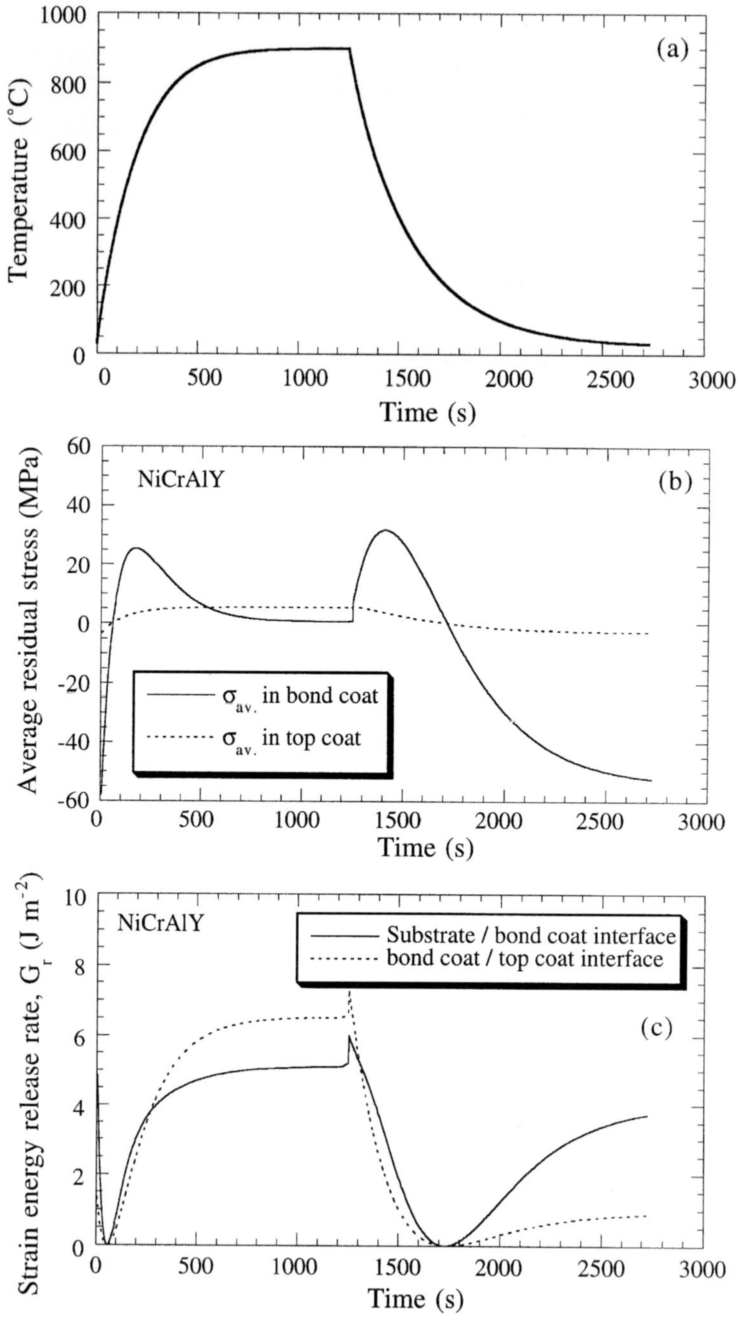

Fig. 8 Changes in parameters during a thermal cycle imposed on a Ni superalloy – NiCrAlY bond coat (100 μm) – ZrO_2 top coat (400 μm) system: (a) temperature, (b) average stresses in bond coat and top coat and (c) strain energy release rate for debonding at substrate/bond coat and bond coat/top coat interfaces.

This is partly a consequence of the lower quenching stress and partly due to the average value of the thermal expansivity over the temperature range concerned being much closer to that of the substrate than for the CoNiCrAlY. The changing stress levels in the top coat, which for this case of a massive substrate are largely dependent on the mismatch in properties between the substrate and the top coat itself, are very similar to those with the CoNiCrAlY bond coat. It is clear that for this system there would be a much greater danger of debonding occurring between top coat and bond coat (or within the top coat close to the interface) than at the substrate/bond coat interface. Note also that the NiCrAlY is in residual compression at room temperature, whereas the CoNiCrAlY is in residual tension. There is thus a danger of through-thickness cracks forming in the CoNiCrAlY, particularly since the stress levels are high, whereas this is not expected to occur with the NiCrAlY bond coat.

6. CONCLUSIONS

The following conclusions may be drawn from this work.

1. Creep data have been obtained at up to 850°C for VPS CoNiCrAlY and NiCrAlY bond coat materials. Measured steady state creep rates conform to power law expressions, with similar activation energies (~350 kJ mole^{-1}) for the two materials but different stress exponents of, respectively, about 2.9 and 4.5. The CoNiCrAlY thus creeps faster than the NiCrAlY at low stresses, but the reverse is true at high stresses.
2. Expansivity values have been obtained at up to 700°C. Measured strain–temperature plots conformed well to second order polynomial expressions. The expansivity of the CoNiCrAlY is consistently higher than that of the NiCrAlY.
3. Comparisons between measured and modelled curvature histories during deposition onto thin substrates allowed estimation of the quenching stress values for CoNiCrAlY and NiCrAlY. These were found to be about 80 MPa and 10 MPa respectively. This difference has been explained in terms of the relative values of the expansivity, creep parameters and modulus.
4. Using these measured properties, predictions have been presented from a numerical process model of the changing stress levels and corresponding strain energy release rates for interfacial debonding during spraying and subsequent thermal cycling of TBC systems. It is shown that significant stress relaxation within the bond coat is expected to occur during a brief hold at 900°C. Differential thermal contraction stresses are high for the CoNiCrAlY case and a typical cycle of slow heating followed by relatively rapid cooling has the effect of enhancing the stress level in the bond coat on returning to room temperature.
5. For both bond coats, spallation is more likely between top coat and bond coat than between bond coat and substrate. This is largely a consequence of the much lower fracture energy (toughness) of the top coat/bond coat interface, which is probably progressively degraded by thermal growth of the oxide in this region.

However, the calculations presented here do indicate that spallation between bond coat and substrate is much more likely for CoNiCrAlY than for NiCrAlY.

ACKNOWLEDGEMENTS

Financial support for this work at Cambridge is being provided by the EPSRC and Rolls-Royce. The authors are grateful for a number of useful discussions with Dr A. Bennett and Dr P. Morell, of Rolls-Royce, Derby.

REFERENCES

I. A. BENNETT, F. C. TORIZ and A. B. THAKKER: 'A Philosophy for Thermal Barrier Coating Design and Its Corroboration by 10000 h Service Experience on RB2JJ Nozzle Guide Vanes', *Surf. Coat. Technol.*, 1987, **32**, 359–375.
2. W. J. BRINDLEY and R. A. MILLER: 'TBCs for Better Engine Efficiency', *Adv. Mater. Processes*, 1989, **136**, 29–33.
3. R. A. MILLER: 'Current Status of Thermal Barrier Coatings - An Overview', *Surf. Coat. Technol.*, 1987, **30**, 1–11.
4. J. R. BRANDON, and R. TAYLOR: 'Phase Stability of Zirconia-Based Thermal Barrier Coatings Part I: Zirconia–Yttria Alloys', *Surf. Coat. Technol.*, 1991, **46**, 75–90.
5. W. J. BRINDLEY and J. D. WHITTENBERGER: 'Stress Relaxation of Low Pressure Plasma-Sprayed NiCrAlY Alloys', *Mater. Sci. Eng.*, 1993, **A163**, 33–41.
6. K. RAHAT, R. R. FLOYD and H. REITER: 'The Effect of Residual Stress on the Formation of Cracks in Plasma Sprayed Zirconia Thermal Barrier Coatings', *12th International Conference on Thermal Spraying*, The Welding Institute, 1989, Paper 98.
7. P. SCARDI, M. LEONI and L. BERTARMINI: 'Influence of Phase Stability on the Residual Stress in Partially Stabilised Zirconia TBC Produced by Plasma Spray', *Surf. Coat . Technol.*, 1995, **76/77**, 106–112.
8. M. ALAYA, G. GRATHWOHL and J. MUSIL: 'A Comparison of Thermal Cycling and Oxidation Behaviour of Graded and Duplex ZrO_2-Thermal Barrier Coatings', *3rd Int. Symp. on Structural and Functional Gradient Materials*, B. Ilschner ed., PPUR, 1994, 405–412.
9. G. J. CHENG and Z. Y. PING: 'Application of Graded Ceramic Coatings for Thermal Barriers', *Surf. Coat. Technol.*, 1994, **63**, 93–96.
10. F. JAMARANI, M. KOROTKIN, R. V. LANG, M. F. GUELLETTE, K. L. YAN, R. W. BERTRAM and V. R. PARAMESWARAN: 'Compositionlly Graded Thermal Barrier Coatings for High Temnperature Aero Gas Turbine Components', *Surf. Coat. Technol.*, 1992, **54/55**, 58–63.
11. R. RAJENDRAN, V. S. RAJA, R. SIVAKUMAR and R. S. SRINIVASA: 'Reduction of Interconnected Porosity in Zirconia-Based Thermal Barrier Coating',

Surf. Coat. Technol., 1995, **73**, 98–200.

12. H. L. TSAI and P. C. TSAI: 'Performance of Laser-Glazed Plasma Sprayed (ZrO_2–12wt.% Y_2O_3)/(Ni–22wt.%Cr–l0wt.%Al–1wt.%Y) Thermal Barrier Coatintgs in Cyclic Oxidation Tests', *Surf. Coat. Technol.*, 1995, **71**, 53–59.
13. B. C. WU, C. H. CHAO, E. CHANG and T. C. CHANG: 'Effects of Bond Coat Pre-alumninising Treatment on the Properties of ZrO_2-8wt.% Y_2O_3/Co-29Cr-6Al-1Y Thermal Barrier Coatings', *Mater. Sci. Eng.*, 1990, **A124**, 2l5–221.
14. P. SAHOO and R. RAGHURAMAN: 'Gator-Gard Applied Bond Coats for Thermal Barrier Coatings', *National Thermal Spray Conference*, C. C. Berndt and T. F. Bernecki eds, ASM International, Materials Park, Ohio, 1993, 369–374.
15. W. LIH, E. CHANG, B. C. WU and C. H. CHAO: 'Effects of Bond Coat Pre-oxidation on the Properties of ZrO_2-8wt.% Y_2O_3/Ni-22Cr-10Al-1Y Thermal Barrier Coatings', *Oxid. Met.*, 1991, **36**, 221–238.
16. J. THORNTON, N. RYAN and G. STOCKS: 'The Production of Stresses in Thermal Barrier Coating Systems by High Temperature Oxidation', *National Thermal Spray Conferenc*, C. C. Berndt and S. Sampath eds, ASM International, Materials Park, Ohio, 1994, 633–638.
17. Y. C. TSUI and T. W. CLYNE: 'Adhesion of Thermal Barrier Coating Systems and Incorporation of an Oxidation Barrier Layer', *Thermal Spray: Practical Solutions for Engineering Problems*, C. C. Berndt ed., ASM International, Materials Park, Ohio, 1996, 275–284.
18. J. A. G. FURNESS and T. W. CLYNE: 'The Application of Scanning Laser Extensometry to Explore Thermal Cycling Creep of Metal Matrix Composites', *Mater. Sci. Eng.*, 1991, **A141**, 199–207.
19. S. C. GILL and T. W. CLYNE: 'Thermomechanical Modelling of the Development of Residual Stress during Thermal Spraying', *2nd Plasma Technik Symposium*, H. Eschenauer, P. Huber, A. R. Nicoll and S. Sandmeier eds, Plasma Technik, 1991, **3**, 227–238.
20. S. C. GILL and T. W. CLYNE: 'Investigation of Residual Stress Generation during Thermal Spraying by Continuous Curvature Measurement', *Thin Solid Films*, 1994, **250**, 172–180.
21. Y. C. TSUI, S. C. GILL and T. W. CLYNE: 'Simulation of the Effect of Creep on Stress Fields During Vacuum Plasma Spraying onto Titanium Substrates', *Surf. Coat. Technol.*, 1994, **64**, 61–68.
22. T. W. CLYNE and S. C. GILL: 'Residual Stresses in Thermally Sprayed Coatings and their Effect on Interfacial Adhesion – A Review of Recent Work, *J. Thermal Spray Technol.*, 1996.
23. S. KURODA and T. W. CLYNE: 'The Quenching Stress in Thermally Sprayed Coatings', *Thin Solid Films*, 1991, **200**, 49–66.
24. S. KURODA, T. FUKUSHIMA and S. KITAHARA: 'Quenching Stress in Plasma Sprayed Coating and Its Correlation with the Deposit Microstructure', *J Thermal Spray Technol.*, 1995, **4**, 75–84.
25. Y. C. TSUI, J. A. THOMPSON, R. C. REED and T. W. CLYNE: 'On the Change

in Stress State Associated With Bond Coat Oxidation During Isothermnal Heat Treatment of a Thermal Barrier Coating System', *Nat. Ther. Spray Conf.*, C.C. Berndt ed., ASM International, Materials Park, Ohio, 1997.

The Spallation of Porous Oxides from Plasma-Sprayed NiCrAlY Overlay Coatings

P. NIRANATLUMPONG, C. B. PONTON and H. E. EVANS
School of Metallurgy and Materials, The University of Birmingham, Birmingham B15 2TT

ABSTRACT

Thermobalance oxidation tests have been performed on plasma-sprayed NiCrAlY coatings in air at 1100°C for exposure periods up to 1000 hours. After an initial transient period, a protective alumina layer develops on the coating surface and parabolic growth kinetics obtain. Throughout the oxidation period, porous Ni-rich oxides exist on the outer surface of the alumina layer but their thickness and the extent of porosity increase after long periods as the gravimetric oxidation kinetics depart from parabolic. The critical temperature drop to initiate the spallation of these porous oxides has been measured for various exposure periods and found to decrease as the oxide-layer thickness increased. Such spallation tended to be initiated above protuberances in the underlying coating and was thought to be due to the propagation of a crack-like void in the Ni-rich oxide as a result of tensile stresses developed across the oxide/metal interface during cooling. Assuming this normal stress decreases linearly to zero through the thickness of the oxide layers, it has been shown that the predicted values of the critical temperature drop to initiate spallation agree reasonably well with those found experimentally.

1. INTRODUCTION

Plasma-sprayed coatings are produced by introducing powder particles into an ionised gas or plasma stream which simultaneously melts or partially melts the particles and transports them onto the substrate forming a coating that is lamellar in nature and having a convoluted outer surface. Such so-called overlay coatings are applied to protect the component from oxidation through forming a continuous protective oxide layer to limit oxygen reaction with the substrate alloy. This allows the component to operate at higher efficiencies, e.g. at higher temperatures or for a greater number of thermal cycles. For high-temperature oxidising environments, e.g. in gas turbines, an MCrAlY-type coating, where M is Ni and/or Co, is usually used. For operation under oxidising conditions, typical overlay coatings contain sufficient aluminium to form an alumina protective layer.

A problem with such coatings is that their aluminium content can become exhausted after a period of time due to the growth of the alumina layer and to interdiffusion with the alloy substrate; in such circumstances, the ability to retain protective conditions becomes limited. A manifestation of this general degradation in NiCrAlY-type coatings is an increasing rate of formation of Ni-rich, porous oxides at the outer oxide surface and, when chemical failure arises, the growth of Ni-rich oxides beneath the alumina layer. In extreme cases, the original alumina layer can cease to exist and truly non-protective conditions will develop. Prior to this stage of chemical breakdown, i.e. after shorter exposure periods or lower-temperature exposures, a porous oxide layer can form above the alumina layer but protective oxidation can still exist. The purpose of the present work is to characterise and, if possible, model the spallation behaviour of such porous oxides in recognition that their release from the coating, whilst not necessarily affecting oxidation kinetics, may produce downstream erosion and increase the dust burden within closed systems.

Results have already been reported[1] for the spallation of Ni-rich oxides after exposure at 1200°C after a sufficiently long exposure period completely to degrade the alumina layer. Under these conditions, the porous oxide was formed directly on the overlay coating and spallation was initiated by the growth of void-like defects adjacent to coating protuberances. Defect growth was caused by the tensile stress generated across the curved oxide/coating interface during cooling. The present study extends this work to tests performed at 1100°C for which the alumina layer has remained intact and the porous Ni-rich oxide is remote from the surface of the coating.

2. SAMPLE PREPARATION AND EXPERIMENTAL PROCEDURE

Nimonic 80A, of nominal composition Ni20Cr2.25Ti1.4Al, was used as a substrate as it is representative of many component materials and is also readily obtainable in wrought sheet form. The oxidation tests were performed on coupons measuring approximately 10 x 20 x 0.5 mm, which were cut from the as-received alloy strip.

Before plasma-spraying, all surfaces of these coupons were grit blasted using 250 μm alumina grit in order to create a sharp peaked topography and a roughness average of at least 4–5 μm, essential to the adhesion of the coating.[2] An overlay coating material, of composition Ni25Cr6Al0.4Y (wt.%), commercially known as Amdry 963, with an average particle diameter of 50 μm, was air plasma-sprayed onto these coupons using recommended spraying parameters[3] to achieve a thickness of about 100 μm.

The as-sprayed samples were then oxidised in air at 1100°C for periods of 50 to 1000 hours in the thermobalances used in the earlier study.[1] The mass of the specimen was measured frequently both during the isothermal oxidation and during subsequent cooling. The critical temperature drop required to initiate spallation could, in principle, be determined as a function of the mass gain of the specimen at the start of cooling but care has to be taken to allow for buoyancy effects on the mass change during cooling and also for random errors in the recorded values of specimen mass. The magnitude of the buoyancy change was determined by monitoring the apparent mass change during cooling of an alumina sample of similar geometry and dimensions as used in the oxidation tests. Correction factors were, thus, generated which, when applied to an oxidised sample, which did not suffer spallation during cooling, resulted in no nett change of mass over the temperature transient. It follows, of course, that over the time period for cooling, the rate of change of specimen mass dW/dt = 0 in the absence of spallation. A negative departure from this condition was used in this work as a criterion for the detection of a spallation event but, in order to ensure that such departures were statistically significant, it was necessary to adopt a 3-standard deviations significance level on the data generated by the balance. Examples of the results obtained are given in he subsequent section.

After oxidation, the specimens were Ni-plated before sectioning to prevent any further damage to the oxide layers. Microstructures of the specimens were observed in both planar and cross-sectional directions using a JEOL 5410 scanning electron microscope.

3. RESULTS

3.1 Oxidation Kinetics and Oxide Morphology

Gravimetric oxidation kinetics for sample A200 (i.e. series A oxidised for 200 hours at 1100°C) are shown in Fig. 1. In the initial stages, transient oxidation rates were rapid until a continuous alumina layer developed; this regime, described as Stage 2 in Fig. 1, commenced within the first 10 hours of exposure. During this stage, the oxidation kinetics were parabolic as shown by the fitted square symbols in Fig. 1. Significant departures from parabolic behaviour occurred after about 150 hours and this third stage is shown in Fig. 1 (diamond symbols) as a region of linear kinetics, i.e. a constant oxidation rate. The trends shown for this particular specimen were typical of most of those tested.

The departure from parabolic kinetics was associated with the formation of Ni-rich oxides above the alumina layer as shown in Fig. 2 at an early stage. The measured thicknesses of these oxides and of the alumina layer are shown as a function of exposure time at 1100°C in Fig. 3. The trend of increasing thickness with time of the outer

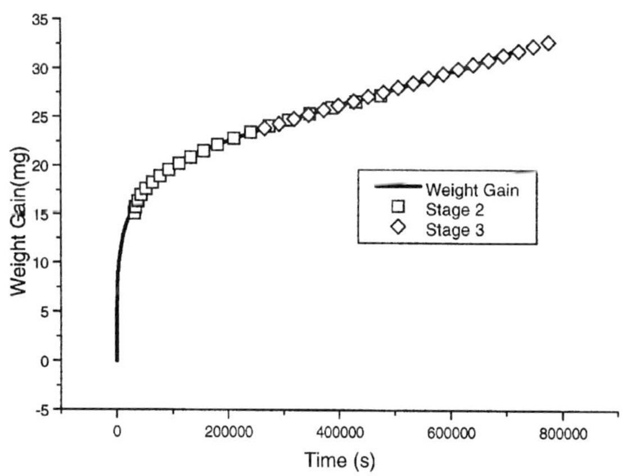

Fig. 1 Mass gain of specimen oxidised for 200 hours at 1100°C. Stage 2 has been fitted to parabolic kinetics but a constant oxidation rate applies in Stage 3.

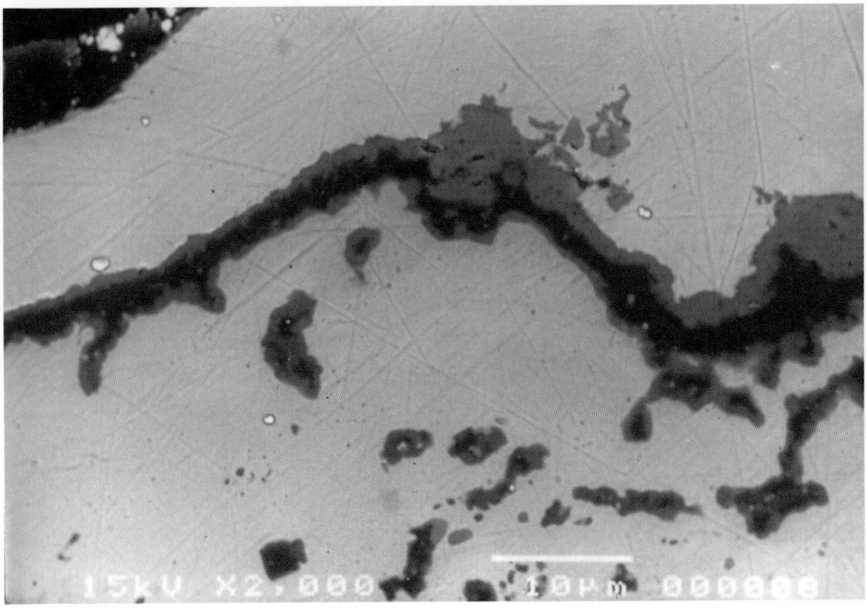

Fig. 2 SEM micrograph of sample oxidised for 100 hours at 1100°C

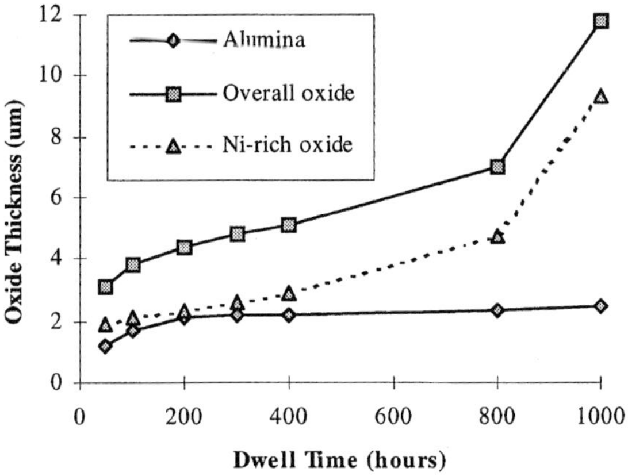

Fig. 3 Oxide thicknesses, measured using the SEM, as a function of dwell time.

oxide is evident but it is also clear that substantial amounts also exist after the initial stage of transient oxidation, i.e. prior to the formation of the alumina layer. For all exposure periods, there was considerable porosity within the outer oxides but the mean of the maximum pore sizes found on a single metallographic section increased approximately linearly with time as shown in Fig. 4. Similar observations on the development of porous oxides with increasing exposure time have also been reported[4] recently for the oxidation of the Ni-base alloy, MA956.

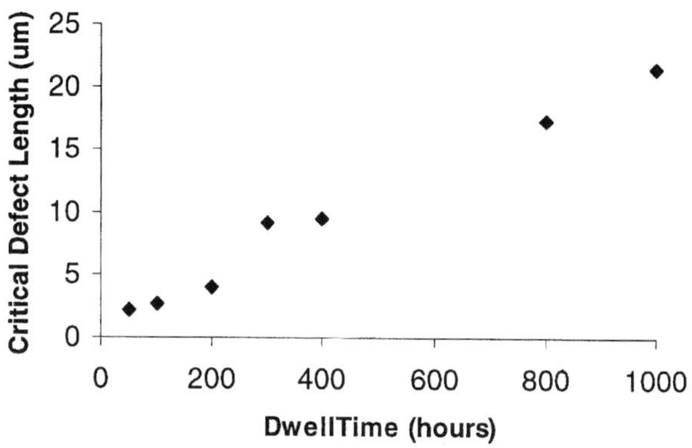

Fig. 4 Increase in the mean-maximum void length within the Ni-rich oxides with dwell time.

It is also clear from Fig. 3 that, whereas the Ni-rich oxide thickness tended to increase with exposure time, the growth rate of the alumina layer decreased. The overall growth kinetics of the alumina layer were sub-parabolic and, indeed, no significant increase in thickness was measured over the last 700 hours of exposure. Prior to this, parabolic kinetics obtained (as implied in Fig. 2) with a constant rate, k_p, of 1.77 x 10^{-17} m^2 s^{-1}. This value of the rate constant is broadly similar to those reported[5,6] for conventional alumina-forming alloys at 1100°C. Associated with the decline in the alumina growth rate was an increase in the quantity of Ni-rich oxides formed *beneath* the alumina layer. This is a classic indication of chemical instability of the alumina, i.e. chemical failure, and is likely to have resulted from the depletion of the aluminium content of the coating; such depletion can also account for the reduction in growth rate of the alumina layer in the longer term (Fig. 3). After long exposure times, the alumina layer was found to be absent in some areas, as shown in Fig. 5, and represented an advanced stage of chemical failure.

3.2 Oxide Spallation

Representative thermogravimetric data, showing the variation of specimen mass during cooling, are given in Fig. 6 for sample A100 (100 hours exposure at 1100°C) and in Fig. 7 for sample A800 (800 hours exposure). All such data sets were corrected for buoyancy effects as described in section 2 and the rate of change of mass calculated throughout the cooling period. These processed data are shown in Figs 8 and 9, respectively, for the above two samples together with the upper and lower 3-standard deviation lines. It can now be appreciated that no spallation was detected for sample A100 and that the decrease in mass during cooling, shown for this specimen in Fig. 6, was due solely to buoyancy changes. By contrast, specimen A800 did show spallation, as indicated in Fig. 9, at a critical temperature drop, ΔT_c, of 234°C. In fact, all specimens other than A100 spalled during cooling. The measured values of ΔT_c are shown in Fig. 10 as a function of oxidation time. The decrease in the critical temperature drop

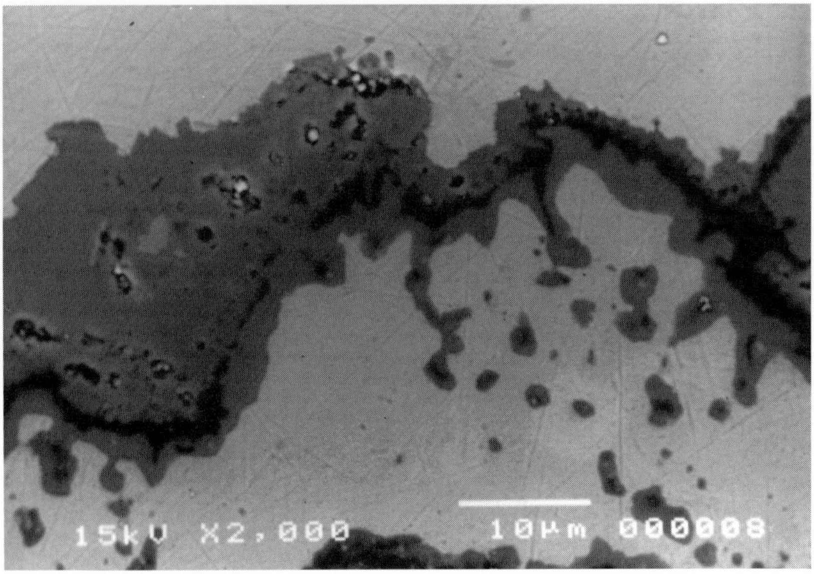

Fig. 5 The oxide morphology after 800 h oxidation at 1100°C.

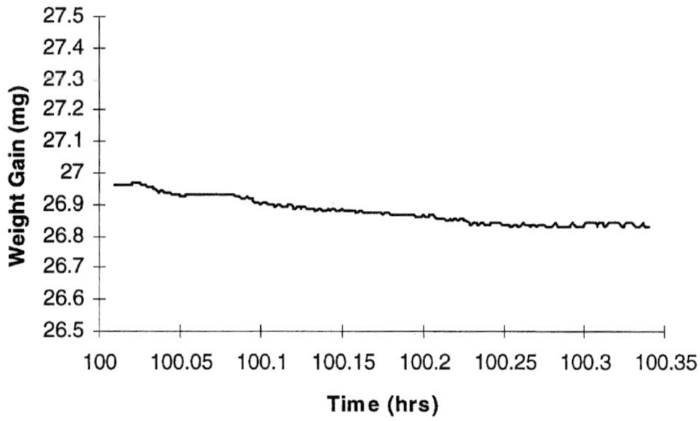

Fig. 6 Mass change during cooling of sample A100 after oxidation for 100 h at 1100°C.

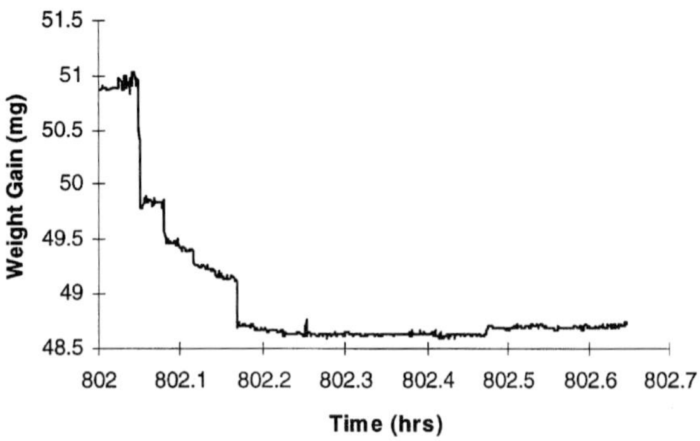

Fig. 7 Mass change during cooling of sample A800 after oxidation for 800 h at 1100°C.

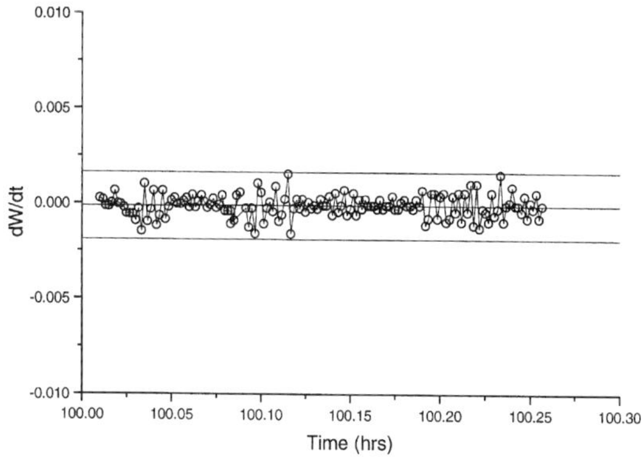

Fig. 8 First differentiation of the cooling data of sample A100.

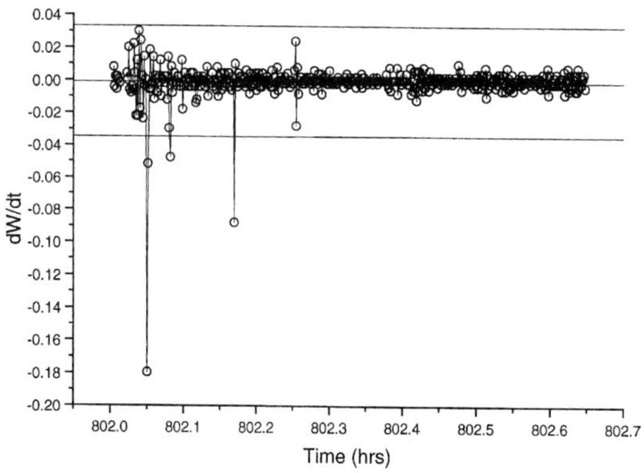

Fig. 9 First differentiation of the cooling data of sample A800.

to cause spallation as exposure time increases is clearly shown. The relationship of ΔT_c to oxide thickness will be discussed below.

SEM examination of the oxidised samples showed that the spallation sites tended to lie above protuberances in the underlying coating. Residual oxide was found in these spalled areas, indicating that cracking had taken place within the oxide layer rather than at the oxide/coating interface. A typical example of the partial propagation of such a crack is given in Fig. 11 and shows that the crack path lay within the porous Ni-rich oxide.

4. DISCUSSION

The present observations that the crack which initiates spallation propagated within the porous, Ni-rich oxide and that spallation tended to occur above protuberances of the underlying plasma-sprayed coating confirm earlier observations by Strawbridge *et al.*[1] on the same coating system oxidised at 1200°C. In their case, chemical break-

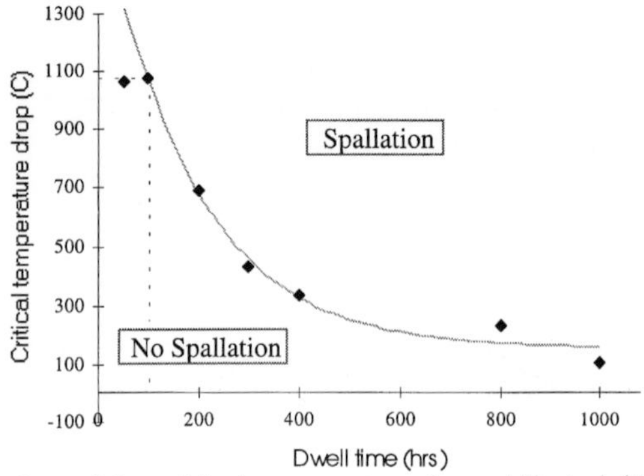

Fig. 10 The variation of the critical temperature drop, ΔT_c, to initiate spallation with dwell time at 1100°C.

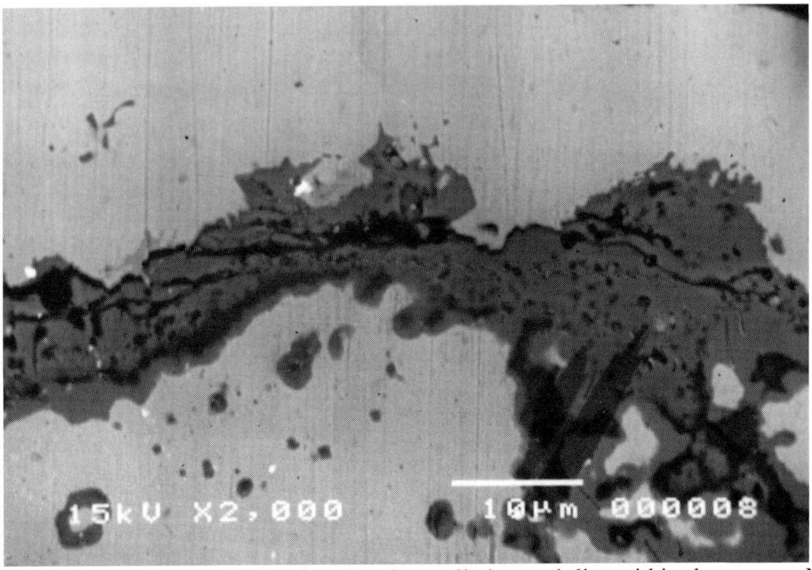

Fig. 11 SEM micrograph showing that the spallation path lies within the porous, Ni-rich oxide.

down of the alumina layer had occurred early during exposure and porous oxides existed directly at the coating interface for much of the exposure period. It was recognised that differential thermal strains during cooling would generate tensile stresses across the oxide/metal interface in regions of convex curvature (Fig. 12a) and that, in principle, these could propagate large pre-existing voids within the oxide (Fig. 12b) leading to oxide spallation (Fig. 12c). Assuming that a large void existed very close to the oxide/coating interface and that its growth was determined by elastic fracture mechanics (i.e. it propagated within the oxide layer rather than in contact with the metallic coating of low creep strength), Strawbridge was able to predict the critical temperature drop to initiate spallation as:

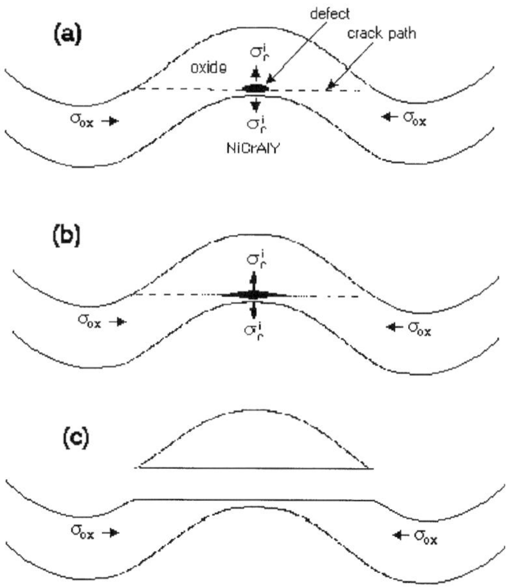

Fig. 12 Spallation of a porous oxide scale, after Strawbridge *et al.*[1]

$$\Delta T_c = \frac{K_{IC}(1-\nu)R}{2E\sqrt{\pi c}\left(\alpha_m - \alpha_{ox}\right)\xi} \tag{1}$$

Here, R is the radius of curvature of the coating protuberance, c is the half-length of the embedded crack-like defect within the oxide, ξ is the total oxide thickness, ν is the Poisson's ratio and E the Young's modulus of the oxide, α_m and α_{ox} are, respectively, the thermal expansion coefficients of metal and oxide (N.B. $\alpha_m > \alpha_{ox}$) and K_{Ic} is the critical stress intensity factor for the oxide. Using values of the elastic and fracture parameters appropriate to NiO together with measured values of R and ξ, Strawbridge *et al.* were able to obtain reasonable agreement with their observed measurements of ΔT_c. Their basic model will also be used to describe the present results but modification is required to account for the dense alumina layer adjacent to the coating and the consequent displacement of voids away from the oxide/coating interface.

The postulated spallation mechanism applying in the present tests is shown schematically in Fig. 12. It is envisaged that a continuous, void-free alumina layer exists over the coating surface and that the porous oxide has formed on the gas side of the alumina; this is representative of most of the samples tested. In order for the large void shown near the interface with the alumina layer, but at a distance a from the coating interface, to propagate, the local tensile stress, σ_a, must equate to $K_{IC}/(\pi c)^{0.5}$ where, again, c is the half-length of the void. Whereas in Strawbridge's approach, it was envisaged that the critical void lay very close to the oxide/coating interface so that rigorous calculations of the oxide/coating interfacial stress could be used, the displacement of voids from that interface in the present tests makes calculation of the local stress σ_a much less exact. In this first approximate approach to the problem it is simply assumed that the tensile stress decreases linearly from its maximum value at the oxide/coating interface to zero at the outer surface of the Ni-rich oxide. Thus, following Strawbridge *et al.*[1] the tensile stress, σ_i, across the oxide/coating interface is taken as:

$$\sigma_i = \frac{2E\Delta T\left(\alpha_m - \alpha_{ox}\right)\xi}{(1-\nu)\,R} \tag{2}$$

where the other parameters have been defined previously. Noting that in the present approximation, the local normal stress, σ_a, varies with distance a from the oxide/metal interface as:

$$\sigma_a = \sigma_i\left[1 - \frac{a}{\xi}\right] \tag{3}$$

it can be shown that the critical temperature drop for propagation of a void of half-length c located at a distance a from the oxide/coating interface is:

$$\Delta T_c = \frac{K_{IC}\,(1-\nu)\,R}{2E\sqrt{\pi c}\left(\alpha_m - \alpha_{ox}\right)\left(\xi - a\right)} \tag{4}$$

This equation is identical to that given by Strawbridge (eqn 1) except that the term (ξ-a) replaces ξ.

It should be noted that approximations other than the assumed linear variation of normal stress are implicit in this derivation. In particular, it is assumed that the duplex oxide layer can be characterised by single values of the thermal expansion coefficient, α_{ox}, the elastic constants ν and E and the critical stress intensity factor, K_{IC}. It would be remarkable if this were the case but, as will be shown, useful progress can be made by sensible selection. Thus, since the maximum normal tensile stress will be located at the alumina/coating interface, it is reasonable to use values of the elastic constants and expansion coefficient for alumina in its evaluation. On the other hand, a value of K_{IC} appropriate to the Ni-rich oxide should be used because crack propagation occurred within this oxide. On this basis, calculations were performed using values of E_{NiO} =

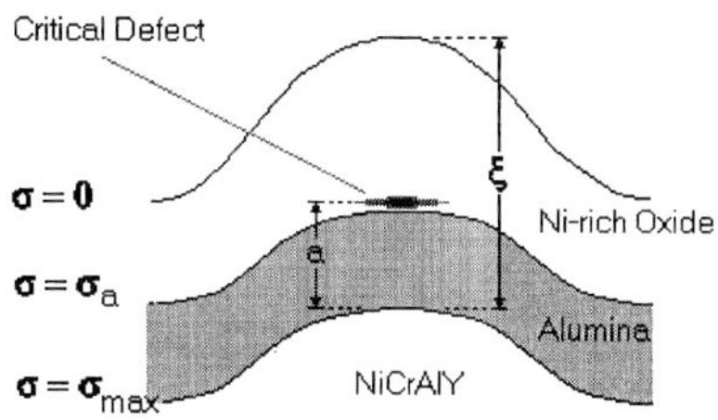

Fig. 13 In the presence of a dense alumina layer adjacent to the coating, a void of critical length, existing within the Ni-rich oxide, will lie at a distance a from the oxide/coating interface. The out-of-plane tensile stress will be at a maximum at this interface but is assumed to decrease linearly to zero at the oxide/gas interface.

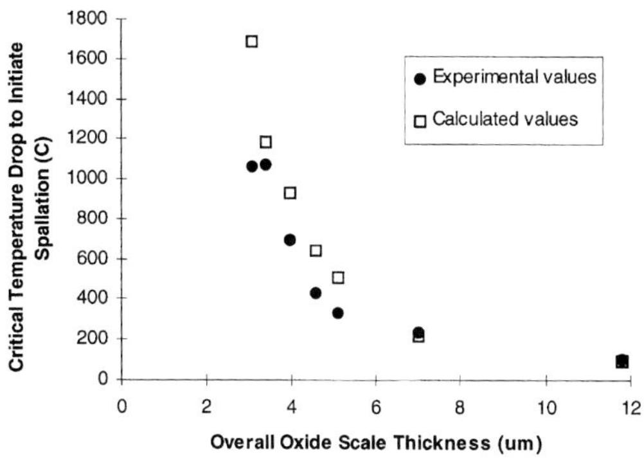

Fig. 14 A comparison of the predicted with the observed values of the critical temperature drop to initiate spallation as a function of total oxide thickness.

1.9 x 10^5 MNm^{-2}, ν_{NiO} = 0.3, $\alpha_{alumina}$ = 7.7 x 10^{-6} K^{-1}, K_{IC} = 1 $MNm^{-3/2}$ [7] and $\alpha_{NiCrAlY}$ = 15.5 x 10^{-6} K^{-1}.[8] The calculational procedure entailed evaluating the critical temperature drop, ΔT_c, for each protuberance of the overlay coating visible in the metallographic section. Not only did this require measurement of its radius, *R*, but also determination of the half-length, *c*, of the largest void above the protuberance and of its distance, *a*, from the coating surface (see Fig. 12). By this method, some 20–30 estimates of ΔT_c were obtained for a single sample and their average was taken as the best estimate.

A comparison of these computed values with those obtained from the thermobalance (TGA) tests is provided in Fig. 13 against the total oxide thickness. There is remarkably close numerical agreement between theory and experiment but, in view of the various approximations made in the model, this is probably fortuitous. Of more significance is that the observed rapid decline in ΔT_c with increasing oxide thickness is predicted, indicating that the basic concepts within the model are sound. In particular, it can be appreciated from eqn 4 that this drop in ΔT_c is due not only to the increasing oxide thickness, ξ, but also to an increase with oxidation time of the void length, *c*, as shown in Fig. 14.

5. CONCLUSIONS

A continuous alumina layer had formed on the present NiCrAlY coatings after approximately 10 hours exposure in air at 1100°C. Its subsequent growth for a further 100 hours or so was parabolic but this growth rate declined over longer exposure periods. This was associated with a depletion of aluminium within the coating, also manifested by the more general formation of porous Ni-rich oxides. The maximum void length within these porous oxides increased with exposure period.

Spallation of these Ni-rich oxides occurred during cooling at a critical temperature drop which decreased rapidly with increasing total oxide thickness. Such oxide loss tended to occur above protuberances in the underlying overlay coating.

A simple approximate model has been developed which recognises that out-of-plane tensile stresses developed above the convex geometry of a coating protuberance can produce spallation of the Ni-rich oxide by propagating a void of critical size. The model agrees well with the experimental observations and shows that the strong dependence of the critical temperature drop to initiate spallation on oxide thickness results not only from the presence of a thicker oxide but also from the larger voids formed within the thick oxide.

REFERENCES

1. A. STRAWBRIDGE, H. E. EVANS and C. B. PONTON: *Mater. Sci. Forum*, in press.
2. B. J. GILL and R. C. TUCKER Jr: *Mater. Sci. Tech.*, 1986, **2**, 207.
3. Plasma-Technik AG, in *Fl Gun, Plasma Spray Parameters*, Switzerland, 1990, 57 and 90.
4. J. P. WILBER, M. J. BENNETT and J. R. NICHOLLS, 3rd Int. Conf. on Microscopy of Oxidation, in press.
5. K. M. N. PRASANNA, A. S. KHANNA, RAMESH CHANDRA and W. J. QUADAKKERS: *Oxidation of Metals*, 1996, **46**(5–6), 465.
6. A. ANDOH, S. TANIGUSHI and T. SHIBATA: *Oxidation of Metals*, 1996, **46**(5–6), 481.
7. M. M. NAGL, S. R. J. SAUNDERS, W. T. EVANS and D. J. HALL: *Corros. Sci.*, 1993, **35**, 965.
8. W. J. BRINDLEY: to be published in *Mater. Sci. Eng.*, 1996.

Test Methods for Evaluating the Mechanical Properties of Coatings

J. P. BANKS, S. OSGERBY and S. R. J. SAUNDERS
Centre for Materials Measurement and Technology, National Physical Laboratory, Teddington, Middlesex TW11 0LW

ABSTRACT

Procedures have been developed to measure ductile–brittle transition temperature, adhesion and thermo-mechanical fatigue behaviour of coatings. Ductile–brittle transition temperature is measured by application of tensile strain to the coating at a series of temperatures: coating ductility may be derived from direct observation of the extent and morphology of cracking or by acoustic emission monitoring. Three methods of adhesion measurement (scratch testing, indentation and four point bend tests) have been investigated. Each test method is capable of producing a quantitative measure for coating adhesion, however a different parameter is derived from each test and correlation of these parameters is not yet possible. The development of a thermo-mechanical fatigue testing machine is well advanced. Control of temperature and load cycles has been established and an extensometer that will allow strain control has been designed.

INTRODUCTION

Most gas turbine components must be optimised for bulk and surface performance to provide the resistance to deformation and surface degradation experienced in an operating engine. In the past this has meant that a compromise between the two, often conflicting, requirements was made. However with the rapid development of coating technology it is common to optimise bulk and surface performance of the component by, for example, coating a creep resistant alloy for use as an HP turbine blade. The usual approach in the first stage of coating development is to consider only corrosion resistance, but there is the recognition that the mechanical performance of the coating is equally important. Clearly, no matter how corrosion resistant the coating, if mechanical failure occurs by, for example, cracking or spalling, the usefulness of the coating will be severely compromised. The nature of coatings is such that it is only possible to measure properties *in situ*, and this presents an added challenge to the materials scientist. The research described in this paper discusses developments in the evaluation of three important properties of coatings: ductile brittle transition temperature (DBTT), adhesion and thermal mechanical fatigue (TMF).

DUCTILE BRITTLE TRANSITION TEMPERATURE

Uniaxial, flexural and small punch[1] loading methods may be used to impose tensile strain on the samples at temperatures of interest up to about 1000°C. There is no intrinsic advantage of using any particular mode of loading and the choice relies upon the geometry of test samples and availability of equipment. However care must be taken to ensure that premature failure of the coating is not induced by the specimen design or at extensometer attachment points. Coating cracking is normally determined by direct observation of the sample after straining to a given level if a simple pass/fail test is sufficient. Examples of the successful use of this test method are shown in Fig. 1 which shows micrographs of a platinum aluminide coating strained to 2% at 600°C (brittle) and 800°C (ductile – no cracks); the DBTT was between 750°C and 800°C.

It has been found that strain to 2% is sufficient to distinguish between ductile and brittle behaviour, but alternatively higher strains can be imposed and then the morphology of the cracks is used to distinguish between ductile and brittle failure.

Alternatively the acoustic emission (AE) signals during deformation can be monitored to detect the onset of cracking if more quantitative information is required. The samples are heated in a furnace with waveguides attached to monitor the acoustic response. Using state of the art acoustic emission (AE) equipment it is possible to locate the AE events and thus avoid confusion from noisy grips etc which are not part of the coating being assessed. This is particularly helpful when high temperature experiments are being carried out and ceramic components are required for the test jig. Care must also be taken to ensure that the AE signals result from cracking events in the coating and are not the result of the substrate itself also showing brittle behaviour as may be the case for some blading alloys. This can easily be checked by straining a blank (uncoated) sample and observing the AE signals produced. When the substrate itself shows brittle behaviour in the temperature of interest a convenient strategy is to apply the coating to a ductile substrate. In this case, of course, only overlay coatings can be assessed and not coatings where interdiffusion with the substrate occurs.

ADHESION

Adhesion is perhaps the most difficult property to measure, and, when values are quoted, interpretation may be controversial. Many existing methods of measuring adhesion may not determine adhesion of the coating to the substrate, but some other property, for example cohesive strength of the coating or fracture stress. Thus the overriding objective of this work has been to develop procedures that will give quantitative information which is unique to the interface. Furthermore, the tests being developed must be relatively easy to perform so that they can be carried out at low cost to industry. This usually invokes the irony that if it is easy to do it is difficult to interpret! The methods currently being investigated viz. scratch, four point bending and indentation tests are no exception. However in each case the analysis procedures have been simplified as much as possible to aid potential users in industry. Generally there is no adhesion problem for diffusion coatings so that methods are only being developed for overlay systems.

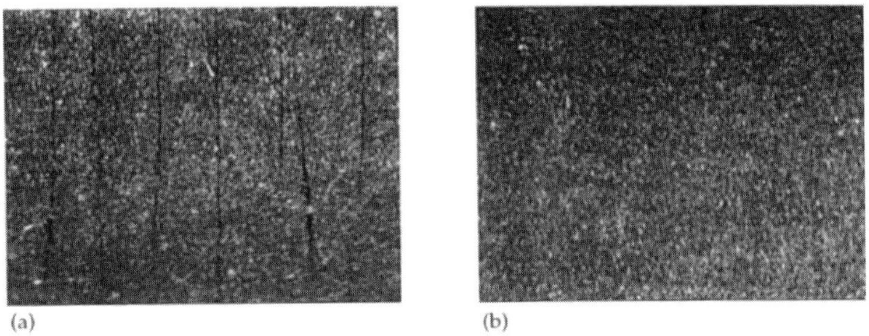

Fig. 1 Optical micrographs of platinum alumidie coating after straining to 2% at (a) 600°C and (b) 800°C.

Scratch Testing

The yardstick for the work is the scratch test which is already widely used in industry, particularly with thin (< 10 μm) thick ceramic coatings. In the case of gas turbine coatings this has been used by the authors' laboratory for the assessment of novel PVD SiAlON coatings. Here ranking of different types of coating was made on the usual basis of the applied normal load required to induce chipping of the coating from the side of the scratch. Concerns about this test are; the lack of standardisation in interpretation and in monitoring the results; and the behaviour of the Rockwell C diamond stylus used in the test. The critical load, as derived by monitoring acoustic emission or friction force measurements (with or without observation of the scratch) is not a material parameter and can be very misleading. Furthermore, indenter wear and geometry greatly affect results, so that it is necessary to monitor these parameters. This can be achieved by performing tests in a standard material, and proposals to produce certified reference materials are now being considered.

Quantitative analysis of the scratch test is in its infancy. The most appropriate analysis is that of Bull,[2] who has related the critical load during scratching to the interfacial fracture energy and the wedge fracture stress of the coating. This analysis relies upon the intercorrelation of the critical load in the scratch test, coating thickness and residual stress. In the original analysis, on $A1_2O_3$:MA956 and TiN:stainless steel systems, it was suggested that a series of tests at constant coating thickness but differing residual stress should be carried out to calibrate the scratch test for imposed stress. This should then be followed by a series of tests, at constant residual stress but varying coating thickness, to derive the interfacial fracture energy. This test matrix may not be possible for coatings of interest for gas turbine components, e.g. it may not be possible to vary residual stress in the coating independently of thickness without changing the processing conditions which may in turn influence the interfacial fracture energy.

A simplified analysis route has been derived where the two steps involved in the Bull analysis are combined to produce a single analytical expression relating residual stress, σ_R, critical load, L_c, and coating thickness, t:

$$\sigma_R = a - bL_c + ct^{-\frac{1}{2}}$$

The coefficients, *a*, *b* and *c*, can be derived easily using commercial curve fitting or spreadsheet software. The interfacial fracture energy, γ_F, can then be calculated from the expression:

$$\gamma_F = \frac{c^2(1-\nu_c)}{E_c}$$

where ν_c and E_c are, respectively, the Poisson's ratio and Young's modulus of the coating.

Using this approach it is possible to calculate the interfacial fracture energy of the coated system from as few as four tests. It is good practice however to carry out more tests than this in order to reduce the influence of measurement scatter from individual points on the result.

Scratch tests have been carried out on MCrAlY/SiAlN coatings on a γ-TiAl substrate and a typical example is shown in Fig. 2. It can be seen that failure has been initiated at the coating/substrate interface and that the critical load can be measured. Insufficient coatings were available to carry out the full quantitative analysis and in this case only the semi-quantitative measure of critical load is possible.

Indentation Tests

Encouraging progress has been made with indentation. For some years an international standard that allows qualitative assessment of adhesion using a Rockwell C indenter has been available.[3] Dependent upon the interfacial strength, radial or circumferential cracking of the coating will occur and, in the limit, delamination is observed. The type and extent of the cracking/delamination is then correlated with interfacial strength. Recently this test has been further developed[4] to allow calculation of the interfacial toughness from the relative size of indentation and the extent of coating delamination. The original analysis required the input of a large number of other input variables and an iterative numerical routine as one of the steps in the calculation. Little can be done to reduce the number of input variables required but it has been possible to generate an analytical expression that provides a good fit to the results of the numerical iteration over the range of values for the variables examined.[5]

Figure 3 shows an indent produced using a Rockwell indenter into a PVD MCrAlY coating where delamination around the indent can be clearly seen; in this case the interfacial toughness has a value of 0.68 J m^{-2}. A direct comparison of scratch testing and indentation procedures to measure coating adhesion is shown in Fig. 4. The results for a series of MCrAlY coatings produced using processing conditions are com-

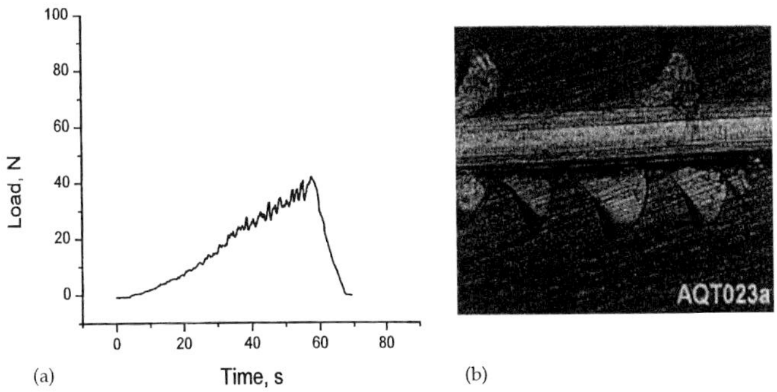

Fig. 2 Scratch testing of MCrAlY/SiAlN coating on γTiAl substrate (a) loading trace, (b) image of scratch test.

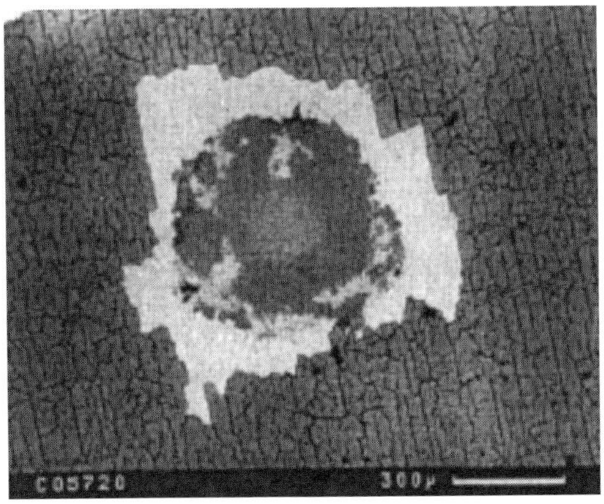

Fig. 3 Backscattered electron micrograph of Rockwell indentation into a PVD MCrAlY coating on a Mar M002 substrate showing delamination of the coating.

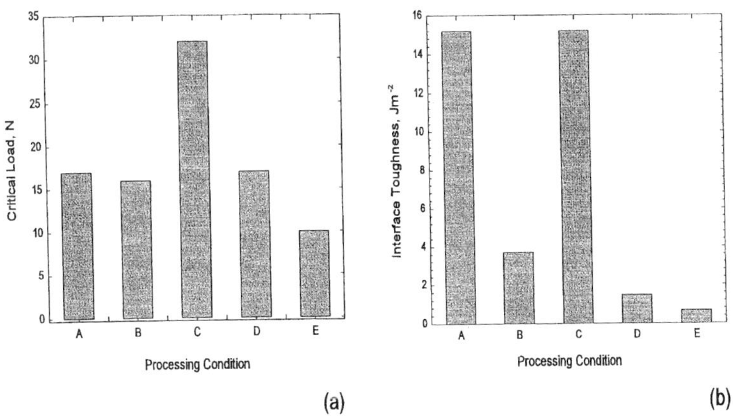

Fig. 4 Comparison of adhesion measurements using (a) scratch testing, (b) indentation.

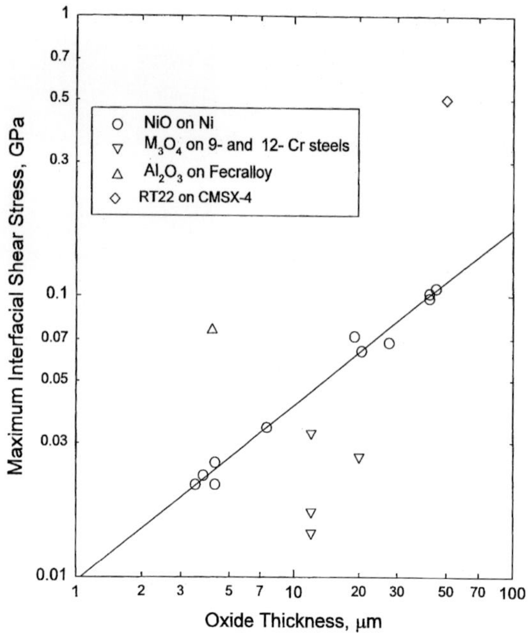

Fig. 5 Adhesion measurements using flexural testing.

pared with the critical load measured during scratch testing of identical specimens. It is clearly demonstrated that the trend in measured adhesion using the two methods is broadly similar with processing condition C producing the highest measure of adhesion and condition E producing the lowest in both cases. The minor discrepancies between the measurements may be due to uncertainties in the input values for some parameters in the indentation model eg no value for residual stress in the coating was available so a constant zero value was used.

Four Point Bend Tests

Use of four-point bending in tension causes cracking of the coating, and as the strain is increased, spallation of the coating is induced. At this point the crack spacing becomes a minimum and this value is used to calculate the shear strength of the interface. Various models have been developed in the literature analysing this phenomenon and these have been compared[6] in a recent review. Differences of up to two orders of magnitude were found in the calculated value of interfacial shear strength. Analysis of these data indicates that results based on a shear lag model are most likely to be correct.

Figure 5 shows results from the four-point bend test for several thermally-grown oxide–substrate combinations and for a RT22 coating on a CMSX-4 substrate. The interfacial stress that can be supported by the coated system is up to 5 times greater

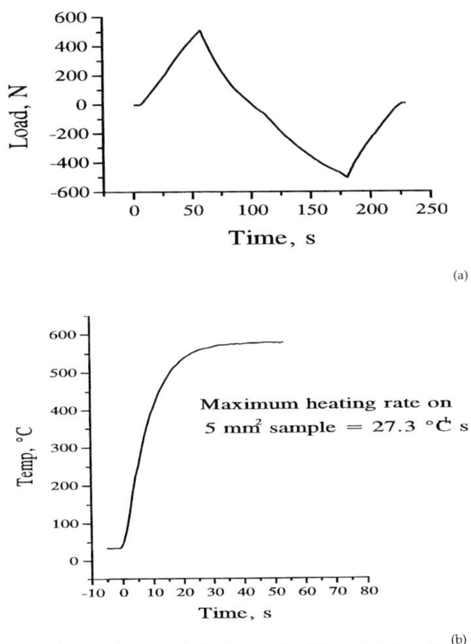

Fig. 6 Control of thermal-mechanical fatigue facility (a) load cycle, (b) temperature cycle.

than that in a thermally grown oxide system. This measurement illustrates the good adhesion of commercial coatings and demonstrates the applicability of the test method at least to the ranking of combinations of coating and substrate.

A more complex model is now being developed. Software available at NPL[7] allows 'virtual testing' to be carried out, and this has been applied to the shear lag model. This intriguing development allows rapid assessment of interfacial shear strength as a function of changes in substrate and/or coating properties.

Overview of Adhesion Tests

The three tests described above have all been shown to generate quantitative measures of coating adhesion. However, each test derives a different parameter for the coating adhesion and it is not yet possible to determine the precise relationship between these parameters. Moreover, currently available models[8] for spallation require yet another different parameter for interfacial properties ie interfacial tensile fracture stress. Further work on measurement methods to derive this parameter or to relate the other adhesion parameters to this quantity is urgently required.

THERMAL MECHANICAL FATIGUE

Thermal mechanical fatigue resistance is a difficult property to measure and there are only a few laboratories able to impose realistic strain and temperature cycles. The objective of the current work is to develop a simplified procedure by modifying an instrument designed at NPL – a miniature thermomechanical test rig which can simultaneously impose and monitor rapid thermal and strain or loading cycles (Fig. 6). The test specimens are simple beams, typically 3 x 2 x 30 mm and are heated directly by passage of a large electrical current. The advantage of the method is that it allows examination of very small samples and thus greatly reduces the costs of testing. Potential disadvantages are the small gauge length and nonuniform temperature distribution which pose problems for the measurement and control of strain during the test. This concern is being addressed through the development of spring loaded extensometer arms, incorporating capacitance transducers which are currently under development. Control of the equipment is enabled by LABVIEW© software.

TMF cracking of coatings can be detected using acoustic emission or by direct observation of the specimen during testing. However, care must again be taken to ensure that the AE signals arise from cracking events in the coating and not from any phenomena occurring in the substrate.

The test rig can be used for the measurement of a number of different mechanical, electrical and thermal properties over a wide temperature range (ambient – about 2000°C depending on specimen size and material).[9] Using coated superalloys a maximum heating rate of ~30°C s^{-1} has been achieved for specimens with a 5 mm^2 cross section.

ACKNOWLEDGEMENTS

The work described in this paper was carried out under the 'Materials Degradation in Agressive Environments Programme': a programme of underpinning research financed by the UK Department of Trade and Industry.

REFERENCES

1. J. CAIMITO and R RANJAN: *Mats. Sci. & Eng.*, 1994, **A183**, 121.
2. S. BULL: *Proc 4th Int. Conf. on Advances in Surface Eng.*, Newcastle upon Tyne, May 1996.
3. W. HEINKE, A. LEYLAND, A. MATTHEWS, G. BERG, C. FRIEDRICH and E. BROSZEIT: *Thin Solid Films,* 1995, **270**, 431.
4. M. D. DRORY and I. W. HUTCHINSON: *Proc. Roy. Soc. A,* 1996, **452**, 2319.
5. S. OSGERBY, S. R. J. SAUNDERS and J. P. BANKS: 'Scale and Coating Adhesion: A Procedure for Laboratory Tests', *NPL Measurement Note CMMT MN25,* January 1998.
6. C. DAMERELL: 'Assessment of Simple Models of the Cracking of Coatings Strained under Tension', *NPL Report CMMT(A)39*, September 1996.
7. N. J. McCORMICK and L. N. McCARTNEY; *Materials World,* May 1997, 257.
8. H. E. EVANS, G. P. MITCHELL, R. C. LOBB and D. R. J. OWEN: *Proc. Roy. Soc. A*, 1993, **440**, 1.
9. B. ROEBUCK and M. GEE: *Mats Sci & Eng.*, 1996, **A209**(1–2), 358.

Influence on Phase Configuration of Protective MCrAlY-Coatings Deposited by Advanced PVD Techniques and Determined by High Temperature XRD

E. LUGSCHEIDER, C. BARIMANI and C. SIRY
RWTH Aachen University of Technology, Materials Science Institute
'Werkstoffwissenschaften, Templergraben 55, D–52056 Aachen, Germany

ABSTRACT

High temperature Ni-base materials require a reliable protection by MCrAlY coatings to prevent detremental oxidation failure at high temperatures. Advanced physical vapour deposition (PVD) produces quality coatings of high density in the as-coated condition. PVD deposition technologies offer a wide range of parameter to vary the growth condition and to influence the microstructure. The deposition conditions of the universal Magnetron-sputter-ion-plating-PVD process is introduced and correlated to its characteristic phase distribution. Also the properties of the deposited coatings are compared by looking at two differently produced targets to determine a reliable and suitable target production process concerning coating quality and cost-performance ratio. The influence of possible gas contamination incorporated during the target production process is described and compared, especially the influence on phase formation during deposition.

The achieved phase distribution of characteristic analysis of PVD MCrAlY coatings is measured by X-ray diffraction in the 'as-coated' condition and compared to the condition after annealing procedures within high temperature X-ray diffraction. Additionally, the measurement facilities allow measurement coatings at high temperature during the annealing procedure and hence enable detection of on-line crystallisation effects.

1. INTRODUCTION

A variety of processes may be used to deposit corrosion and oxidation resistant coatings. The increasing importance of the physical vapour deposition techniques can also be established for protective coatings that are applied under high temperature conditions. Several PVD techniques can deposit high quality MCrAlY coatings. The processing parameters allow controlled deposition of coatings with dense homogeneous compositions. All of these techniques have a special profile which can influence the coatings. The complex MCrAlY alloys require special attention to be payed to the melting point of the containing elements.[1] Due to its suitable and advantageous nature the magnetron sputter ion plating (MSIP) PVD technique was selected and investigated. One of the most important key questions is the operational behaviour of the sputter-target (cathode) during the sputtering process. The results presented here compare two types of targets. Besides conventionally sintered HIP targets HVOF-sprayed MCrAlY targets have been produced and tested for their application in the MSIP-PVD process. A thermal spray process technology like HVOF has been investigated as to whether it may offer a new way of producing sputtering targets.

Moreover, the characteristic morphology of sputtered MCrAlY-coatings is described. In this paper recent results on their characteristic microstructure are presented also containing crystallisation effects during vacuum annealing.

The investigations presented herer are focussing on these two aspects: the target constitution including the potential and limits of the process when using a HVOF-target and the coatings dependence on the negative subtrate bias voltage.

2. EXPERIMENTAL

2.1 Selection of a Suitable PVD-Deposition Technique

At our laboratories three different PYD techniques have been systematically investigated for protective high temperature coatings: arc-ion plating-PVD (ALP), electron beam (EB)-PVD and magnetron sputter ion plating (MSIP) PVD. The application of these PVD techniques leads to respectively typical coating conditions like composition and structure. Principially both techniques, electron beam and arc ion plating, exploit the effect of melting the material by cathodic electrical arc and electron beam respectively during evaporation. In contradiction to this the MSIIP process enables a material transfer into the gas phase by the impingement of the process gas that allows it to avoid passing through melting state. This effect is of special advantage when depositing materials containing components with significantly different melting temperatures. The MSIP process allows the sputtering and deposition of almost every solid material and produces a composition very close to those of the target materials. This process has also the potential to deposite at a relatively low temperature, for example on heat-treated metal substrates or even extremely temperature-sensitive plastic materials. The MSLP process has been selected here for the investigations presented due to its many advantages, including defined composition and homogenity.

2.2. Target Production

The condition of the cathode target mounted inside the vacuum chamber authoritatively determines the operation behaviour and by that the restrictions for the process parameters during PVD. Besides a conventionally sintered HIP target for comparison purposes, target cathodes have been prepared by the HVOF thermal spray procedure. The background for these investigations is the promising coating quality of the HVOF-sprayed MCrAlY-coatings. The main features of these, in-house experiments, are the relatively low porosity and contamination of the sprayed coatings[2] that makes this way of target production attractive. A very important fact is the potential to achieve a sufficiently low oxygen content and coating thickness. Typical thicknesses of sinter

Fig. 1 HVOF sprayed MCrAlY target for application within the MISP process (target diameter 75 mm on copper cathode holder).

and melt metallurgical targets used in the sputter facility are in the range 5–6 mm. The trials were performed directly onto the copper target holder system of the applied sputter facility. The successfully achieved thickness of the NiCoCrAlY targets were 1 mm. From that, a basic evaluation of the potential of target production potential by HVOF spraying could be performed. The production of the sprayed targets has been carried out with a Sulzer MetCo Diamond Jet Hybride HVOF facility. The HVOF-sprayed sputter target is depicted in Fig. 1.

2.3 Deposition

Austenitic steel, Ni-base and cemented carbide THM08 substrates have been prepared by a five step grinding and 6 μm-diamond polishing process followed by ultrasonic cleaning in isopropanol. Before the deposition, of the coating, an *in situ* r.f. ion etching cleaning procedure has been performed.

The investigations were carried out with a Leybold Hereaus Z400 sputter facility using Argon as the process gas. The negative substrate bias voltage has been varied from 0 V to -150 V, with the d.c. power beening kept at 18.24W cm^{-2}. The deposition temperature was fixed at 300°C. The keyplan of the sputter process is given in Fig. 2.

2.4 Evaluation

The coatings were evaluated by metallographic investigations. One aim is to describe the typical structure of the MSIP-MCrAlY coatings after deposition. Rupture structures have have been generated applying cemented carbide samples. Moreover, the phase formation of these coatings during a defined annealing procedure has been investigated by high temperature X-ray diffraction. The substrate material for these investigations was Ni-base MAR M002. The measurements were performed with a Seifert X3000 diffractometer using Cobalt Kα radiation ($\lambda = 0.17889$ nm) and a Fe-filter in

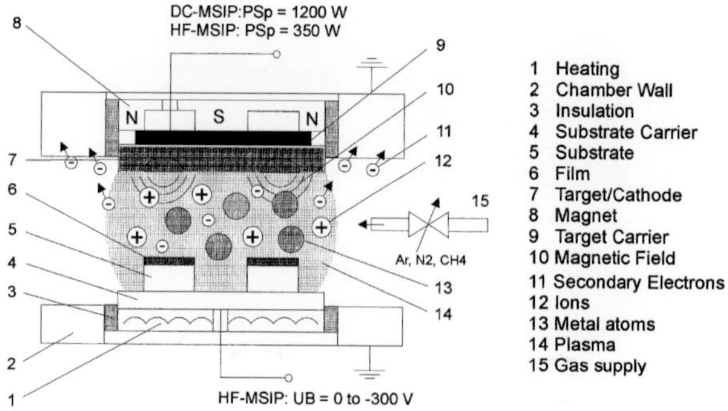

Fig. 2 Key plan of the 'magnetron sputter ion plating' (MSIP-PVD) process.

the 1:1 Omega/theta mode for a theta angle range from 10° to 70°. This system is equipped with a Buehler/Otto high temperature vessel HDK 1.4. The measurements were performed on samples under vacuum. The remaining base pressure was kept below 1 x 10^{-3} Pa. Temperature measurement was performed applying a Pt/RhPt thermoelement (thermocouple?) spotwelded on the rear side of the Ta heater and sample holder element. Diffraction patterns were recorded at 30°C, 500°C, 850°C and 1200°C. Before each recording a holding sequence of 30 minutes has been projected. The heating and measurement procedure selected is shown in Fig. 3. The hardness has been tested with a Vickers Microhardness equipment. To confirm these results and collect information on the Youngs modulus of PVD MCrAlY coatings, cross checking investigations have been performed with a indentation XP nanoindenter of nanoinstruments at a depth of 800 nm using a Berkovich diamond tip.

3. RESULTS AND DISCUSSION

3.1 Target Operation

Beside the application of a conventionally Sinter/HIP-produced target, a HVOF-sprayed MCrAlY-target was tested. The d.c. power was kept constant for all tests. The trials with the sintered HIP target were performed at a standard process gas pressure level of 1 Pa. The according plasma current amounts to 1.569 mA at a line drop of 510 V.

After mounting this target the sputtering process could be started using the d.c. deposition mode. During the run-in of the newly sprayed target, in the d.c. deposition mode, the change in the deposition parameters had to be established. When reducing the argon process gas pressure level to a typical level, the behaviour during the sputtering process was restricted. During the run-in tests the argon pressure level during operation, with this target, could not be below 4 Pa. By that, the plasma conditions are shifted towards a higher voltage that has been determined by the on-line operation condition control. The characteristic of the current-voltage dependencies detected for the HVOF-sprayed target are given in Table 1 in comparison to the status when using the sintered HIP target.

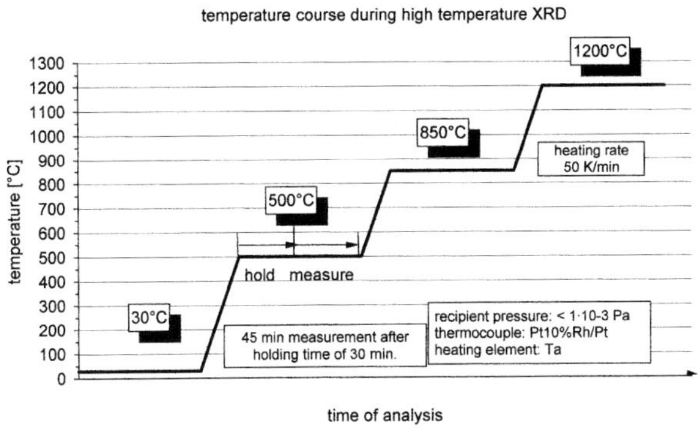

Fig. 3 Temperature course during high temperature X-ray diffraction.

Table 1 Current voltage characteristics during the run-in operation of the NiCoCrAlY sputter target produced by HVOF spraying.

Ar pressure (Pa)	10	8	6	4
Current (mA)	595	650	798	1085
Voltage (V)	1272	1163	982	677

The targets after operation already demonstrate the formation of the circular sputter traces due to the effects of the magnetron. The for comparison used targets are depicted in Fig. 4. The sinter/HIP target (left) with a origin thickness of 6 mm is given as a typical target at a largely exploited state, in comparison to the thinner HVOF-sprayed target on the (right). Due to the circular material carry-off the overruns of the single sprayed layer become visible and can be detected due to its darker oxigen enriched top zone formed on top of each layer immediately after being sprayed.

3.2 Coating Morphology

In all cases the MSIP-coatings are of a dense homogeneous quality. SEM micrographs show the homogenity of the coating and the surface of sputtered MCrAlY coatings that have been deposited to a thickness of 8 μm. Figure 5 shows a coating deposited from the sintered HIP MCrAlY target. The graph is produced with a tilt angle of 30° to obtain a view of the coating's surface. The coating deposited from the HVOF-sprayed target is depicted in Fig. 6 showing the cross section of a rupture. Also the deposites of the sprayed target fullfill the requirements of high homogenity and low porosity to ensure successful protection.

The coating composition was measured by electron micro probe analysis. Generally, the results for the sputtered coatings show a high correlation to the reference powder composition. Coatings deposited from the sinter/HJiP target contain amounts of oxygen in the range of 0.01 wt%, the lower limit of measurement resolution. Coat-

Fig. 4 Operated MCrAlY sputter targets (left) sintered HIP and (right) HVOF sprayed in comparison (target diameter Ø 75 mm).

ings deposited from a HVOF-sprayed target yield a higher amount of oxygen, in the range of 1.1 wt%, and increases by that two orders of magnitude. An examplarily composition is given in Table 2. For both coatings deposited from the sintered HIP target and the HVOF-sprayed target very closely meet the metal element distribution of the source material.

Table 2 Composition of PVD coatings deposited from differently produced sputter target determined by EMPA.

Electron microprobe analysis (wt%)	Ni	Co	Cr	Al	Y	O
Powder reference	47.8	21.7	17.2	12.6	0.65	<0.01
Coating deposited from sintered HIP target at 300° C, -80V substrate bias	46.8	22.1	17.9	11.9	0.52	0.017
Coating deposited from HVOF sprayed target at 300° C, -80V substrate bias	46.7	21.5	17.4	11.5	0.65	1.13

3.3 Phase Formation

The measured phase of sputtered coatings in the 'as-deposited' state yield diffraction patterns with broad reflections of low intensity. In general this kind of reflection is known for a low order X- ray amorphous state of materials and that is typical for rapidly cooled materials and vapour deposites.[3,4]

The main result from the X-ray diffraction measurements confirmed that sputtering

Fig. 5 Rupture structure of a sputtered MCrAlY PVD coating deposited on cemented carbide from a sintered HIP target (tilt angle 30°C).

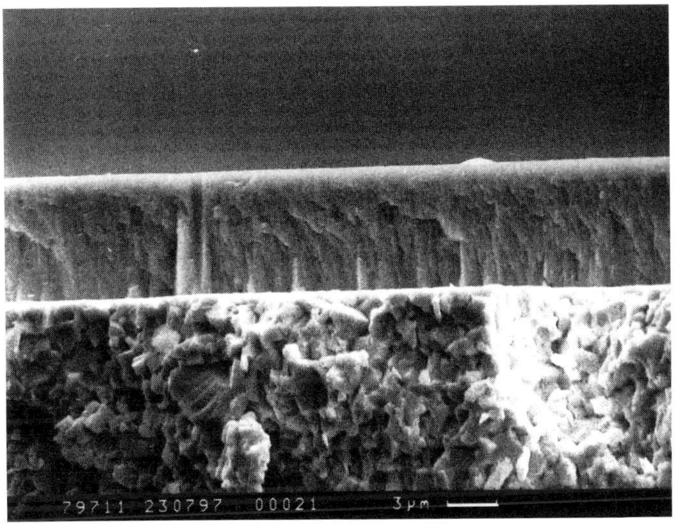

Fig. 6 SEM picture of a MCrAlY PVD coating deposited from HVOF sprayed target.

of the pure MCrAlY material, without additional influences on the growth, resulted in a broad and less sharp reflection of the sputtered MCrAlY close to Nickel (111). The phase formation during the annealing procedure generally led to an increase of the crystalline order that is established by the intensity and sharpness of the reflections. All *in-situ* measurements at higher temperatures are influenced by thermally induced lattice expansion. The widened lattice structures result in a generally reduced inflexion that led to a peak shift towards lower angles. The main reference lattice reflections like Ni, Co, Ni_3Al are marked in the patterns for information. In some cases a final interpretation allocation is limited due to the overlapping effects of the different phases and characteristic growth properties of magnetron sputtered coatings.

The diffraction patterns of a MCrAlY-coating deposited on a substrate at 0 V substrate bias, and investigated during a four-step vacuum annealing procedure are given in Fig. 7. Crystallisation effects have occurred primarily before recording the 850°C pattern and have become increasingly sharp and intense with the 1200°C pattern. The increased bias voltage during deposition tends to reduce this reflection in favour of some smaller peaks at higher theta values. In the 'as-coated' condition a small but intense peak appears that corresponds to the γ (200) (NiCo) reflection, and respectively to the Ni_3Al γ' phase. The patterns given in the Figs 8 and 9 for coatings deposited under -50 V and -80 V respectively shows the phase formation measured at 30°. At the higher bias voltage of -150 V an additional broad reflection around theta = 45.44° can be observed that can be interpreted as the γ-(NiCo) phase in the (220) orientiation, whereas the (111) reflection almost disappears as shown in Fig. 10. A comparison of the patterns recorded at 30°C is shown in Fig. 11. This confirms the influence of the bias voltages effect on the preferential crystal structure already in the as-deposited condition. The increasing impact energy of the argon ion impingement causes an increased degree of crystallisation.

The dependance of the substrate bias on crystalline order after this annealing procedure results in a MCrAlY strong reflection of the main peak allocated at the (111) orientation as well as the (222) reflection at 62.1° being increasingly reduced in favour of the (200) and (220) reflections at 30° and 45.6°. The corresponding patterns of coatings deposited at different bias voltages after crystallisation are given in Fig. 12.

The coatings deposited using the HVOF-sprayed target besides showing the low order reflection close to (NiCo) (111), already in the as deposited state, also show a characteristically strong line at 30.5° corresponding to d = 1.77. The reflection cannot be defined by an oxide or nitride phase. The high incorporation of oxygen that had been established by high resolution EPMA strongly supports the formation of the γ-(200) reflection. The annealing procedure results in a crystallisation that especially forms the (200) reflection an example is given in the patterns depicted in Fig. 13, which gives no indication for oxide phase formation.

3.4 Microhardness as Coated and After Annealing Procedure

Basic mechanical evaluation, performed with a Vickers Microhardness equipment, shows a high hardness value for all the investigated sputtered MCrAlY coatings. The level of hardness has been found to be around 800 HY 0.05. The Vickers hardness results show neither a significant dependence on substrate bias voltage nor on the differently produced targets with its deviating deposition condition. These results show that the substrate bias voltage, causing the argon ion impact on top of the growing films, has no significant potential to increase the hardness of the sputtered MCrAlY coatings, as it has been found for example for ultrahard light element coatings in the B–N–C triangle.[5]

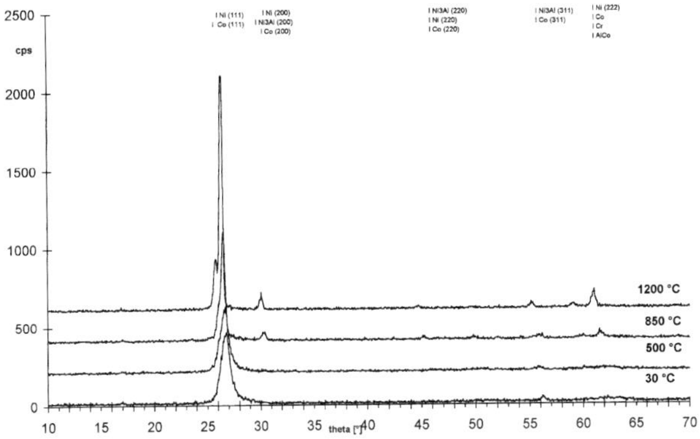

Fig. 7 Diffraction pattern for a 8 μm thick NiCoCrAlY coating at T = 30°C, 500°C, 850°C and 1200°C sputtered at 0 V substrate bias from a sintered HIP target.

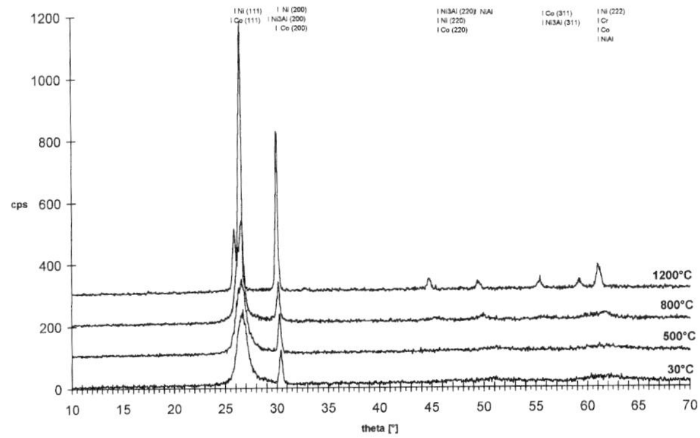

Fig. 8 Diffraction pattern for a 8 μm thick NiCoCrAlY coating at T = 30°C, 500°C, 850°C and 1200°C sputtered at -50 V substrate bias from a sintered HIP target.

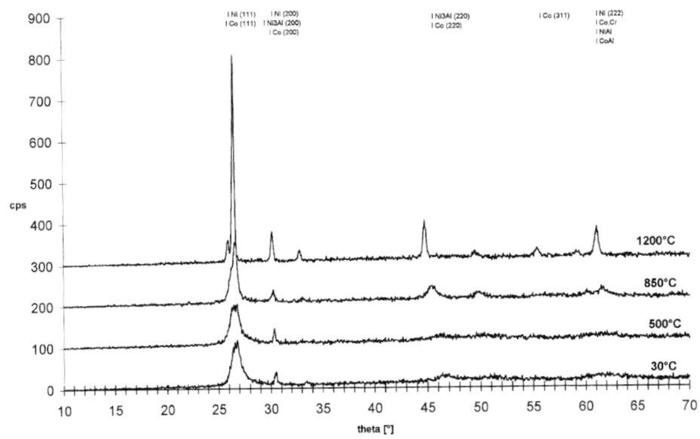

Fig. 9 Diffraction pattern for a 8 μm thick NiCoCrAlY coating at T = 30°C, 500°C, 850°C and 1200°C sputtered at -80 V substrate bias from a sintered HIP target.

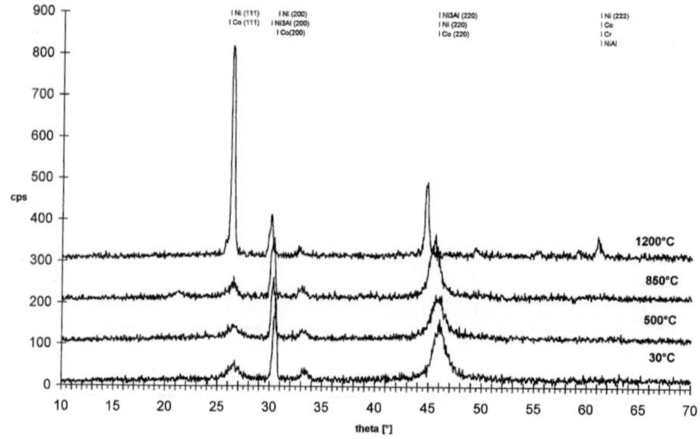

Fig. 10 Diffraction pattern for a 8 μm thick NiCoCrAlY coating at T = 30°C, 500°C, 850°C and 1200°C sputtered at -150 V substrate bias from a sintered HIP target.

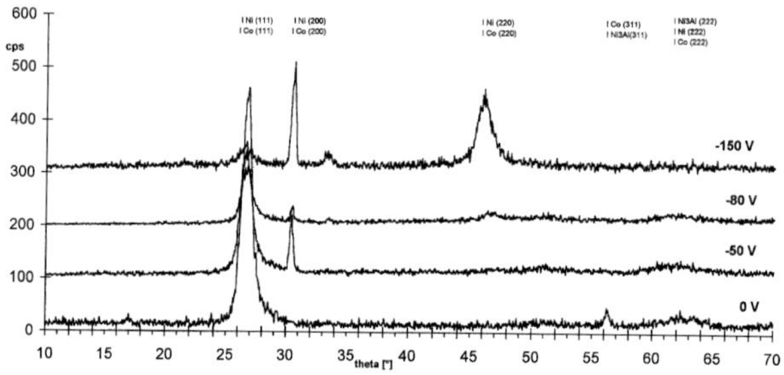

Fig. 11 Diffraction pattern of NiCoCrAlY coatings at T = 30°C sputtered at 0 V, -50°C, -80°C and -150°C substrate bias from a sintered HIP target.

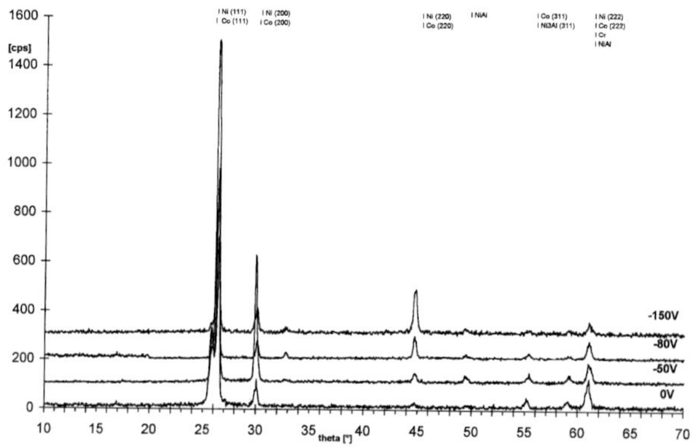

Fig. 12 Diffraction pattern of NiCoCrAlY coatings at T = 1200°C sputtered at 0 V, -50°C, -80°C and -150°C substrate bias from a sintered HIP target.

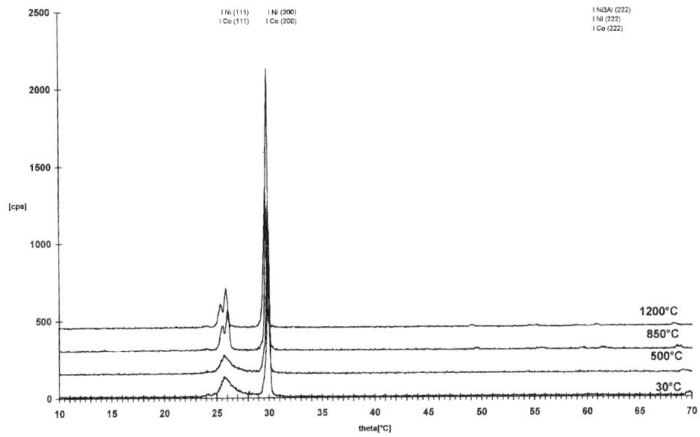

Fig. 13 Diffraction pattern for a 8 μm thick NiCoCrAlY coating at T = 30°C, 500°C, 850°C and 1200°C sputtered at -80 V substrate bias from a HVOF sprayed target.

Hardness measurements performed after the annealing procedure showed that in all cases a reduction in hardness to a level around 550 HV0.05 was observed. This significant effect coincides with the phase formation and can be followed back to the mechanism releasing a state of distortion. Usually this effect of distortion in the 'as-coated' state but not the different phase formation is due to the varied bias substrate voltage, but does cause differences in hardness. It can be concluded the bias does clearly influence the growth condition of the as-coated condition and after annealing but does not significantly influence the basic coating Vickers hardness values. Figure 14 shows the details for coatings deposited from the sintered HIP-MCrAlY target applying different substrate bias voltages. The 'as-coated' condition is compared to that after the vacuum annealing procedure as given previously. The high hardness values for the as-coated condition could also be found for the coatings that are sputtered from the HYOF-sprayed target.

To confirm the high hardness of the metal coatings in the as-coated condition, and to determine a modulus of elasticity of these coatings, nanoindentation tests have been performed. To establish data and determine the range of Young's modulus for MCrAlY PVD-deposited coatings typical data is given in Table 3. These result also show the high hardness for all the investigated MCrAlY coatings and confirm that the conclusion of hardening by distortion is characteristic of the 'as-coated' condition.

4. SUMMARY

Magnetron sputter ion plating (MSIP-PVD) permits the deposition of very homogenous MCrAlY coatings that are dense and non-porous. Coatings deposited from a sintered HIP target tend to have characteristic broad mono-reflection in the 'as-coated' condition indicating a low crystalline order after the deposition. Changes in the substrate bias voltage not only result in shifts of the main reflections but also shows broadening

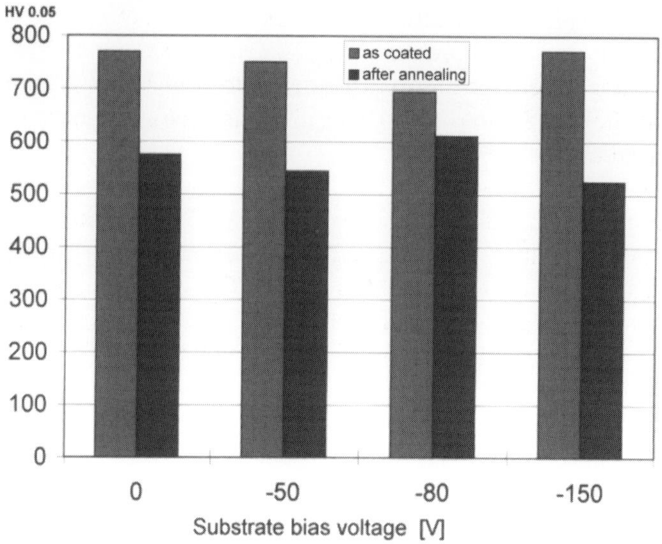

Fig. 14 Microhardness of NiCoCrAlY coatings deposited by sputtering from the sintered HIP target applying different substrate bias voltages, measured in the as-coated condition and after annealing.

Table 3 Nanoindentation hardness and determination of Youmgs modulus (XP Nanoindenter, depth 800 nm, Berkovich tip).

Coating	Hardness Nanoindenter (GPa)	Youngs modulus (GPa)
NiCoCrAlY sputtered coating Target sintered HIP -80V bias	11.70	216
NiCoCrAlY sputtered coating Target HVOF sprayed -80V bias	11.30	215

of the low order refications. By means of high-temperature X-ray investigations the crystallisation during annealing could be established. The dependance of substrate bias on crystalline order after this annealing procedure results in the (Ni,Co) γ–phase in the (111) orientation as well as the (222) orientation being reduced in favour of the (200) and (220) reflections. Significant steps could be detected *in situ* during crystallisation annealing up to 1200°C.

Microhardness could determined at a high level of about 800 HV0.05 in the as-coated condition. After the performed annealing procedure a reduction of hardness could be observed. It can be concluded that this is caused by the release of the distortion during the annealing procedure and confirmed by the detected crystallisation.

MCrAlY targets have been produced by sitnered HIP and the HVOF Diamond Jet facility. This latter target type has been produced to a thickness of 1 mm on top of a copper cathode holder. Sputter PYD Deposition trials have been performed with this HVOF produced target, operating in the d.c. mode. The argon pressure level was reduced to about 4 Pa during run-in tests. The quality of the coatings density and homogeneties meet the requirements; whereas the oxygen contamination is increased by two orders of magnitude. Hardness and Youngs modulus are in the same range for both coatings sputtered from sintered HIP and HVOF sprayed targets.

The coatings produced by means of sputter PVD generally showed high density and homogenenity and are recommend especially for applications where low porous protective coatings are required and a defined complex composition has to be produced and observed.

5. REFERENCES

1. U. SCHULZ, K. FRITSCHER, M. PETERS and W. A. KAYSSER: 'Em Mantel für die Turbinenschaufel', *DLR-Nachrichten,* 1993, **73**(11), 27–31.
2. E. LUGSCHEIDER, C. HERBST and L. ZHAO: 'Charactensation of MCrAlY coatings sprayed by HVOF and reactive plasma spraying', *this publication*, ?–?
3. N. MATTERN, W. PISCHKE, A. DANZIG and S. DOYLE: 'X-Ray diffraction at high temperatures', *Fresenius Journal of Analytical chemistry.* 1994, **349**, 91–96.
4. N. MATTERN, W. PISCHXE, G. WEISE, D. KRAUT and H. D. BAUER: 'X-ray diffraction studies of the structure of molybdenum sulphide thin films', *Thin Solid Films,* 1994, **245**, 255–259.
5. O. KNOTEK, E. LUGSCHEIDER and C.W. SIRY: 'Tribological properties of B–C thin films deposited by magnetron-sputter-ion plating method', *Surface and Coatings Technology,* 1997, **91**,167–173.

The Effect of Creep on the Residual Stress in Vapour Deposited Thin Films and Thermally Grown Oxide Scales

S. J. BULL

Materials Division, Department of Mechanical, Materials and Manufacturing Engineering, University of Newcastle, Newcastle upon Tyne, NE1 7RU

ABSTRACT

The residual stress state of coatings or oxide scales can be critically important in dictating their performance. When the residual stress in the coating is compressive the substrate is put into tension and, as the temperature increases, it can reduce this tensile stress by creep processes which in turn will reduce the stress in the coating. A model has been developed to characterise this process by predicting how stress is generated and relaxed during coating deposition or scale growth. The output from the model has been validated by experimental measurements on vapour deposited TiN on stainless steel and alumina scales on a weak alloy (FeCrAlY) and a strong alloy (MA956) substrate and shows excellent agreement to experimental data measured by X-ray diffraction. The amount of stress relaxation by creep depends on the thickness of the substrate and coating, the time of exposure at temperature and the cooling rate as well as on the creep properties of the substrate materials.

1. INTRODUCTION

Failure of oxide scales by spallation, buckling or through-thickness cracking is controlled by the levels of stress in the scale and substrate.[1] The main stress generation and relaxation processes are listed in Table 1. Of the stress generation mechanisms thermal expansion mismatch is well understood and expressions exist for calculating the variation of thermal stress with temperature assuming that elastic behaviour is maintained.[2] Growth stresses are less well understood, but have been measured experimentally and appear to be related to the detailed mechanism of oxide growth.[3]

Table 1 Processes affecting stress generation and relaxation in the scale/substrate system.

Stress generation	Stress relaxation
Growth	Scale creep
Thermal expansion mismatch	Substrate creep
Applied stresses	Scale cracking
Component geometry	Substrate cracking
	Substrate yield
	Scale yield

Stress relaxation mechanisms have attracted much less research interest, although the effects of substrate creep have been shown to affect the measured growth stresses in some cases.[4] Although the stresses in the substrate are usually much less than the stresses in the scale, the effect of creep in the substrate is generally far more important than creep in the scale at a given temperature since, in the majority of cases the creep rates of alloy substrates are much greater than those of ceramic materials. Microfracture in the scale can also be very important if the scale is weak.

A model which predicts the variation of stress with temperature during thermal cycling has recently been developed which quantifies the generation of growth stresses during oxidation and the subsequent relaxation of both these and thermal stresses by creep in the substrate.[5] This paper describes this model and how its predictions compare with measurements of alumina scales on FeCrAlY and MA956 and demonstrates that it can be equally well applied to sputtered TiN coatings on 304 stainless steel.

THE MODEL

2.1 Modelling Approach

The model is based on the calculation of stress increments and relaxation in small time increments, *dt*. At the start of the calculation process the original stress in the scale (or coating), σ_{ox} is entered as an input parameter. This stress is then incremented by the growth and thermal stresses which would have developed in the small time period. The stress in the substrate, σ_s, which will be of opposite sign, is then calculated by a simple force balance. For a flat plate sample with a scale of equal thickness on each side so that no bending occurs we have

$$h_s\sigma_s + 2h_{ox}\sigma_{ox} = 0 \qquad (1)$$

where h_s and h_{ox} are the thickness of substrate and scale respectively. The stress in the substrate is then allowed to relax by creep in the time, *dt*. Finally the stress in the scale is recalculated using eqn (1). At temperatures where creep relaxation of the substrate is significant the stress in the scale is relaxed, but as the temperature is reduced the stress in the scale will tend to the value predicted by an elastic analysis. The process is repeated until a well defined end point is reached.

There are two main model approaches adopted to determine this time:

1. Isothermal oxidation or coating. In this approach the model is run until a well-defined time has passed. There are no thermal stresses generated, but the thickness of the scale or coating is updated in each time step and a new growth stress increment is incorporated. As the scale or coating thickness increases the stresses on the substrate, and hence the amount of stress relaxation, also increase.
2. Cooling. In this approach the scale/substrate system is allowed to cool from the oxidation or coating temperature to room temperature. The cooling rate is an input parameter which is used to determine the temperature drop that has oc-

curred within the timestep. As with the isothermal oxidation model the coating thickness and growth stress are updated at each time step if necessary and in addition the thermal stress increment is calculated.

2.2 Stress Generation Components

2.2.1 Thermal stress

Thermal stress increments, σ_{th}, are calculated using the well-known expression for the elastic stresses generated due to thermal expansion mismatch where the coating or scale may be a significant fraction of the substrate thickness:[2]

$$\sigma_{th} = \frac{\dfrac{E_{ox}(T)}{1-\nu_{ox}}\left[\alpha_{ox}(T)-\alpha_s(T)\right]dT}{1+\dfrac{h_{ox}}{h_s}\dfrac{E_{ox}(T)}{E_s(T)}\dfrac{(1-\nu_s)}{(1-\nu_{ox})}} \tag{2}$$

where E and ν are the Young's Modulus and Poisson's ratio, α the thermal expansion coefficient and dT the temperature increment. The subscripts s and ox refer to the substrate and scale (or coating) respectively and the bracketed T indicates that temperature variant mechanical and thermal properties are used where available (see later).

2.2.2 Growth stress

A constant growth stress, σ_g, is assumed in this model to be generated in all new scale or coating as it forms. However, the stress at the oxidation or deposition temperature can be relaxed by substrate creep during the process so the stress that is in other parts of the coating or scale which have not just been formed, σ_{gr}, may be somewhat reduced. The average stress in the scale, σ_{gav}, on addition of a layer of thickness, dh, is thus given by

$$\sigma_{gav} = \frac{h_{ox}\sigma_{gr} + \sigma_g dh}{h_{ox} + dh} \tag{3}$$

For a scale or coating where no stress relaxation occurs $\sigma_{gav} = \sigma_g$. In all other cases the average stress will be lower.

2.3 Stress Relaxation in the Substrate

In general the stresses in the substrate are small and tensile since the scale is thinner than the substrate and both growth stresses and thermal stresses induced on cooling a scale from the oxidation temperature are compressive (i.e. the thermal expansion coefficient of the scale is less than that of the underlying alloy). A similar situation occurs for hard coatings on stainless steel. The substrate can relieve these stresses by creep at high temperature leading to a permanent elongation.

If the creep process can be described by a power law of the form[6]

$$\frac{d\varepsilon_p}{dt} = A\exp\left(-\frac{Q}{RT}\right)\sigma^n \tag{4}$$

where A is a constant, Q is the activation energy for creep and n is the stress exponent, the stress relaxation is then given by[5]

$$\frac{1}{\sigma^{n-1}} = \frac{1}{\sigma_i^{n-1}} + A\exp\left(-\frac{Q}{RT}\right)E_s(n-1)t \tag{5}$$

where σ is the stress after relaxation and σ_i is the initial stress. Equation (5) can be used to determine the change in stress at each timestep as all the constants and the temperature are known. The same expression could be used to determine the creep relaxation of stresses in the scales or coatings. However, preliminary calculations indicated that this was much less than the relaxation in the scale due to creep in the substrate so this has been ignored in the model.

2.4 Scale and Substrate Properties

2.4.1 Scale thickness

The thickness of the oxide scale was calculated at each timestep using a parabolic growth law which was fitted to experimental data. The fitted curve used was

$$h(t) = \left(h_0 \exp\left(-\frac{Q_{ox}}{RT}\right)t\right)^{\frac{1}{2}} \tag{6}$$

where h_0 is a constant and Q_{ox} is the activation energy for scale growth. For convenience of calculation this was used in the differentiated form such that

$$h(t+dt) = h(t) + \frac{h_0 \exp\left(-\frac{Q_{ox}}{RT}\right)}{2h(t)}dt \tag{7}$$

The fitted parameters h_0 and Q_{ox} for alumina scales on FeCrAlY and MA956 are presented in Table 2. The scales were grown by isothermal oxidation of coupons in alumina boats at a range of temperatures between 1100 and 1350°C for times up to 1000h. The scale thickness was an average value determined from metallographic cross sections.

For the TiN coatings produced by sputtering a much simpler version of eqn (7) can be used because the growth rate is fixed by the sputtering parameters used. Thus

Table 2 Properties of substrate and scale used in the modelling.

Property	FeCrAlY	Alumina scale on FeCrAlY	MA 956	Alumina scale on MA 956	304 Stainless steel	TiN
Young's Modulus, E (GPa)	210	380	210	380	210	600
dE/dT, k (GPa/K)	0.09	0.056	0.09	0.056	0.09	0
Poisson's ration, ν	0.3	0.26	0.3	0.26	0.3	0.28
Thermal expansion coefficient, a (K^{-1})	14.8×10^{-6}	8×10^{-6}	14.8×10^{-6}	8×10^{-6}	16.9×10^{-6}	8.9×10^{-6}
Creep constant, A ($Pa^{-n}s^{-1}$)	5.96×10^{-27}	–	4.2×10^{-87}	–	2.1×10^{-24}	–
Creep exponent, n	5.5	–	11.0	–	4.0	–
Creep activation energy, Q (kJ/mol)	392	–	191	–	366	–
Oxide growth rate constant, h_0 (m^2/s)	–	19284	–	53174	–	–
Oxide growth activation energy, Q_{ox} (kJ/mol)	–	580	–	250	–	–
Coating rate (μm/h)	–	–	–	–	–	0.1

$$h(t + dt) = h(t) + Kdt \tag{8}$$

where K is the growth rate measured from calibration experiments for a given set of processing conditions. In this study coatings were produced by sputter ion plating at a coating rate of 0. 1 μm/h (see later). The deposition temperature was 500°C.

2.4.2 Elastic constants

The Young's Modulus and Poisson's ratio used in the calculations are presented in Table 2. The Poisson's ratio was assumed to be temperature invariant whereas the temperature variation of Young's Modulus was given by:

$$E(T) = E(T - 300K) - k(T - 300) \tag{9}$$

The room temperature elastic moduli of alumina scales and substrates were measured by Nanoindentation and were found to be very similar to previously reported values.[7] The temperature coefficients for Young's Modulus were determined by fitting to literature data.[7–9] For the TiN coatings Nanoindentation was used to measure the room temperature Young's Modulus as with the alumina scales. Again similar values were found to those reported in the literature previously.[10] No data on the temperature variation of this property was available so the Modulus was assumed to be independent of temperature.

2.4.3 Thermal expansion coefficients

Although the model has been developed to incorporate the temperature variation of the thermal expansion coefficient of both scale and substrate it proved impossible to obtain high quality data for both and thus all model predictions have been run with constant values obtained from the literature[3,7,11] which are listed in Table 2.

2.4.4 Substrate creep equation

The power law creep equation for FeCrAlY was determined by Evans[12] by fitting to a large range of experimental creep data. The creep equation for MA956 was determined by fitting to the experimental data of Wittenburger[13] with some extra data provided by Wilber and Nicholls.[14] The quality of the fit to this data is not very good, partly because of the variation in material from the two sources and partly due to the very high creep exponent shown by this material. By fixing the creep exponent at 11, which is a value obtained from fitting to the Wittenberger data at two different temperatures, a much better fit to the rest of the data to obtain the activation energy and pre-exponential constant was possible. However, this creep equation must be still treated with caution. For the 304 stainless steel substrate a values for the creep exponent at low applied stress and activation energy were found in the literature[15] and the pre-exponential constant was determined by a limited number of experiments performed on the material used in this study. All creep parameters used are listed in Table 2.

Clearly the use of this equation makes the assumption of secondary creep occurring in the substrate throughout the test. However, the length of the primary creep segment for FeCrAl alloys at the high oxidation temperatures used in this study is quite short (<lh) and since a period of scale growth and substrate creep will have occurred at temperature before cooling it is anticipated that the assumption of secondary creep processes operating is reasonable. The assumption of secondary creep for the stainless steel substrate is less reasonable due to the relatively low deposition temperature. However, the long coating times required to produce thick layers at the low coating rate used are expected to lead to steady state creep conditions for this material.

3. EXPERIMENTAL

1.8mm FeCrAlY sheet (Resistalloy) and MA956 sheet (INCO) was cut into 20 mm x 10 mm coupons and polished to a 600 grit finish by standard metallographic methods.

Table 3 Compositions of all the alloys investigated in this study.

Composition (wt%)												
Alloy	Fe	Cr	Ni	Al	Si	Mn	Y	Y_2O_3	Zr	Ti	C	P
FeCrAlY	Bal	20.3	–	5.4	0.26	0.32	0.045	–	0.08	0.084	0.024	0.018
MA 956	Bal	19.7	–	4.62	0.09	–	–	0.5	–	–	0.01	–
304	Bal	18.2	9.3	–	1.0	1.6	–	–	–	–	0.08	–

The compositions of the alloys are reported in Table 3. Coupons were weighed and placed in a crucible to catch any spalled oxide. Crucibles were then placed in a carrier and loaded into a cold furnace. A programmed temperature cycle was used for all oxidations involving heating in flowing laboratory air to the oxidation temperature at 10°C/min, isothermal at the oxidation temperature for a defined period, and cooling to room temperature in the furnace. In the early stages of cooling, the cooling rate was determined to be about 20°C/min, but this rate was rapidly reduced as the temperature dropped giving an average cooling rate of around 2°C/min. Oxidations were performed at between 1100 and 1350°C for periods up to 1000h. After oxidation both samples and crucibles were weighed to determine the total weight gain and the amount of spall.

Scales where excessive spallation had taken place were not investigated further as in these cases some stress relaxation could have occurred in the coating which is not included in the model.

TiN coatings were deposited by sputter ion plating[16] onto 304 stainless steel coupons which were prepared in an identical fashion to the coupons for oxidation. After ion cleaning the samples were deposited with a -35V substrate bias at a deposition temperature of 500°C and a coating rate of 0.1 μm/h. After deposition the coating thickness was determined by metallographic methods and varied from 1 to 10μm.

The residual stress in the alumina scales and coatings was measured at room temperature using the X-ray diffraction $\sin^2\psi$ method.[17] Stress measurements on the scales were made using the reflections from the (4.0.10) planes of α-alumina whereas the (422) planes of TiN were used for the coatings. X-ray elastic constants were calculated from the data in Table 2.

4. RESULTS AND DISCUSSION

4.1 Alumina Scales

4.1.1 Isothermal oxidation

The model was previously used to predict the effect of isothermal oxidation at high temperature on any growth stresses introduced into the alumina scales during the oxidation process.[5] It was shown that for the alumina scales on a weak substrate (FeCrAlY) the creep relaxation of the substrate reduces any growth stresses to a very low value. In this study it was found that there was no measurable creep relaxation in the scales grown on the strong alloy (MA956).

Some high temperature measurements of growth stresses are available for alumina scales on FeCrAlY type alloys which can be as high as 1.4GPa.[18] However, these data are usually generated for very short exposures and previous predictions of the stress in these layers after the longer exposures used in this study show very low stress values (~200MPa[5]). Furthermore, the room temperature stress data collected for the scales on MA956 is not consistent with such high growth stresses in the scales on this alloy. For this reason it has been assumed that there are no growth stresses in the scales present at the end of the oxidation process in all that follows.

4.1.2 Cooling to room temperature

As the sample cools from the oxidation temperature thermal expansion mismatch stresses are generated as well as some extra growth stresses due to continued oxidation. At high temperatures there is some creep relaxation in the substrate and the scale stresses are reduced compared to the simple elastic thermal stress calculation, but once sufficient cooling has been achieved the creep processes are effectively halted and the linear relationship between stress and temperature expected from elastic behaviour is achieved. For 8 μm alumina scales on a 1.8mm thick FeCrAlY substrate cooled from 1250°C substrate creep is important for temperatures above ~750°C (Fig. 1). The form of the curve with a gradual transition to the elastic behaviour is similar to that measured for residual strain for chrome oxide on chromium during cooling previously.[3] There is no creep effect for the scales on MA956.

For FeCrAlY the room temperature stress also depends on cooling rate as the slower the sample is cooled, the more time there is for creep relaxation. Over a range of practical cooling rates, the difference in residual stress at room temperature is about 0.5 GPa (Figs 1 and 2) which is small compared to the 1.3 GPa difference between the elastic thermal stress and the modelled stress incorporating creep. Cooling rates of 1000 K/min or greater are needed to achieve the full elastic stress component which is achieved for scales on MA956 at all practical cooling rates (Fig. 2).

For FeCrAlY, changing the oxidation temperature also dramatically affects the development of stress in the scale (Fig. 3). For oxidation above 950°C substrate creep processes limit the stress in the scale and all scales of the same thickness produced at all temperatures have the same residual stress at room temperature. At lower oxidation temperatures elastic thermal stress behaviour is maintained. The experimentally measured room temperature residual stresses for samples oxidised at temperatures between 1100°C and 1350°C show very good agreement with the predicted values (Fig. 4a).

For MA956 no creep relaxation occurs on cooling and the predicted room temperature residual stress increases with oxidation temperature as expected from the thermal expansion mismatch stress eqn (2). Again there is very good agreement with experimentally measured stress values (Fig.

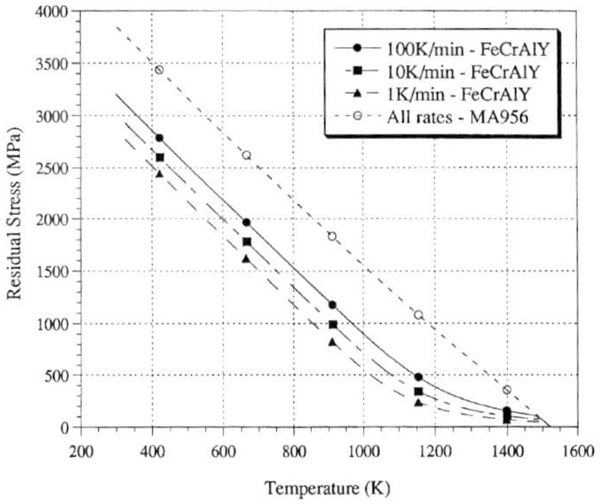

Fig. 1 Predicted variation of scale stress with temperature during cooling of an 8 μm thick alumina scale on a 1.8 mm thick FeCrAlY or MA 956 substrate as a function of cooling rate.

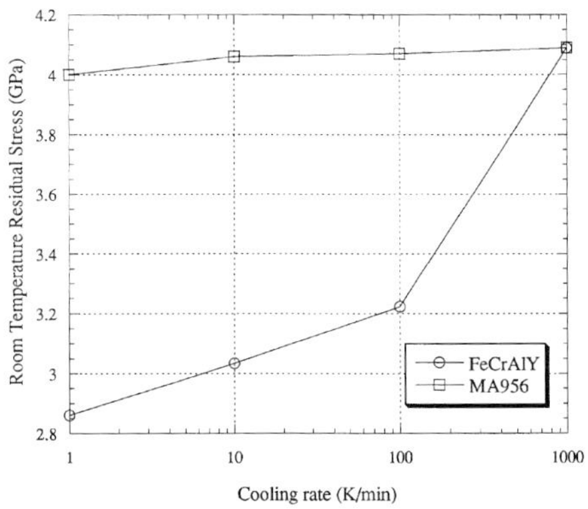

Fig. 2 Variation of room temperature scale stress with cooling rate for an 8 μm alumina scale on 1.8 mm thick FeCrAlY or MA 956 substrates.

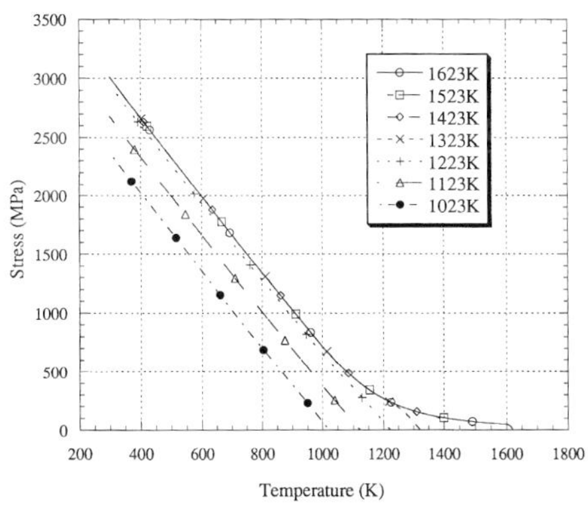

Fig. 3 Predicted variation of scale stress with temperature during cooling of an 8 μm thick alumina scale on a 1.8 mm thick FeCrAlY substrate as a function of oxidation temperature.

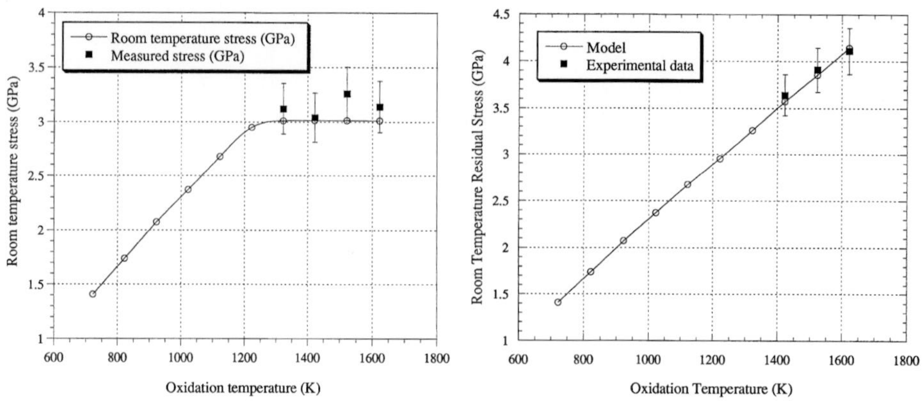

Fig. 4 Comparison between predicted and measured room temperature scale residual stress values for an 8 μm thick alumina scale on an (a) 1.8 mm thick FeCrAlY substrate and (b) a 1.8 mm thick MA 956 substrate.

4.1.3 Effect of substrate thickness

For a given scale thickness, oxidation temperature and cooling rate the residual stress in the scale will increase as the substrate thickness increases (Fig. 5). This is because the stress in the substrate is reduced as its thickness increases and the driving force for creep is also reduced. For the FeCrAlY substrate investigated here, there is no creep relaxation in the substrate from the stresses in an 8 μm alumina scale if the substrate thickness exceeds 20 mm. Clearly there will be significant stress relaxation for thin substrates such as the 50 μm foils being developed for automotive catalyst support and there will be a reduced chance of scale spallation on these thin foils compared to on thicker artefacts. For the MA956 substrate the thickness below which creep becomes significant is reduced to less than 2mm.

4.1.4 Effect of scale thickness

As the scale thickness increases the stress induced in the substrate to balance the scale stresses will increase and the propensity for substrate creep will also increase. The amount of stress relaxation in the scale increases with scale thickness (Fig. 6). Thus for longer oxidation times where the scale thicknesses are larger, there will be considerable stress relaxation on cooling and the driving force for scale spallation will be reduced in cases where substrate creep is occurring. This has been observed for nickel oxide scales on nickel[19] and is confirmed for alumina scales by the experimental measurements made in this study (Fig. 6).

4.2 Titanium Nitride Coatings

4.2.1 Effect of coating rate

The ion bombardment of the coating during deposition in the sputter ion plating process leads to a growth stress which is around 2 GPa compressive. Creep relaxation of

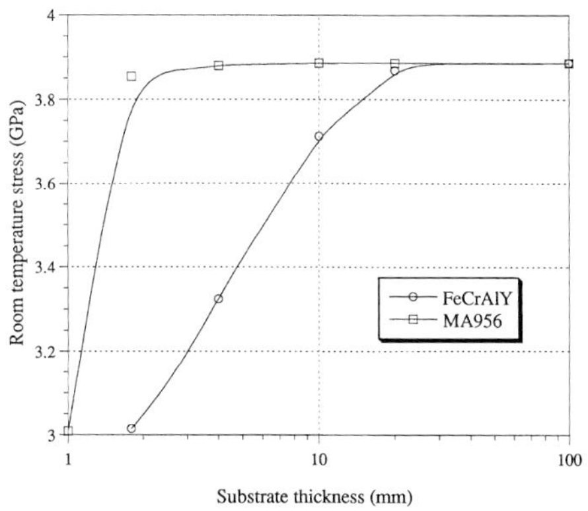

Fig. 5 Variation of room temperature scale stress with substrate thickness for an 8 μm thick alumina scale on FeCrAlY and MA 9565.

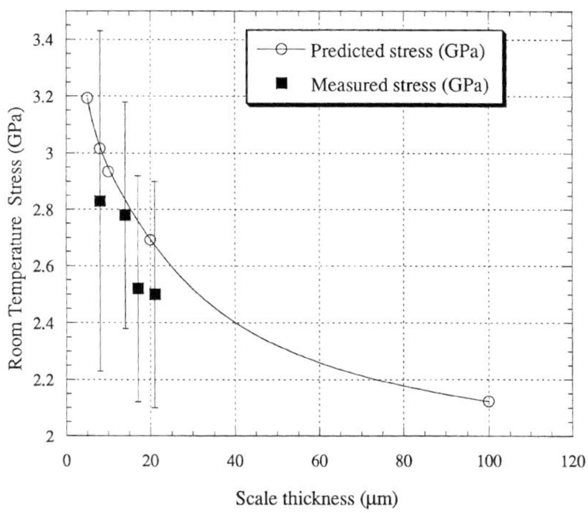

Fig. 6 Variation of room temperature scale stress with scale thickness for alumina scales on a 1.8 mm thick FeCrAlY substrate.

this stress becomes significant as the coating rate is reduced and the coating thickness increases since the time that the substrate can creep is increased under these conditions (Fig. 7). Most industrial coating systems work with coating rates which are of the order of 1 μm/h under which conditions creep relaxation is not significant. For the coatings investigated in this study, the relatively low deposition rate of 0.1 μm/h means that relaxation of stresses in the coating due to substrate creep is very important.

4.2.2 Effect of coating thickness

A combined prediction of the growth stress generation and relaxation during coating and the generation and relaxation of growth and thermal stresses on cooling has been

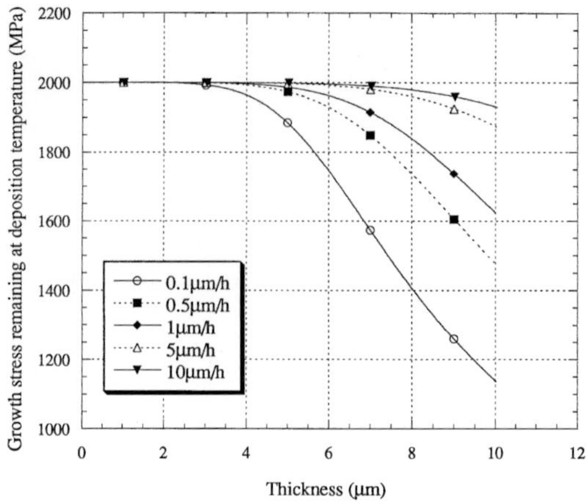

Fig. 7 Variation of growth stress with coating thickness as a function of deposition rate for sputter ion plated TiN on 304 stainless steeel. The deposition temperature was 500°C.

performed to predict the stresses which are expected to remain in the TiN coatings at room temperature. The prediction shows that little or no stress relaxation occurs on cooling and that the room temperature compressive stress in the TiN is reduced as coating thickness increases by an amount the corresponds to the relaxation in growth stresses at the deposition temperature (Fig. 8). Experimental data shows the same trends though there is considerable scatter in the measurements.

The reduction in stress with coating thickness for PVD TiN has been reported previously and was attributed to the increase in grain size of the material as the coating thickness increases and the greater likelihood that large-grained material can relieve stress by plastic deformation.[11] However, the results from this study indicate that a much more likely explanation is that creep relaxation in the substrate leads to stress relaxation in the coating.

5. CONCLUSIONS

A model developed to predict the development of residual stress in oxide scales during isothermal oxidation and on subsequent cooling to room temperature has been tested on a range of alloy systems and shows excellent agreement with experimental data. The modelling approach can also be applied to coatings deposited by physical vapour deposition. The model shows that the interpretation of growth stress measurements made on scales at the oxidation temperature is very difficult since these stresses will be continually changing if the substrate can creep at the test temperature. However, for long exposures at high temperatures growth stresses can be discounted when substrate creep occurs. The importance of substrate and scale/coating thickness on creep relaxation and stress generation cannot be overemphasised. Room temperature

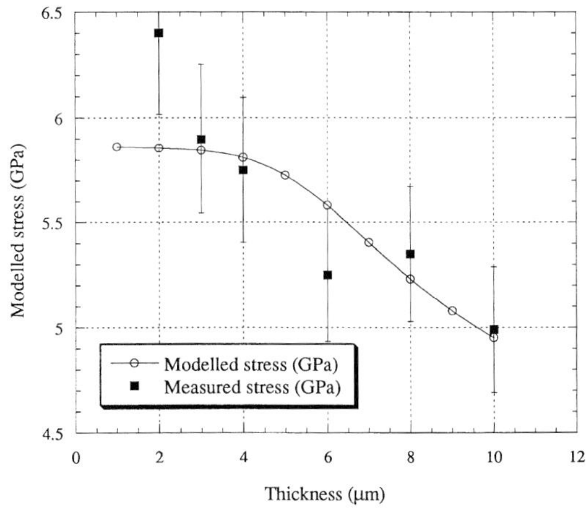

Fig. 8 Variation of predicted and measured room temperature residual stress with coating thickness for sputter ion plated TiN on 304 stainless steel deposited at 500°C with a coating rate of 0.1 μm/h.

scale or coating stresses will be lower on alloys where creep can occur compared to their creep resistant counterparts. In addition, the propensity for scale or coating spallation is reduced as the scale thickness increases or the substrate thickness is reduced. In both cases this is due to the increasing creep relaxation in the substrate.

ACKNOWLEDGEMENTS

The author would like to thank H. E. Evans, J. R. Nicholls and M. J. Bennett for the discussions which instigated this work. Some of the experimental work was supported by the CEC BRITE-EURAM programme on 'How to improve alumina scales'. The TiN coatings were deposited at AEA Technology and the author would like to thank Phil Warrington and Jennie Cullen for their provision.

REFERENCES

1. H. E. EVANS: 'Stress effects in high temperature oxidation of metals', *Int. Mater. Reviews,* 1995, **40**, 1–40.
2. A. M. HUNTZ and M. SCHUTZE: 'Stresses generated during oxidation sequences and high temperature fracture', *Materials at High Temperatures,* 1994, **12**, 151–161.
3. A. M. HUNTZ: 'Stresses in NiO, Cr_2O_3 and $A1_2O_3$ oxide scales', *Mat. Sci. Eng.,* 1995, **A201**, 211–228.
4. H. E. EVANS: 'A creep analysis of the deflection test for evaluating oxide growth stresses', *Mat. Sci. Eng.,* 1995, **A203**, 117–127.
5. S. J. BULL: 'Modelling of residual stress in oxide scales', *Oxidation of Metals,* 1998, **49**, 1–17.

6. G. E. DIETER: *Mechanical Metallurgy,* McGraw-Hill, London, 1981, 453.
7. J. R. NICHOLLS, D. J. HALL and P. F. TORTORELLI: 'Hardness and modulus measurements on oxide scales', *Materials at High Temperatures,* 1994, **12**, 141–150.
8. H. J. FROST and M. F. ASHBY: *Deformation Mechanism Maps, 1st Edition,* Pergamon Press, Oxford, 1982, 24–27.
9. S. C. GILL and T. W. CLYNE: 'Stress distribution and material response in thermal spraying of metallic and ceramic deposits', *Met. Trans.,* 1990, **B21,** 377–385.
10. M. E. O'HEARN, W. C. OLIVER, C. J. MCHARGUE, D. S. RICKERBY and S. J. BULL: 'The hardness and elastic modulus of TiN films as determined by ultra-low load indentation', *Mat. Res. Soc. Symp. Proc.,* 1990, **188**, 139–143.
11. D. S. RICKERBY: 'Internal stress and adherence of titanium nitride coatings,' *J. Vac. Sci. Technol.*', 1986, **A4**, 2809–2814.
13. J. D. WITTENBERGER: 'Elevated temperature mechanical properties of the iron base oxide dispersion strengthened alloy MA956 bar', *Met. Trans.,* 1981, **12A,** 845–851.
14. J. WLLBER and J. R. NICHOLLS: *private communication*, 1997.
15. V. SUSTEK, M. PAHUTOVA and J. CADEK: 'Investigation of creep behaviour of a low carbon 18Cr-12Ni (304L type) stainless steel at 873-1173K', *Mat. Sci. Eng.,* 1994, **A177**, 75–81.
16. D. S. RICKERBY and R. NEWBERY: 'Structure, properties and applications of TiN coatings produced by sputter ion plating,' *Vacuum,* 1988, **38**, 161–166.
17. J. A. SUE: 'Stress determination for coatings,' *ASM Handbook Volume 5: Surface Engineering,* ASM International, Metals Park, Ohio, 1994, 647–653.
18. C. SARIOGLU, J. R. BLACHERE, F. S. PETTIT and G. H. MEIER: 'Room temperature and *in-situ* high temperature strain or stress measurements by XRD techniques', *Microscopy of Oxidation 3,* S. B. Newcombe and J. A. Little eds, Institute of Materials, London, 1997, 41–51.
19. C. LIU, A. M. HUNTZ and J. L. LEBRIN: 'Origin and development of residual stress in Ni-NiO system - in situ studies at high temperatures by X-ray diffraction', *Mat. Sci. Eng.,* 1993, **A160**, 113–126.

Properties of Electron Bombardment Assisted EBPVD PYSZ TBCs

A. MATTHEWS, S. J. YOUNG, S. J. DOWEY, M. JOSEPH,
A. LEYLAND, G. FISHER*, P. K. DATTA*
and J. BURNELL-GRAY*

Research Centre in Surface Engineering, University of Hull, Hull, HU6 7RX
**Surface Engineering Research Group, University of Northumbria, Newcastle upon Tyne, NEI 8ST*

ABSTRACT

Studies of a novel EBPVD PYSZ coating process are described. This uses electron bombardment to pre-heat the substrate and to modify the coating structure. The films produced are dense and well adhered, and possess good cohesive strength and toughness. Hot corrosion tests also show the films to offer better protection to substrate attack than coatings produced by conventional EBPVD or plasma assisted PVD methods.

INTRODUCTION

Over the past ten years considerable progress has been made on the development of plasma-assisted electron beam physical vapour deposition (EBPVD) methods for producing thermal barrier coatings Fancey *et al.*,[1] James *et al.*,[2] James *et al.*,[3] Fancey *et al.*[4] However a specific difficulty with the method has been in achieving adequate substrate heating prior to deposition. Such heating can be beneficial in improving the coating structure and in producing an appropriate residual stress state such that the stresses and strains caused by differential thermal expansion between coating and substrate during in-service thermal cycles are minimised.

Recently a new technique has been reported for achieving the necessary substrate heating and enhancing the structural morphology of coatings, Matthews *et al.*[5] This is based on the electron bombardment of the substrate prior to and during coating. In this paper details are given of the properties of coatings produced by this new method. These are compared with those achieved by the previously used plasma-assistance technique, that is the radio frequency biasing method, and with another recent innovation, bipolar pulsed-bias deposition.

EXPERIMENTAL

The main deposition parameters for the three types of coatings reported here are summarised below:

I RF plasma assisted EBPVD
Frequency 13.56 MHz
DC offset bias voltage -200 V
Chamber pressure 5 mTorr (Argon)
Thickness 40 μm

II Bipolar plasma assisted EBPVD
Frequency 100 kHz
Negative bias voltage -250 V
Positive bias voltage + 75 V
(-ve for 30% of period)
Chamber pressure 30 mTorr (Argon)
Filament emission current 35 A
Thickness 30 μm

III Electron bombardment deposition
Bias voltage +140 V
Chamber pressure 3 x 10^{-2} mTorr
Filament emission current 15 A
Thickness 20 μm

Note that both II and III utilised electron enhancement; in the case of II this was to increase the degree of ionisation occurring, and in III it was to directly bombard the substrate surface. In the coatings reported here the substrate temperature during deposition was 500±20°C. The source material was 8 wt% yttria stabilised zirconia, and the substrate material was N75 nimonic alloy, without any bond coat or aluminide layer.

All coatings were evaluated for phase composition by X-ray diffraction and for structure by SEM studies of fracture sections. In addition, various mechanical and corrosion evaluations were carried out.

RESULTS

All coatings exhibited a mixed (tetragonal plus cubic) phase composition. The fracture sections were dense columnar (Zone 1)[6] in the case of I and II; but in the case of III the columnar morphology was much less evident. They were nearer to the Zone T or Zone 2 morphologies described by Thornton.[6]

The strain tolerance and adhesion was evaluated by bending samples through a 90 degree angle and observing the cracking behaviour. The electron bombarded samples showed no sign of debonding, which indicated both their good strain tolerance and excellent adhesion, whilst coatings in group I showed significant cracking and detachment; those from II showed some detachment, but less than I.

The elastic moduli (E values) of all coatings were evaluated by a laser-induced

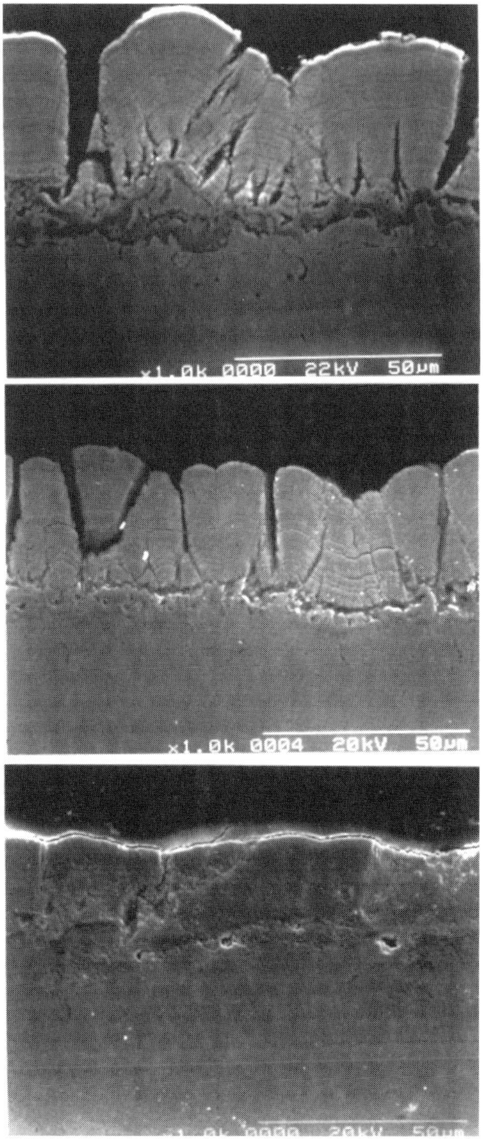

Fig. 1 Coating sections after 24 hour hot corrosion tests at 1000°C, (a) RF plasma PVD, (b) pulsed plasma PVD and (c) electron bombardment assisted deposition.

surface acoustic wave (SAW) method, Schneider *et al.*[7] Coatings from group I gave values of 7–8 GPa. There was considerable scatter in the group II deposits, from 3–57 GPa, and the values for group III were between 14 and 20 GPa. For comparison, plasma sprayed zirconia coatings typically give values of 11–46 GPa, Schneider *et al.*[8] All of these values are much lower than the accepted bulk value for dense zirconia, which is about 240 GPa.

The lower E values for the coatings have previously been attributed to porosity (particularly with regard to plasma sprayed coatings, Schneider *et al.*[8]). However it is significant that the electron bombarded coatings (group III) exhibited relatively low porosity, and thus the low elastic modulus can be seen to suggest a tougher, more resilient coating; this was confirmed by impact tests. In this case the low E values are believed to be due to wave scattering effects from the dense (yet columnar and strain tolerant) coating microstructure.

The hot corrosion protection offered by the different coatings was assessed using isothermal soaks for 24 and 48 hours at 1000°C in laboratory air. The aged coatings were weighed before and after exposure and subsequently mounted in resin and prepared in section. The samples were then analysed using optical and SEM techniques. Typical gravimetric weight gains were 0.1 to 0.2% after 24 hours and 0.2 to 0.3% after 48 hours. This was typical for all the coating investigated, and represented substrate oxide growth thickness of 8 to 10 μm respectively. It should be emphasised that these coatings were relatively thin TBCs, with no bond coat or protective aluminide layer. Nevertheless, it was encouraging that the dense, electron bombarded coatings, gave better corrosion protection, as well as improved mechanical properties. The results are illustrated in Figures l(a)–1(c).

CONCLUSIONS

It is shown that biased DC plasma pulse deposition offers a flexible alternative to radio frequency plasma assisted PVD for TBC. However a new coating process, based on electron bombardment pre-heating and concurrent electron bombardment during coating growth can produce TBCs of higher coating density and thus with enhanced hot corrosion and mechanical properties at surprisingly low substrate temperatures.

REFERENCES

1. K. S. Fancey and A. Matthews: *J. Vac. Sci. Technol.*, 1986, **A4**(6), 2656.
2. A. S. James, K. S. Fancey and A. Matthews: *Surf. Coat. Technol.*, 1987, **32**, 377.
3. A. S. James and A. Matthews: *Surf. Coat. Technol.*, 1990, **43/44**, 436.
4. K. S. Fancey, S. J. Young, A. S. James and A. Matthews: *Mat. Sci. and Eng.*, 1993, **A163**, 171.
5. A. Matthews, S. J. Young, M. Joseph, C. Rebholz, J. M. Schneider and S. J. Dowey: Proc. Int. Conf. Metallurgical Coatings and Thin Films, San Diego, USA, April 1997. To be published in *Surf. Coat. Technol.*
6. J. A. Thornton: *J. Vac. Sci. Technol.*, 1974, **11**, 666.
7. D. Schneider and Th. Schwarz: *Surf. Coat. Technol.*, 1997, **91**, 136–146.
8. D. Schneider, Th. Schwarz, H.-P. Buchkerner and D. Stover: *Thin Solid Films*, 1993, **224**, 177–183.

The Hot Corrosion Resistance of Platinum–Rhodium Modified Diffusion Coating on Directionally Solidified MAR M002 Superalloy at 900°C

W. Y. CHAN, P. K. DATTA, G. FISHER and J. S. BURNELL-GRAY

Surface Engineering Research Centre, School of Engineering, University of Northumbria at Newcastle, Newcastle upon Tyne, NEl 8ST

ABSTRACT

Hot corrosion is a major problem commonly encountered in most gas turbines. Conventional diffusion aluminide coatings on nickel superalloys are generally used to prevent degradation due to oxidation and hot corrosion. However, there is a continuing interest to modify the aluminide coating system with a view to obtaining enhanced performance. This paper discusses the influence of precious metal additions on the stage I hot corrosion performance of modified aluminide coatings on directionally solidified MAR M002 nickel-base superalloy at 900°C in sodium sulphate and sodium chloride containing environments. The scale morphologies and the hot corrosion products formed on the coatings have been characterised using optical and scanning electron microscopy (SEM) and energy dispersive analysis by X-ray (EDX) techniques.

Precious metal additions have been observed to improve hot corrosion resistance by increasing the stability of aluminide phases through the formation of a protective oxide layer. The role of the precious element additions has been discussed in relation to their effects upon the various aspects of hot corrosion.

INTRODUCTION

The development of new aluminide coatings is a continuing process, motivated primarily by the demand for materials with enhanced oxidation and hot corrosion resistance, and high temperature stability for use in modern gas turbine engines. Aluminide coatings developed by the pack-cementation process have been used extensively against hot corrosion of nickel and cobalt superalloys in gas turbine application.[1–3] Corrosion of material in the presence of liquid sodium sulphate, either by itself or in combination with sodium chloride, has been a problem in gas turbines; this corrosion process has been termed 'hot corrosion' which is often described as an accelerated form of corrosion in the presence of oxygen and sulphur.[4] Two types of hot corrosion have been identified: Type I, high temperature hot corrosion (HTHC), operative at 800–

950°C or higher, and Type II, low temperature hot corrosion (LTHC), operative at 600–750°C.

The hot corrosion resistance of platinum-aluminide coatings on IN-738LC has been reported by several authors.[5–7] These studies have been concentrated on HTLA (high temperature low activity) and LTHA (low temperature high activity) platinum-aluminides on IN-738 at temperatures ranging from 700°C (973K) to 900°C (1173K). These temperatures correspond to Type I and Type II conditions. The Type I hot corrosion properties of the platinum-aluminide coatings on IN-738 have been found to have better properties under HTHC conditions, as a more protective scale is formed, than under the LTHC condition. Malush *et al.* have reported that the rate of attack in Type II conditions is 8 times that of Type I for a LTHA basic aluminide, 2 times for the HTLA aluminide.[9]

Currently, there is significant interest in the high temperature hot corrosion performance of precious metal modified aluminide coatings. It has been reported by this laboratory that both high activity Pt-aluminide and Pt–Rh-aluminide systems have excellent cyclic oxidation performance, compared with a basic aluminide.[10] However, as there is little information available on the hot corrosion mechanism of Pt–Rh-aluminide in the open literature, it is the aim of the present study to evaluate the Type I hot corrosion performance of such a coating in relation to a Pt-aluminide coating.

EXPERIMENTAL

The nominal composition of the directionally solidified MAR M002 alloy substrate used in the present investigation is presented in Table 1.

Tabe 1 Nominal chemical composition of MAR M002 substrate material (wt%)

Cr	Al	Ti	Co	W	Ta	Mo	Hf	Fe	Zr	B	C	Ni
9	5.5	1.5	10	10	2.5	0.5	1.25	0.5*	0.055	0.015	0.015	Bal.

*denotes maximum

The materials used in the current studies were supplied by Rolls-Royce and Johnson Matthey. The platinum-modified coatings were prepared by Chromalloy (UK) Ltd. Rod-shaped (25mm in length and 6mm diameter) specimens were coated with a platinum-aluminide of the RT22LT-type which nominally contains 35–55% Pt. First, a layer of platinum was electroplated on the alloy surface. The platinum–rhodium-aluminide coating was prepared by a very similar method, initially a layer of rhodium was electroplated on the MAR M002 superalloy substrate which was followed by the electroplating of a thin layer of platinum prior to diffusion treatment. The diffusion is as described for platinum-aluminide coating, (i.e. heat treat at 1100°C for 1 h in argon). Following a diffusion treatment, the surface was aluminised in a high activity pack. After coating, all specimens were thermally aged for 16 h at 870°C to precipitate the strengthening gamma prime phase. Following heat treatment the test specimens were washed in distilled water and then ultrasonically cleaned in acetone, followed by hot air drying, the pin samples were then coated with 1mg cm^{-2} Na_2SO_4 and dried prior to testing.

The hot corrosion behaviour of the platinum-aluminide, platinum–rhodium-aluminide coating on MAR M002, and the uncoated MAR M002 superalloy substrate were conducted at 50 h intervals up to 200 h in an Na_2SO_4 and NaCl mixture in the presence of 0.2% SO_2 balance O_2 gas mixture at 900°C.

Detailed microstructural evaluations were conducted using optical, scanning electron microscopy and XRD techniques. Quantitative EDX analyses of elements were performed at an accelerated voltage (20kV) using a spot size of 100 nm. The results were quantified using the ZAF-4 programme. To identify all phase constituents X-ray diffraction was carried out using Cu-K_α radiation on a Siemens D-5000 diffractometer. Phase identification was accomplished with the aid of the Joint Committee Powder Diffraction Standards (JCPDS) file.[11]

RESULTS AND DISCUSSION

Kinetics of Hot Corrosion

The uncoated MAR M002, Pt-aluminide and Pt–Rh-aluminide coated MAR M002 superalloy substrate produced by a high aluminum activity process were exposed for 200 h at 50 h intervals in a 0.2% SO_2/balance oxygen environment for hot corrosion studies as described earlier. The surface degradation, as measured by corrosion penetration, of the two coatings and the uncoated MAR M002 at 900°C is shown in Fig. 1. It is clearly evident that the incubation period before the propagation of hot corrosion attack in the uncoated MAR M002 superalloy occurred between 50–100 h of exposure. In contrast under similar conditions of temperature and hot corrosive media, the Pt-aluminide and Pt–Rh-aluminide coatings showed greatly improved hot corrosion resistance. According to the depth of penetration measurements, these precious metal aluminides exhibited relatively long incubation periods up to 200 h. Visual examina-

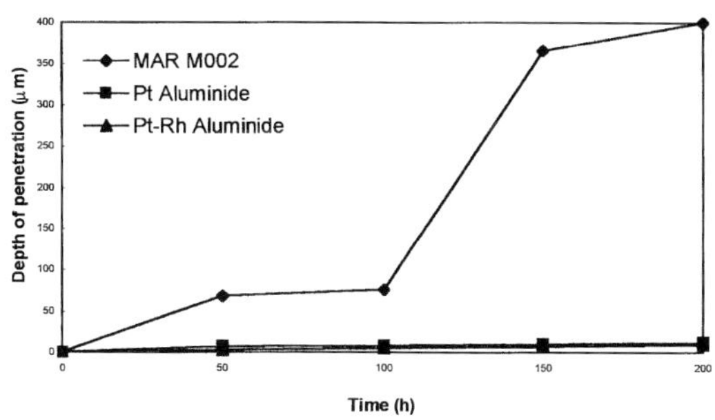

Fig. 1 Depth of penetration versus time after hot corrosion test for the uncoated MAR M002, Pt-aluminide and Pt–Rh-aluminide coatings at 900°C.

tion of the surface morphology of the uncoated MAR M002 indicated a noticeable surface spallation after 100 h exposure and more spallation was evident after 200 h exposure to the hot corrosive media. Negligible surface scale spallation was observed in the Pt-aluminide and Pt-Rh-aluminide coatings.

Hot Corrosion of the Uncoated MAR M002

The hot corrosion of nickel-based alloys[1] and uncoated MAR M002 is charactererised by an 'incubation period' of the order of 100 hours, as shown in Fig. 1. The morphologies of the hot corrosion products associated with the uncoated MAR M002 superalloy have been examined using scanning electron microscopy techniques. A typical morphology includes an outer layer oxide composed of NiO with some $NiCr_2O_4$ spinel which is rather porous. Under the porous layer, a nickel-rich phase is dispersed in a Cr_2O_3 matrix. In the metal matrix beneath this layer, particles of chromium-rich, and sometimes nickel-rich sulphides are surrounded by an alloy depleted in chromium as shown in Fig. 2(a).

It is generally believed that the hot corrosion of uncoated superalloy and the modified coatings involves the following key steps:

(1) formation of oxide scale by oxygen uptake from the Na_2SO_4,
(2) consequent rise in sulphur activity at the salt/coating interface, and
(3) sulphidation of the coating (or alloy matrix in the case of uncoated MAR M002).

In the case of the Pt- and Pt–Rh-modified aluminide coatings, the oxide and sulphide of interest are Al_2O_3 and Al_2S_3, respectively. The Al_2O_3 formed on these alloys can, however dissolve in Na_2SO_4 as $NaAlO_2$,[12,13] i.e. basic fluxing. To make the matter worse, spallation can occur during thermal cycling. All these factors together lead to a gradual Al depletion (i.e. Ni-enrichment) at the scale/coating interface, so that finally NiO begins to appear instead of Al_2O_3.

The uncoated superalloy, consisting mainly of Ni_3Al, will reach this stage earlier than (β-Ni,Pt–Al phase) as observed in the precious element-modified coatings. Once NiO begins to form, corrosion of Ni_3Al continues, as the Al level at the scale/Ni_3Al interface is depleted. This finally leads to a situation where the zone immediately ahead of the advancing corrosion front would be virtually Al-free. On reaching this point, the growing NiO consumes oxygen below the molten Na_2SO_4, thereby the $p(O_2)$ in the melt is decreasing and $p(S_2)$ is increasing. This leads to sulphur transportation through the thin oxide (probably by diffusion of SO_2 through pores and microcracks) and chromium, and nickel sulphide formation beneath the scale. The SO_2 stems from the reaction:

$$SO_4^{2-} = SO_2 + O^{2-} + O \text{ (in NiO)}$$

Since the SO_2 and oxygen are consummed, the activity of oxygen ions O^{2-} in the melt is increasing and becomes more basic. The increase in basicity is greatest over the areas where sulphides form, in these regions NiO will react to form soluble nickelate

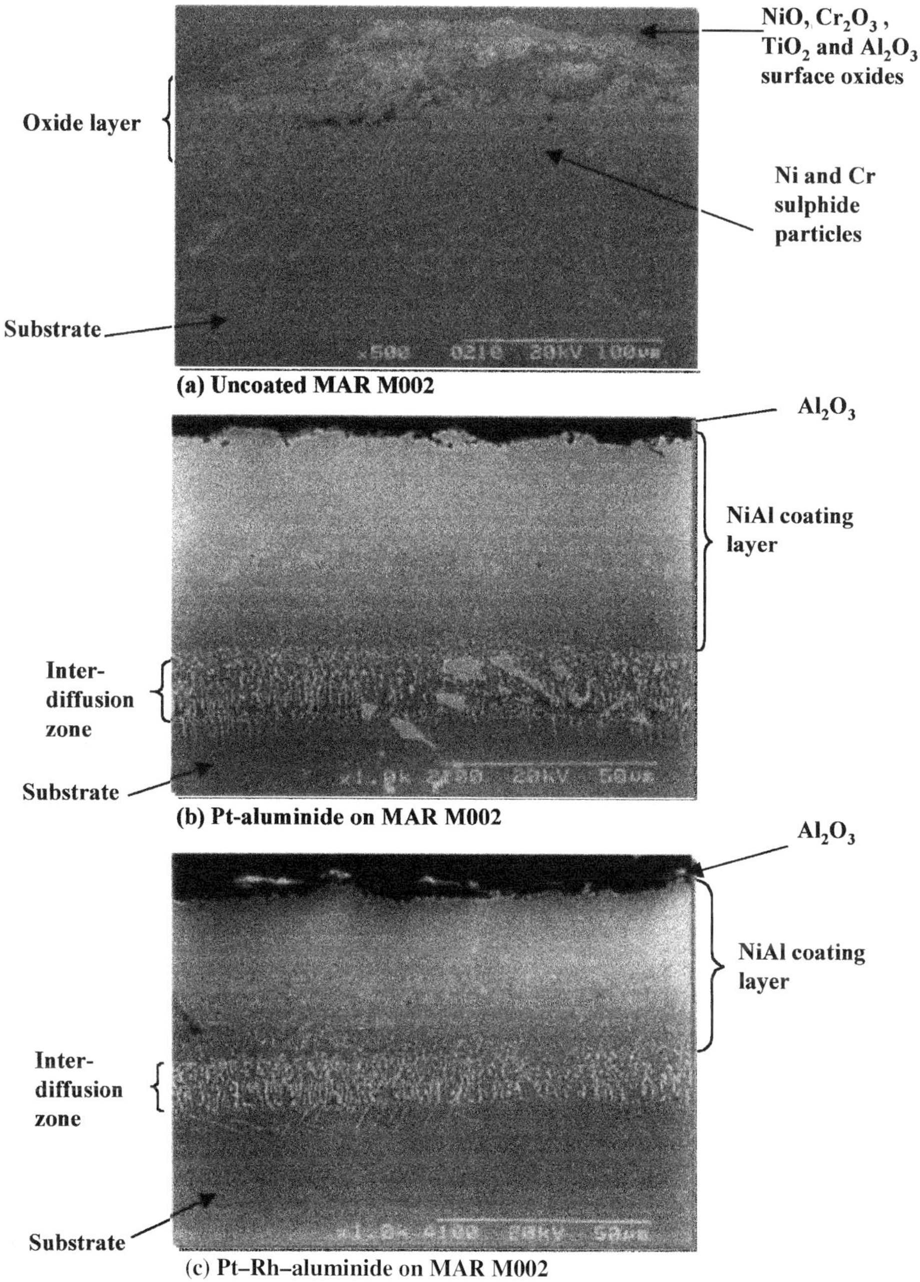

Fig. 2 SEM backscatterd electron images of (a) MAR M002 uncoated, (b) Pt-aluminide and (c) Pt–Rh-aluminide coating on MAR M002 after 100 h hot corrosion test at 900°C.

ions NiO^{2-}. The dissolution of the scale eventually allows the melt to penetrate and to spread along the scale/metal interface lifting and cracking the scale. After cracking the scale, the melt having a high $p(O_2)$, has access to the sulphides which are then oxidised. The sulphur freed by this reaction diffuses further into the metal. Liquid Ni–Ni_3S_2 eutectic could be formed instead of the usual Al_2S_3 or Cr_2S_3. After renewed oxide growth and internal sulphide formation, the reaction sequence occurs repeatedly – a periodic porous structure of the scale results. A schematic diagram depicting the reaction mechanism is shown in Fig. 3. Gradually the Na_2SO_4 is occluded in the porous scale and the corrosion rate diminishes. Because the eutectic temperature in the Ni–Ni_3S_2 system is 645°C, the formation of liquid sulphides at grain boundaries can sometimes impair the alloy's mechanical properties before surface recession would otherwise cause failure. An unusal 'split-open' state could have only resulted from the presence of a liquid phase deep within the alloy, Fig. 2(a).

Hot Corrosion of Pt- and Pt–Rh- Modified Aluminide Coatings

The cross-section profiles illustrating the distribution of platinum, nickel, aluminum, rhodium and tungsten in the coating as well as in the matrix of the interdiffusion zone for each of the coatings studied are presented in Figs 4 and 5. Of particular significance to the protective nature of the coating is the absence of tantalum and tungsten in the outer layers. Extended (200 h) exposure at 900°C was found to have no marked effect on both the structure and of the composition of the outer coating layers.

Detailed analysis of the depth of penetration attack for Pt- and Pt–Rh-modified aluminide coatings revealed that both coatings sustained only very light attack (~9 μm) even after 200 h exposure, Figs 1, 2(a) and (c), i. e. at the incubation stage. This enhanced hot corrosion behaviour is believed to be attributed to the presence of a $PtAl_2$-rich layer which permits the selective formation and healing of an adherent, compact α-Al_2O_3 scale. The addition of Pt and Pt–Rh was demonstrated to have significant influence on the stability of the protective α-Al_2O_3.

The $PtAl_2$ and β-NiAl, are alumina formers, and the preferential formation of Al_2O_3 led to the dissolution of the oxide in the salt melt as an ion.

At the coating/oxide interface, aluminum in the coating would react with sulphide ions passing through the Al_2O_3 scale so that the total reaction increases the oxygen ion activity at the oxide/salt interface, which may lead to basic fluxing of the Al_2O_3:

$$2/3Al + SO_4^{2-} = O^{2-} + 1/3Al_2S_3 + 3/2O_2$$

Consequently, the increase in oxygen ion activity at the oxide/salt interface can lead to the negative solubility gradient as shown in Fig. 6. Hence, alumina is dissolved at the coating/salt interface, and precipitates near the gas/salt interface to form a porous layer. In the case of coatings containing sufficient aluminum to form both oxide and sulphide, not only the fluxing of oxide, but also the oxidation of the sulphide seem important. Chromium is also known to have beneficial influence on the hot corrosion resistance of the coatings, when the activity of chromium is sufficiently high, chro-

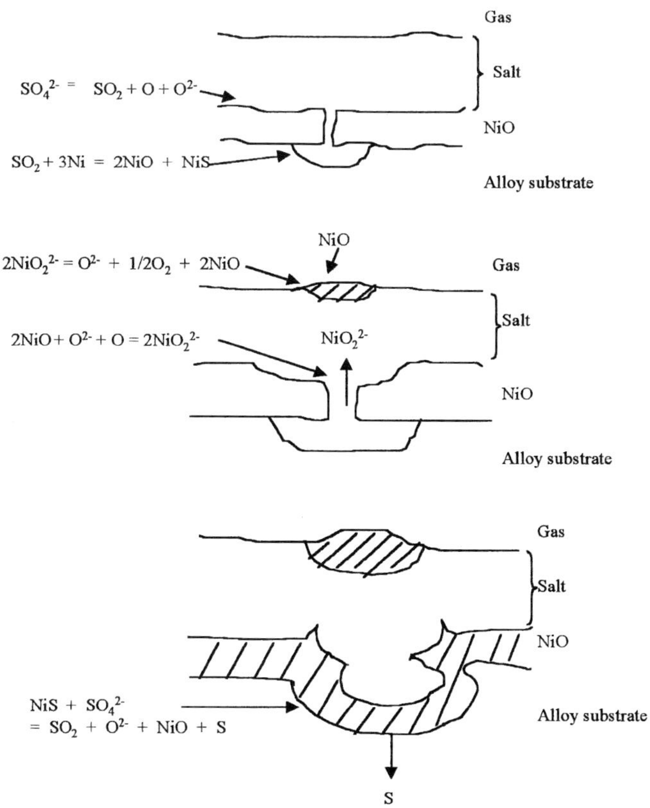

Fig. 3 Schematic reaction sequence describing the hot corrosion of uncoated MAR M002 at 900°C.

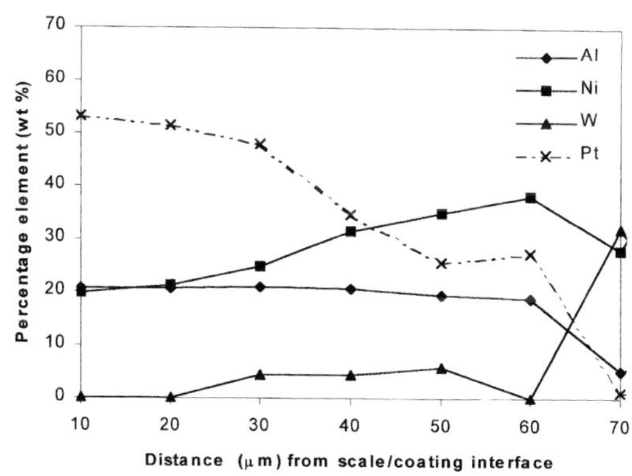

Fig. 4 Elemental profile for Pt-aluminide after 100 h hot corrosion test at 900°C.

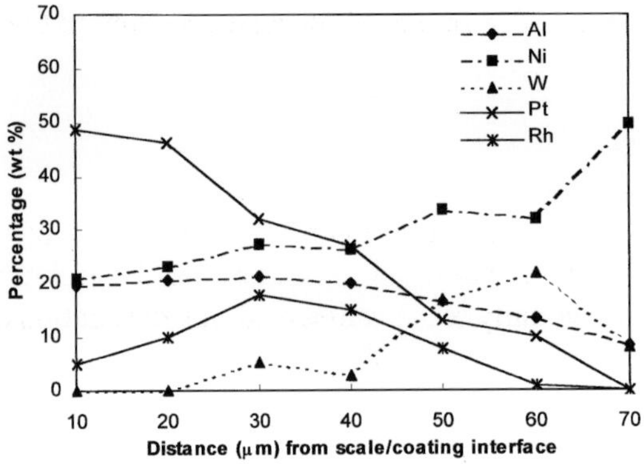

Fig. 5 Elemental profile for Pt–Rh-aluminide after 100 h hot corrosion test at 900°C.

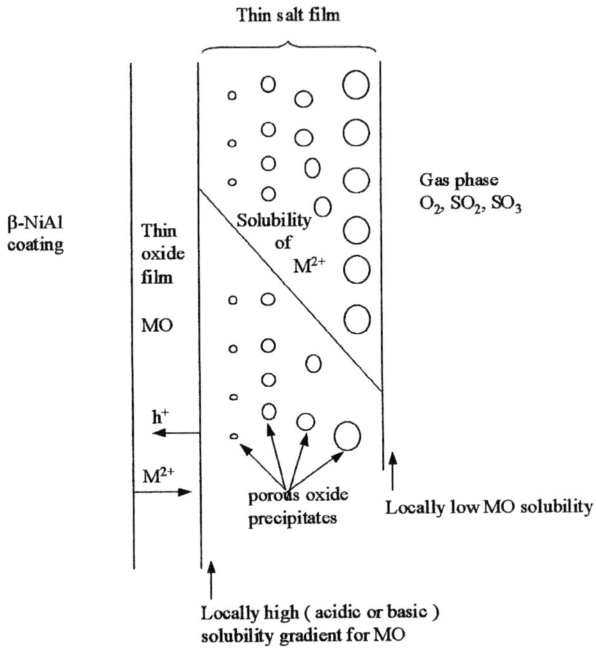

Fig. 6 Schematic diagram showing precipitation of porous MO oxide supported by solubility gradient in fused salt film.[14]

mium will dissolve in the Na_2SO_4 and form chromate ions at the coating/salt interface. This type of fluxing to form aluminate and chromate ions is commonly referred to as 'basic fluxing'. The accelerated oxidation in Na_2SO_4 by basic fluxing is not self-sustaining, since the decomposition of Na_2SO_4 is inevitable in the mechanism, and the amount of Na_2SO_4 is limited.

The second mechanism which can occur in hot corrosion is called 'acidic fluxing'. In this case, acidic oxide, such as WO_3 formed by oxidation of refractory metals contained in the alloy can dissolve in the salt deposits, leading to their acidification (the O^{2-} ion content is decreased):

$$W0_3 + O^{2-} = WO_4^{2-}$$

The concentration of acidic oxide in the condensate generally increases as a function of time. Thus acidic fluxing can often be set up after basic fluxing. The acidity of condensates can thus eventually reach values enabling dissolution of protective oxides as cations:

$$Cr_2O_3 = 2Cr^{3+} + 3O^{2-}$$

$$Al_2O_3 = 2Al^{3+} + 3O^{2-}$$

The role of Rh and Rh-rich phases in influencing the hot corrosion behaviour of aluminide coatings is uncertain at present. It can not be said with certainty whether Rh has a generic role or it performs a metal specific function.

SUMMARY AND CONCLUSIONS

The hot corrosion of MAR M002 is characterised by an ‘incubation period’ of the order of 100 h.[12] For extended exposures (200 h) the Pt- and Pt–Rh-modified aluminide coatings are superior to the uncoated MAR M002 superalloy. The important characteristics and properties of platinum- and platinum–rhodium-modified diffusion aluminide coatings are:

1. Platinum-modified aluminide coatings possess improved cyclic hot corrosion resistance[4,12,15] because the platinum-enriched zone of the Pt- and Pt–Rh aluminide coatings promote selective oxidation of aluminium and minimises the effects of nickel and tungsten in the coatings. The purer α-alumina scales have slower growth rates, which lead to longer exposure times before certain critical thicknesses are reached or where spallation of the oxide occurs.
2. Precious elements rhodium and platinum acted synergistically in maintaining the stability of α-alumina and the β-NiAl after long (200 h) exposures which reduced the detrimental effects of the substrate elements on the coating performance.

ACKNOWLEGEMENTS

This work was supported under EPSRC Contract No. GR/K02367. The provision of materials and coatings from Rolls-Royce plc, Chromalloy (UK) Limited and Johnson Matthey are greatly appreciated.

REFERENCES

1. M. D. EI-DAHSHAN, J. STRINGER and D. P. WHITTLE: *Cobalt,* 1972, **57**, 182.
2. A. J. B. CULTER and C. J. GRANT: *Deposition and Corrosion in Gas Turbines,* A. B. Hart and A. J. B. Culter eds, Halsted Press, 1973, 178.
3. R. G. WING and E. R. McGILL: *Plat. Metals Review,* 1981, **3**, 94.
4. J. STRINGER: *Ann. Rev. Mater. Sci.,* 1977, **7**, 477.
5. W. T. WU, A. RAHIVIEL and M. SCHORR: *Oxidation of Metals,* 1984, **22**(1/2), 59.
6. P. DEB, D. H. BOONE and R. J. STREIFF: *J. Vac. Sci. and Technology,* 1985, **A3**(6), 2578.
7. R. J. STREIFF, O. CERCLIER and D. H. BOONE: *Surface and Coating Technology,* 1987, **32**, 111.
8. R. STREIFF and D. H. BOONE: *J. Materials Eng.,* 1988, **10**, 15.
9. R. E. MALUSH, P. DEB and D. H. BOONE: *Surface and Coatings Technology,* 1988, **36**, 13.
10. W. Y. CHAN, P. K. DATTA and J. S. BURNELL-GRAY: 'High Temperature Air Oxidation Performance of Modified Aluminide Coatings on Nickel-Based Superalloy', *Advances in Surface Engineering*, P. K. Datta and J. S. Burnell-Gray eds, Royal Society of Chemistry, 1997, **Vol. III**, 78–98.
11. Joint Committee on Powder Diffraction Standards, International Center for Diffraction Data, Swartmore, PA, 1995.
12. S. J. GRIAFFE, *The Superalloys,* C. T. Sims and W. C. Hagel eds, John Wiley & Sons, New York, 1972, 342.
13. D. CHATTERJI, D. W. McKEE, G. ROMEO and H. S. SPACIL: *J. Electrochem. Soc.,* 1975, **122**, 941.
14. R. A. RAPP and K. S. GOTO: Proc. '2nd International Symposium on Molten Salts', J. Braunstein ed., *The Electrochem. Soc.,* 1979, 159.
15. T. N. RHYS-JONES: *Materials Development in Turbo-Machinery Design*, D. M. R. Taplin. T. F. Knott and M. H. Lewis, Institute of Metals, 1989, 218.

The Structure of an Interlayer used to Improve the Adhesion of Plasma Sprayed Al_2O_3 Coatings on to a Steel Substrate

J. M. GUILEMANY, J. NUTTING, J. PORTILLO and M. URBAN

Thermal Spraying Center (CPT), Dept. Enginyeria Química i Metal·lurgia, Ceiència dels Materials i Enginyeria Metal·lúrgica, Facultat de Químiques, Universitat de Barcelona, Spain

SCT Serveis Científico-Tècnics, Universitat de Barcelona, Spain

ABSTRACT

In the present work, transmission electron microscopy (TEM) and scanning Auger microscopy (SAM) techniques have been used to study the structure and composition of an interlayer of 95% Ni–5% Al plasma sprayed between a low-alloy steel substrate and a plasma sprayed alumina layer. TEM images of this interlayer reveal a microstructure formed by a discontinuous stacking of curved layers, each one corresponding to a splat. These splats consist on different crystalline zones confined between thin amorphous layers. Small columnar crystals (height 0.1 μm, width 25 nm) form, as well as larger columnar crystals (height 1 μm, width 0.5 μm) and equiaxed ones (mean diameter of 0.5 μm), depending on the distance from the initially cooled surface. These microstructures are argued in terms of the solidification processes that do occur during the formation of the interlayer. SAM experiments reveal that at the interface Ni/Al–steel substrate, there is no diffusion of the signal of Ni, neither Al inside the steel.

INTRODUCTION

Plasma spraying is a thermal spraying technology that is widely used to produce ceramic coatings used in a variety of applications; to provide electrical insulation, increase wear resistance, or chemically unreactive surface. Plasma spraying differs from HVOF spraying in employing higher temperatures (~10 000 compared with ~3000K) and lower particle velocities.

Sprayed coatings of alumina are used as thermal barriers to give enhanced oxidation and wear resistance, and can be applied to a wide variety of substrates.

Although the plasma spraying technique gives good adhesion between the coating and the substrate, when the nature of coating and substrate is different, the presence of an intermediate layer is widely used as a means of increasing the coating–substrate adhesion. In the case of alumina onto a steel substrate, the adhesion of the coating can be greatly improved by using a bond of 95% Ni–5% Al. As the difference of thermal expansion coefficient (TEC) between ceramic coatings and metal substrates might be as high as $\Delta TEC = 22.6 \times 10^{-6}$ 1/K (cordierite sprayed on aluminium substrate) this leads to the generation of residual stresses. That is the main reason for the application of either, bond coatings or particulate composites containing the material of top and bond coating.[1] On the other hand, the roughness that the interlayer offers to the alumina coating, could be high enough to assure a good adhesion between layers.

Experiments have been carried out in order to describe the microstructure of the Ni–Al interlayer and the interfacial zone between Ni–Al and the steel substrate. Transmission electron microscopy (TEM) at 200 and 300 KV, and scanning auger microscopy (SAM) on thin foils prepared through the Ni–Al interlayer and the interfacial regions have been the techniques used to characterise this system.

EXPERIMENTAL

Samples of a low-alloy steel 34CrMo4 (UNS–G41350) containing 0.37 wt%C and 0.97 wt%Cr, were coated by plasma spraying with 95%Ni–5%Al powder (Amdry 380), with a particle size of -95 +45 μm and with alpha alumina (Amdry 6062) powder, with a particle size of -45 +22 μm. The surface of the substrate piece was grit blasted with a 24 grade corundum for one minute, to giver a fresh and rough surface and to improve the coating–surface adhesion. Plasma Technik A3000 equipment with an F4 torch was employed with (Ar + H) as plasma gas and the spraying distances were of 120 mm for the alumina powder and of 140 mm for the Ni/Al powder. The total thickness of the layers were 225 μm for the Ni/Al and 50 μm for the alumina.

The Ni–Al interlayer was studied with a transmission electron microscope (H–800, 200 kV), HRTEM (philips CM–30, 300 kV) and scanning auger microscope (Perkin Elmer PHI Model 670 Scanning Auger Nanoprobe System).

Cross-section specimens for transmission electron microscope studies were prepared using methods described elsewhere.[2] The same specimens prepared for TEM where examined by SAM, in order to study the steel–Ni/Al interfacial region and to establish the profiles of Fe, Ni and Al in order to determine if some diffusion of these elements had occurred.

RESULTS AND DISCUSSION

Within the Ni–Al interlayer, low magnification TEM images reveal a microstructure consisting of a series of splats, Fig. 1. These splats are typical of the thermally sprayed coatings, that are believed to be deposited through the build-up of splats quenched from a series of molten droplets, The coating is then the result of the deposition of successive layers, and although not all the powder particles do melt during their flight from the spray gun to the substrate, the solidification processes are the major responsible of the final microstructures which are present in every splat.

High magnification TEM observations of each individual splat show an ordered structure composed of different crystalline zones confined between thin amorphous-like layers, Fig. 2. Selected area diffraction pattern (SADP) of these layers confirm

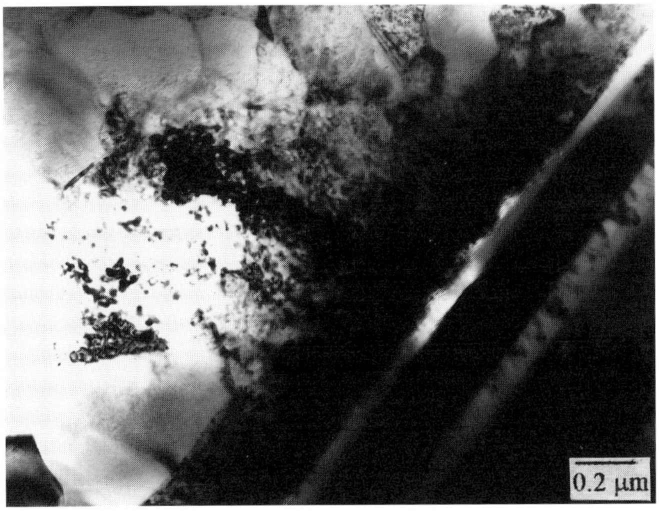

Fig. 1 TEM micrograph showing the different structures present in a splat.

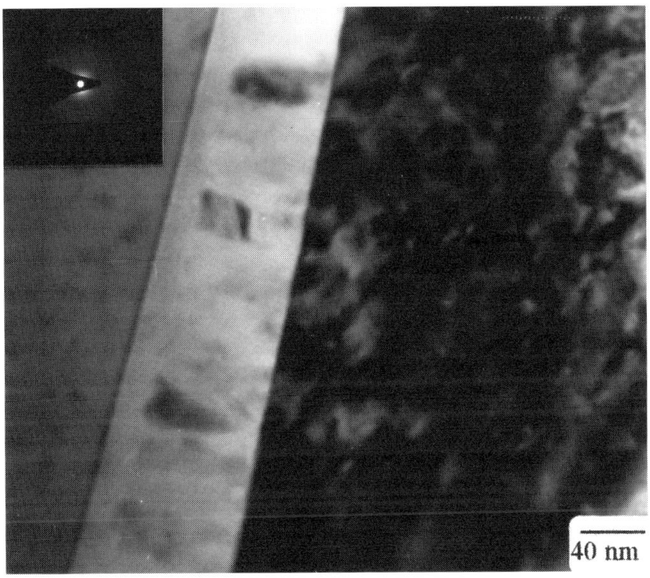

Fig. 2 TEM micrograph showing amorphous layers and small growing columnar crystals.

their amorphous nature, in showing the typical diffuse rings corresponding to a non crystalline materials.

The structural sequence of crystalline and non crystalline layers is repeated from the base of the Ni–Al splats for a number of times. Small columnar crystals (height 0.1

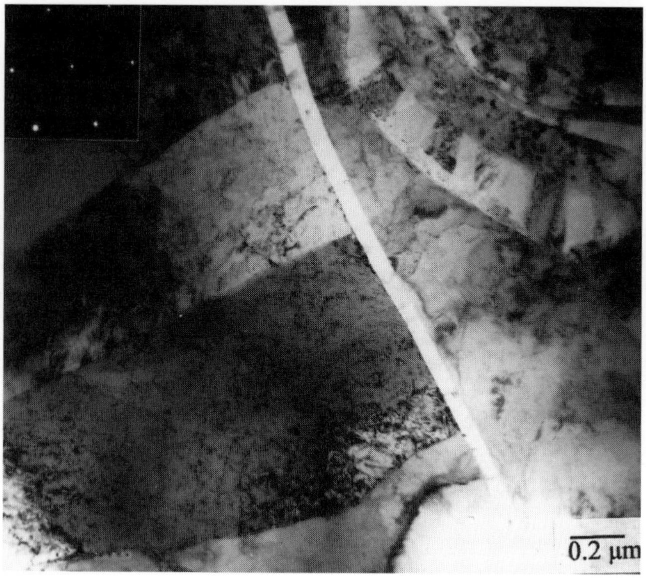

Fig. 3 TEM micrograph showing large columnar crystals. SADP with zone axis [011].

μm, width 25 nm) are usually found close to the amorphous zones. As the distance to the amorphous zone increases, larger columnar crystals are formed (height 1 μm, width 0.5 μm), Fig. 3. SADP of this columnar crystals correspond to a FCC phase, with lattice parameters very close to those of the pure nickel. Eventually, equiaxed crystals may form (mean diameter of 0.5 μm). Inside these crystals, small precipitates are frequently found, Fig. 4. It is interesting to note that, although these structures are present in every splat, the proportion of each one may change depending on the distance of the spalts from the steel substrate.

SAM analysis at the interfacial zone between the steel substrate and the, Ni/Al interlayer, shows the distribution of Fe and Ni along the interfacial zone. Also indicated is a sharp compositional change from Ni to Fe, without any coexistence zone of both elements, Fig. 5. Chemical distribution maps of Fe, Ni and Al were obtained, Fig. 6a–c. In these maps, signal Fe was not found inside the Ni/Al layer, neither was a signal of Ni or Al found inside the steel substrate.

In relation to the microstructure observed by TEM and HRTEM in the Ni–Al interlayer looking at the grain size and crystal structure, can be explained in terms of the extremely high cooling rates after a droplet flattened. When a droplet impacts on a substrate, the droplet generally spreads radially to form a flattened particle or splat or flat disc. The structure of such flattened particles, including morphology and size, depends basically on the state of the droplet (melting state, temperature and velocity).[3]

According to the Ni–Al phase diagram, Fig. 7, and due to the high temperatures reached during the plasma spraying process, the particles of the starting powder might

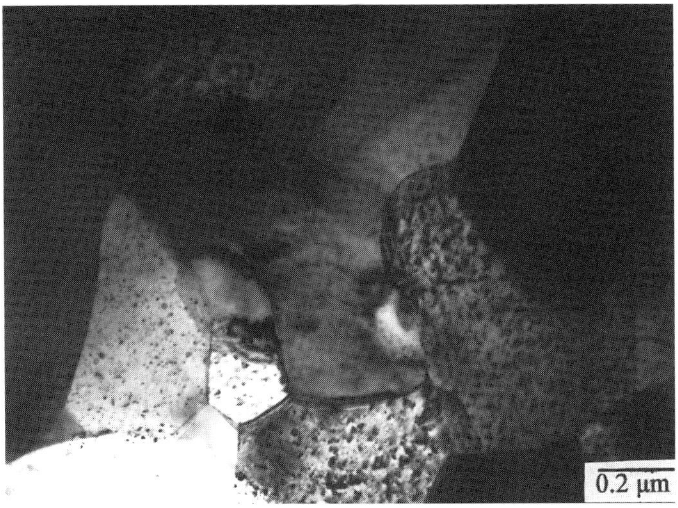

Fig. 4 TEM micrograph showing equiaxed crystals with small precipitates.

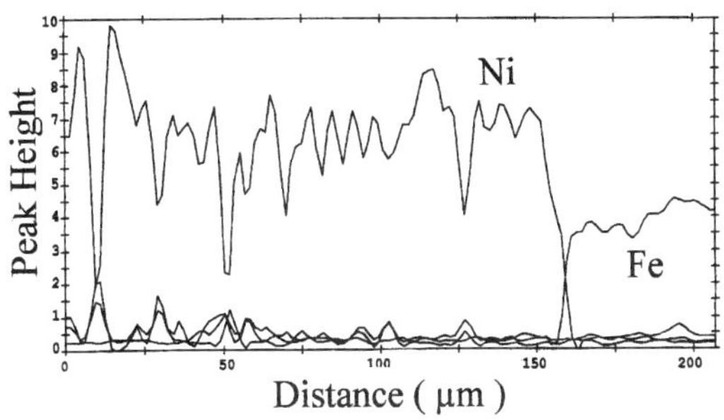

Fig. 5 SAM spectra showing the Fe and Ni distribution at the interfacial zone.

be molten when they reach the substrate. Due to the steep temperature gradient between the steel and the newly arriving particles, a high velocity of cooling results, producing an amorphous phase. As the solidification process continues, the cooling rate decreases, so allowing the system to produce small crystals of nickel-rich phase which is supersaturated in aluminium. If the cooling rate were decreased further, larger crystals would form. Small ($AlNi_3$) precipitates appeared in the last steps of the solidification process, as a direct consequence of the low solubility of the aluminium in the nickel matrix, as it is indicated in the Ni–Al equilibrium diagram. Due to the small size of the ($AlNi_3$) precipitates, SADP of them has not been possible to obtain. Though as the lattice parameters obtained with SADP for the precipitate-free crystals are in very good agreement with those corresponding to pure nickel, it is believed that these precipitates might correspond to the $AlNi_3$ phase.

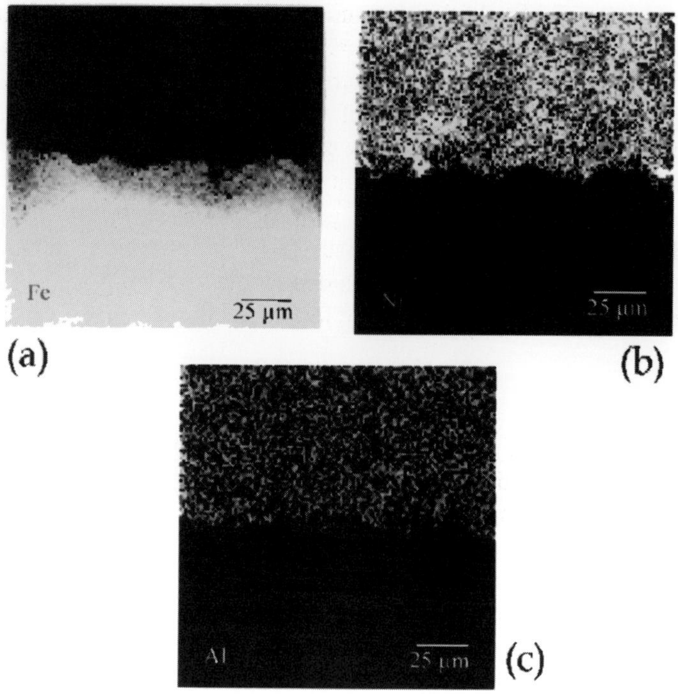

Fig. 6 SAM distribution maps of (a) Fe, (b) Ni and (c) Al.

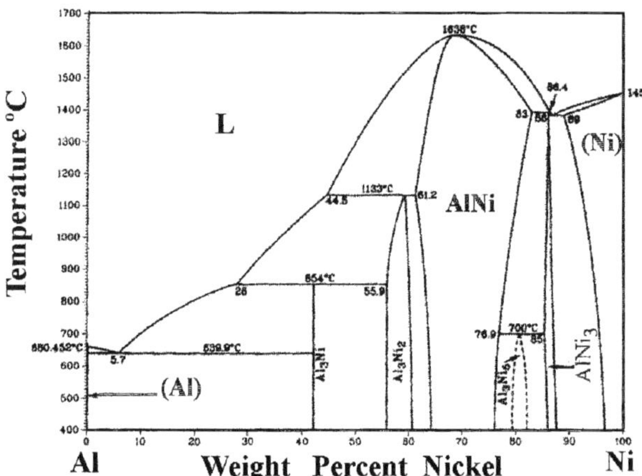

Fig. 7 Ni–Al phase diagram.

It is also possible that the cooling rates were high enough to assume that the matrix was aluminium supersaturated at any time during the solidification process. However, it is believed that as the spraying process continues, the deposition of further splats heats the immediately adjacent splats. This brings about crystallisation of the amorphous phase or precipitation of the aluminium supersaturated nickel grains, so allowing the formation of the $AlNi_3$ phase and giving rise to structural differences.

SAM experiments in the Ni/Al–steel substrate interfacial zone, have demonstrated that up to 0.1μm there is no diffusion of Ni into the steel. Therefore the adhesion between the substrate and the Ni/Al layer seems to be a result of mechanical interactions. It has not to be forgotten that the grit blasting operation, prior to plasma spraying, is a widespread method of assuring a good adhesion between the coating and the substrate. However, it should not to be discarded that in the first atomic layers some diffusion processes might have occurred aided by the high temperatures reached during the solidification of the droplets at the interface.

CONCLUSIONS

1. The microstructure of an interlayer of Ni/Al has been described. It consists on splats, each one containing amorphous zones confined between crystalline layers. In this layer columnar crystals – of different size – and equiaxed grains can be found.
2. Inside these equiaxed grains, precipitates are usually found, and according to the Ni/Al phase diagram, they might correspond to the $AlNi_3$ phase.
3. There is no diffusion of Ni up to 0.1 μm into the steel substrate, so the adhesion between the Ni/Al layer and the steel is mainly due to mechanical interactions.

ACKNOWLEDGEMENTS

The authors would like to thank the CICYT and the Generalitat de Catalunya for finance under the projects MAT 96–0426 and SGR 97–15 respectively.

REFERENCES

1. PAWLOWSKI: *The Science and Engineering of Thermal Spray Coatings*, John Wiley & Sons Ltd, West Sussex, England, 1995.
2. J. NUTTING, J. M. GUILEMANY and Z. DONG: 'Substrate/coating Interface Structure of WC–Co High Velocity Oxygen Fuel Sprayed Coating on Low Alloy Steel', *Materials Science and Technology*, 1995, **11**, 961.
3. AKIRA OHMORI and CHANG-JIU LI: 'The structure of thermally sprayed ceramic coatings and its dominant effect on the coating properties', *Plasma Spraying: Theory and Applications*, R. Suryanarayanan ed., World Scientific Publishing Co., London, 1993.

A Mass Spectroscopy Study of the Electron Beam Evaporation of Yttria Partially Stabilised Zirconia (PYSZ)

A. S. JAMES and A. MATTHEWS
Research Centre in Surface Engineering,
University of Hull, Hull HU6 7RX

ABSTRACT

Yttria partially stabilised zirconia (PYSZ) is commonly deposited by electron beam evaporation for use as a thermal barrier coating in aero-engines. This paper considers how the material dissociates on evaporation and whether the degree of dissociation depends upon the operating conditions and the state of the melt. It is found that the zirconia does dissociate on evaporation with the liberation of oxygen-based species. especially at 32 amu, the most significant part of the evaporation occurs in the early stages of the run, as thereafter tunnelling effects into the melt reduce the overall evaporation rate. It was also observed that the evaporation tends to proceed as a result of individual granules evaporating separately, and the effect of thermionic support was also investigated. In this work on zirconia based coatings there is no indication of substoichiometry in the coatings as a result of the dissociation of the material.

1. INTRODUCTION

Although in our previous experiments on thermal barrier coatings[1-9] there was no evidence of substoichiometry in the coatings produced, it seems likely that the zirconia based source material, (ZrO_2–8Y_2O_3–PYSZ), will dissociate on evaporation, and this is the phenomenon under investigation in this paper. It has been suggested that zirconia will dissociate to produce ZrO and O_2.[10] The objective of this set of experiments was to investigate using mass spectroscopy the evaporation of PYSZ and in particular to determine if and how the material dissociates upon evaporation, and whether the degree of dissociation depends upon the operating conditions and the initial state of the melt.

2. EXPERIMENTAL DETAILS

For these experiments the Tecvac IP35L rig was utilised fitted with a Micromass 200 quadropole mass spectrometer, which can detect ions of mass upto 100 amu. The spectrometer employs an electron impact type ion source, with an electron energy of 65V. Gas evaporation and dc triode conditions were investigated, employing electron beam evaporation techniques. No substrates were coated during this work, although it was necessary to have a substrate holder in position so that the power could be applied for the dc triode runs and for sputter cleaning.

The general run procedure was as follows, the zirconia source material was weighed prior to each experiment and the vacuum chamber pumped down to 10^{-6} Torr and then backfilled to the desired pressure with argon, and the initial species' partial pressures being monitored were recorded. The filaments and power were then switched on, where appropriate, and the partial pressures recorded again. When the chamber conditions had stabilised the electron gun was switched on and after 60 seconds the crucible cover was opened and evaporation was typically continued for 40 minutes, with the partial pressures being recorded every 60 seconds. On completion of the evaporation cycle, everything was turned off and after a further three minutes a final record of the partial pressures was made. In addition one run was performed to monitor the gas phases evolved on heating a coated filament.

The filament assembly consisted of tungsten wire 0.7 mm in diameter, and positioned in the form of a flat loop around the crucible, the loop having a diameter of 29.0 ± 0.5 cm at a height of 6.5 ± 0.5 cm above the base of the chamber.

The analysis chamber of the mass spectrometer was mounted off the side of the vacuum chamber, and was differentially pumped by means of a turbomolecular pump. Therefore the actual partial pressures recorded were lower than those in the main chamber, and it is assumed that any trends observed are similar to any occurring in the actual deposition volume. The species monitored in this work were:

i)	8 amu (O^{++})	ii)	16 amu (O^{+})
iii)	32 amu (O_2^{+})	iv)	40 amu (Ar^{+})

In addition, when evaporation was performed in vacuo, 28 amu (N_2^{+}) and 18 amu (H_2O^{+}) were monitored. Attempts were also made to monitor 91 amu (Zr^{+}) but these were not successful, possibly because such species would condense before reaching the analysis chamber. The limit of the analyser was 100 amu, so species such as ZrO^{+} could not be monitored.

A series of experiments was also carried out to investigate the way in which evaporation changes with time. The evaporation runs were performed in vacuo, at constant electron gun powers for a series of different run times. A static beam position was used and the source material was weighed before and after each evaporation. In order to try to eliminate particle size effects, it was ensured that all of the particles used in the crucible were of a similar size and a similar weight of source material was used for each run.

3. RESULTS

The factors which may affect the dissociation investigated in this work include the origin of the source material (supplied by either Mussel Shoals or Universal Abrasives), changes in electron gun power (7kV x 400mA or 10kV x 400mA), whether the electron beam is moved during evaporation (in order to prevent the beam drilling so deeply into the source material), the deposition conditions, i.e. vacuum evaporation, gas evaporation or dc thermionic triode ion plating. Table 1 details the run conditions which were used in this work.

Table 1 Description of the runs used in this work.

Run	Argon pressure (mTorr)	Material	State	Run type	Gun power (kV x mA)	Weight loss (g)
M1	10	UA	New	GE	10 x 400	5.4
M2	10	UA	Used	GE	10 x 400	7.4
M3	10	UA	New	GE	7 x 400	7.1
M4	10	UA	Used	GE	7 x 400	U/A
M5	10	MS	New	GE	7 x 400	3.8
M6	10	MS	Used	GE	7 x 400	3.8
M7	10	UA	New	GE+F	10 x 400	8.2
M8	10	UA	Used	GE+F	7 x 400	3.4
M9	10	MS	New	GE+F	7 x 400	5.1
M10	10	UA	New	DCT	7 x 400	5.0
M11	10	MS	New	DCT	7 x 400	5.2
M12	10	MS	Used	DCT	7 x 400	4.1

UA - Universal Abrasives MS - Mussel Shoals
GE - Gas Evaporation GE+F - Gas evaporation with filamants DCT - DC triode

Before discussing the effects each of the parameter changes has upon the species detected, it is worthwhile initially to consider a typical trace and indicate some of the main features observed. Figure 1 shows the traces of partial pressure against time for run M3. The points to be noted are:

(i) The trace of 32 amu (O_2^+) is the most complex, showing several maxima and minima, in particular note the reaction on the trace when the beam position is moved after 20–25 minutes, and when the gun power is increased after 2.5 minutes. It seems likely that each peak corresponds to the evaporation of a fresh granule of source material, implying that the material dissociates to form free oxygen species.

(ii) The trace at 8 amu (O^{++}) generally follows the trends in the 32 amu trace for the

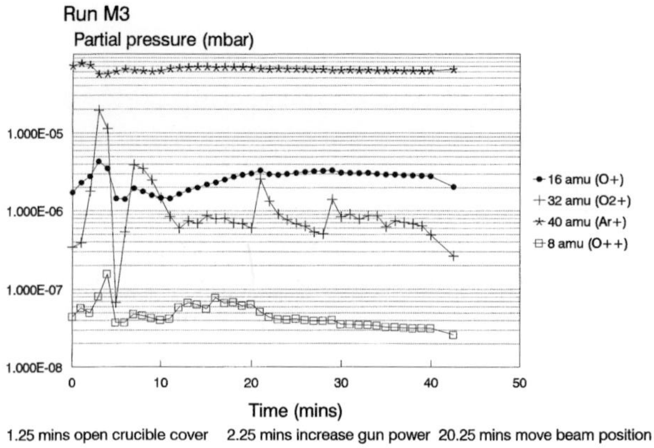

Fig. 1 Typical partial pressure traces observed in these experiments. The traces shown are for 8, 16, 32 and 40 amu and are for run M3.

first half of the run, and similar smaller peaks in the 16 amu trace (O^+) correspond to the major peaks of the 32 amu trace, but does not show much response otherwise.

(iii) After each peak in the 32 amu trace, the partial pressure decays very rapidly. which is consistent with the liberation of a large amount of oxygen very quickly, resulting in an increase in the total pressure. and hence a dilution of the argon. The pumping down of the mixture with flowing argon gives rise to the exponential decay observed.

(iv) The 40 amu trace stays reasonably constant as expected, but does show some reaction from the initial conditions when the crucible cover is opened and the gun power is increased to initiate evaporation.

The two different sources of melt material are considered initially. The Mussel Shoals (MS) material the Universal Abrasives (UA) appear similar but the UA has a slightly smaller particle size and appears slightly less dense. The two sources of material and the difference between using fresh melt material and material used once previously were initially considered. The following points were noted:

(i) For the MS material, the 32 amu trace has much fewer maxima than observed for the UA material, although the individual peaks seem generally to be larger. Similarly the traces for the oxygen species show much fewer variations. This would appear to correlate to the initial granule size.

(ii) There is a greater partial pressure of all three oxygen species monitored with the new melt, for both the MS and the UA materials, and generally there is less activity in the traces for the used melt, particularly towards the end of the runs. The ratio of the partial pressures P16/P32 is generally similar for the two source materials, but tends to be higher for the used material compared to the new material in both cases.

The effects of changing the deposition process will now be considered. The main observations are:

(i) The effect of biased filaments alone was:
(a) To reduce the number of maximna observed in the 32 amu trace when the beam position was moved, compared to the gas evaporation case, but to reduce the rate of decay of the peaks so that each maximum gradually decays until the next peak occurs (see Fig. 2).
(b) The partial pressures of both the 16 amu and 8 amu traces increase compared to the gas evaporation runs. Both of these traces generally follow the trends in the 32 amu trace. which itself shows a similar level of partial pressure for the MS material but is decreased for the UA muaterial.
(ii) The effect of the dc triode system on the evaporation process was
(a) To drastically reduce both the number and size of the peaks on the 32 amu trace, in fact the greatest variation for this system was in the 16 amu trace (see Fig. 2), and in the 8 amu trace towards the end of the run.
(b) The partial pressure of the 8 amu trace appears to decrease for the UA material, but increase for the MS PYSZ.
(c) The partial pressure at 16 amu increases in both cases.
(d) In all cases the dc triode system results in a larger P16/P32 ratio than the gas evaporated runs.
(e) In general for the dc triode the P16/P32 ratio decreases with time, whereas for gas evaporation (with or without biased filament support) the ratio tends to increase, more rapidly when filaments are used.

The effect of increasing the EB gun power is generally to increase the number and magnitude of the maxima observed on the 16 amu trace, and to decrease the partial pressure of this species towards the end of the evaporation cycle. Therefore the P16/P32 ratio is significantly larger at lower gun powers.

4. DISCUSSION

The general observations will be discussed initially, and then the more specific observations will be considered. It seems likely that the peaks in the 32 amu trace correspond to individual granules of zirconia in the crucible beginning to melt, and at least partially dissociating to produce O_2. This is as stated in Ref. 10. The fact that the other traces relating to other oxygen species also follow the general trends observed at 32 amu, however, seems to indicate that the other species are also produced in the dissociation process, or possibly the O_2 is broken down in the mass spectrometer to produce the other species.

The observation that the peaks decay very rapidly seems to suggest that the evapo-

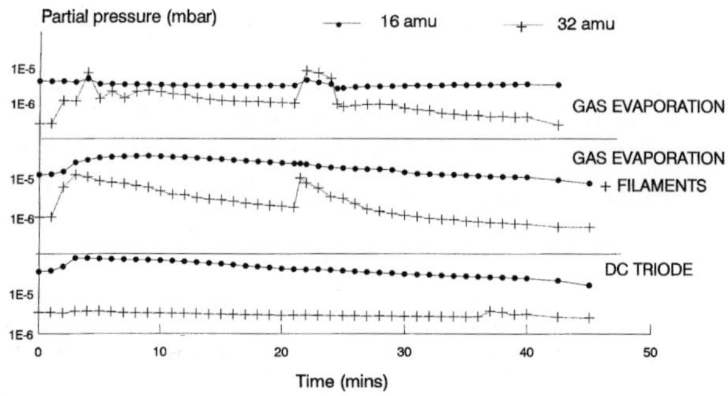

Fig. 2 This illustrates the changes in the traces of the major oxygen species with changing deposition conditions.

ration rate from a given particle is very rapid initially when the material first begins to melt, but this evaporation is quickly reduced. This effect may occur because as the beam starts to burrow or tunnel into the source material it becomes more difficult for the evaporation species to leave the molten surface, as it is likely to condense onto the walls of the tunnel rather than enter the deposition volume. A second possibility is that as soon as the particle begins to heat up all of the oxygen is liberated quickly and as the evaporation progresses the material becomes more and more zirconium rich. This however would seem to be unlikely in view of the XRD studies performed on the coating samples, and observations regarding evaporation rates.

It is not immediately evident why the 40 amu trace (Ar^+) should react to the opening of the crucible cover, and to a lesser extent the movement of the beam. One possibility may be changes in the radiant heat emanating from the crucible causing a local rarefaction of the gas and/or increased outgassing from the chamber walls, and a short-term increase in the effective argon pressure.

The effects due to the different source materials will now be addressed. The UA material exhibits more peaks but they tend to be smaller in magnitude. This is probably just a function of the initial particle size, which was smaller for the UA PYSZ. When the electron beam strikes particles of a smaller size, with the beam in a fixed position, more individual pieces will be evaporated although naturally there is less material in each granule. Due to the tunnelling effects the oxygen peak heights might be expected to reduce with time as observed. With larger size pieces however, the peaks will be larger because more material will be evaporated when the beam strikes the granule. The difference in material density may also enhance this effect, as it has been observed that it is extremely difficult to evaporate very dense zirconia, presumably due to the poor heat transfer characteristics.

The reason why a used zirconia melt results in lower partial pressures for all of the oxygen species monitored, is that the material in this state has already lost some oxygen during its previous evaporation. Despite thorough mixing of the used melt prior to its second usage, and some oxygen being reabsorbed into the material on exposure to air after venting, the overall oxygen level in the melt will be reduced. This should be paarticularly evident in the partial pressures towards the end of the deposition cycle. Further evidence is provided by the P16/P32 partial pressure ratio, which is greater in the case of the used material. The implication is that O_2 is the species preferentially generated on dissociation, but where less oxygen is available the monatomic species may be evolved.

As implied above, it was noticed that the activity in the partial pressure traces decreased towards the end of the run, with the used melt. This implies that less oxygen is evolved due to a combination of there being less oxygen available for dissociation, and the tunnelling of the beam into the melt material. It was observed in the coating experiments that extending the run times had very little effect on the weight of material evaporated if the beam was kept stationary, i.e. the overall evaporation rate was observed to decrease with increasing run time. This confirms the results observed by Fancey.[11] This is illustrated in Fig. 3.

For a system to be used in a production environment, it is obvious that these problems which occur in the evaporation of zirconia are a major concern. The most likely route to achieve controllable and predictable evaporation rates would be to utilise material in a dense, solid form rather the granular material used in this work, probably to be used in a rod-fed crucible arrangement to enable long deposition run times to acquire the required coating thicknesses in a single cycle. Furthermore in order to reduce the effects due to the beam tunnelling into the source material, it would be necessary to employ a high power, scanning EB gun in order that the beam does not remain stationary for long periods and would be able to evaporate from the entire surface of the material. It has been shown in previous experiments that moving the beam during deposition can greatly increase the weight of material evaporated.[12]

The effects due to the changes in the operating conditions are now discussed. Firstly the effects observed as a result of the use of biased filament support will be considered. The observations showed a reduction in the number of peaks after the beam was moved, a decrease in the rate of decay of the peaks and an increase in the partial pressures of the 8 and 16 amu species, compared to the gas evaporation runs. The first two of these effects are difficult to explain, but several possibilities exist:

(i) The electron emission from the filaments will result in a small degree of bombardment over the entire surface of the melt which may result in some additional dissociation of the zirconia which could explain the reduced rate of decay of the peaks.

(ii) The magnetic field generated by the current in the filaments may affect the electron beam in some manner, possibly broadening the beam spot which could lead to a reduction in the number of peaks as the beam power density would be reduced. However, the rate of decay could be reduced as a larger area of the melt surface is being heated and the evaporation rate will be slightly reduced.

(iii) Similarly there is a possibility that the electron emission from the filaments may

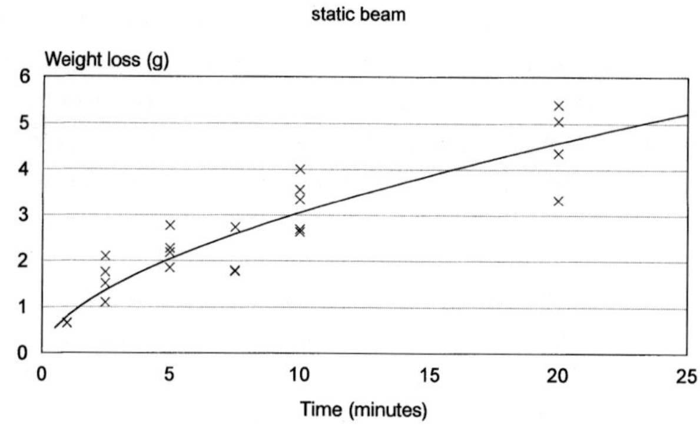

Fig. 3 Graph illustrating the change in evaporation rate with time, when using a static electron beam in evaporation.

interact with the electron beam to cause a broadening effect. It should however be pointed out that there is no evidence for this change in the electron beam spot size, although direct observation is difficult due to the intensity of the light produced by zirconia when it is evaporating.

The increase in the partial pressures of the 8 and 16 amu traces suggests that the oxygen species evolved in gas evaporation (O_2) is further dissociated as a result of electron impact ionisation by the electrons emitted from the filaments, which would cause an increase in the partial pressures of the lower mass species.

The effects of operating a dc triode system upon the evaporation characteristics are now considered. The most obvious effect is the severe change in shape of the partial pressure traces, and in particular the fact that the traces showing the greatest variations are the lower mass oxygen species. Presumably the reason is further ionisation or dissociation of the O_2 species by a combination of electron impact ionisation as discussed above, together with a significant effect due to the presence of the glow discharge. In addition it is not implausible that species other than those monitored are created in this case, for instance some form of combination between the argon and the oxygen species, or even the formation of negative oxygen ions is possible. All that can readily be inferred is that much less of the O_2^+ species is present as a result of evaporation in a dc triode situation, with at least some of the evaporated oxygen being converted to lower effective mass species as a result of further ionisation processes occurring within the plasma.

It is noted that the partial pressure at 16 amu is increased, which tends to support this explanation, as do the observations regarding the P16/P32 partial pressure ratio. It seems likely that towards the end of the run, where the greatest variation occurs in the 8 amu trace, that some of the evaporating oxygen species are monatomic and more readily further ionised, rather than given off in molecular form. Possibly this is simply due to the oxygen deficiency in the source material, which may be exacerbated in this case due to the additional heating of the source material from the plasma together with the radiant heating from the filaments. The fact that the proportion of O_2^+ is dependent upon the source material is probably due to the physical nature of the two alternatives.

The effects due to increasing the EB gun power can be explained in terms of the higher gun power causing increased evaporation, in terms of the number of granules consumed. There is, however, virtually no difference in terms of the actual weight loss from the crucible, implying that although the initial evaporation rate is probably higher, after a few minutes tunnelling becomes the dominant effect and in the case of higher gun powers the tunnelling will probably be deeper, so the effect will be exaggerated.

Many of the arguments presented implicitly presume that there is a relationship between the peak height observed in the 32 amu trace and the evaporation rate. This is verified in Fig. 4, where the first peak on the trace was measured and normalised, to account for the differences in the initial baseline partial pressure recorded, and is plotted against the overall evaporation rate. The justification for using only the first peak is that this is where the evaporation is not occluded by tunnelling effects, and this peak is likely to contain the largest proportion of material actually depositing upon the substrate. A linear relationship is observed which seems to confirm both of the above points.

5. CONCLUSIONS

The main points to arise from this part of the work are:

(i) Zirconia is dissociated upon evaporation to produce several oxygen-based species.

(ii) The most significant part of the evaporation occurs in the early stages of a run, at least with granular source material, as thereafter tunnelling effects will reduce the overall evaporation rate.

(iii) The use of biased thermionic support tends to prolong the useful evaporation time as it reduces the rate of decay of the observed oxygen peaks.

(iv) The use of a dc triode system results in further dissociation of the liberated oxygen species, and there is a large difference in the nature of the mass spectromneter results. It may be that the melting of individual granules becomnes less important in this system.

(v) The evaporation of zirconia tends to proceed as a result of individual granules evaporating separately, with the material currently being used.

ACKNOWLEDGEMENTS

We acknowledge the financial support of the UK Engineering and Physical Sciences Research Council and Ion Coat Ltd, and wish to thank our co-workers at the RCSE for helpful discussions.

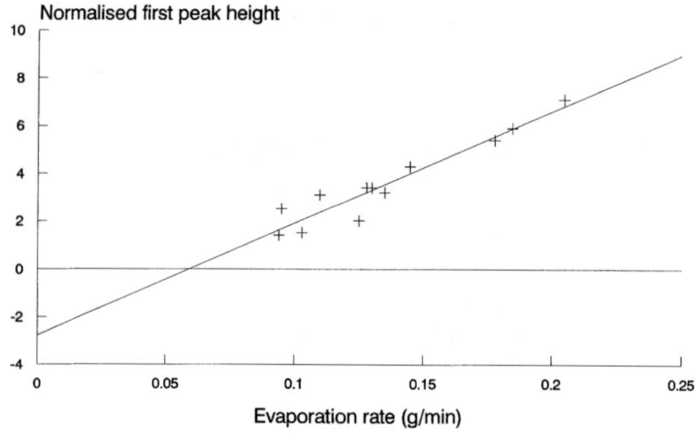

Fig. 4 Graph illustrating the relationship between the normalised height of the initial 32 amu peak and the overall evaporation rate.

REFERENCES

1. K. S. FANCEY, A. S. JAMES and A. MATTHEWS: Proceedings of the International *Conference Engineering the Surface*, London 1986, Institute of Metals, 1987.
2. A. S. JAMES, K. S. FANCEY and A. MATTHEWS: Proceedings of the International Symuposiumn on *Trends and New Applications in Thin Films*, Strasbourg, March 1987, Societe Francaise du Vide, Paris, 1987.
3. A. S. JAMES, K. S. FANCEY and A. MATTHEWS: *Surf. Coat. Technol.*, 1987, **32**, 377.
4. A. S. JAMES and A. MATTHEWS: Proc of the 1st International Conference on *Plasma Surface Engineering (PSE 88)*, Garmisch-Partenkirchen, Sept 1988, DGM, Oberusel, FRG, 1989.
5. A. S. JAMES and A. MATTHEWS: *Surf. Coat. Technol.*, 1990, **41**, 305.
6. A. S. JAMES and A. MATTHEWS: *Surf. Coat. Technol.*, 1990, **43/44**, 436.
7. A. S. JAMES and A. MATTHEWS: *Materials Science Forum*, 1992, **102–104**, 633.
8. K. S. FANCEY, S. J. YOUNG, A. S. JAMES and A. MATTHEWS: *Mat. Sci. Eng.*, 1993, **A163**, 171.
9. J. ROBSON, A. S. JAMES, B. LEWIS and A. MATTHEWS: *Surf. Coat. Technol.*, 1993, **60**, 536.
10. R. GLANG: *Handbook of Thin Film Technology*, L. I. Maissel and R. Glang eds, McGraw Hill, New York, USA, 1970.
11. K. S. FANCEY: *MSc thesis*, University of Hull, 1986.
12. A. S. JAMES: *PhD thesis*, University of Hull, 1991.

INDEX